Interdisciplinary Applied Mathematics
Volume 6

Springer-Science+Business Media, LLC

Interdisciplinary Applied Mathematics

Ulrich Hornung

Editor

Homogenization and Porous Media

With 35 Illustrations

 Springer

Ulrich Hornung
Department of Computer Science
University of Federal Armed Forces, Munich
D-85577 Neubiberg
Germany

Editors

L. Kadanoff
Department of Physics
James Franck Institute
University of Chicago
Chicago, IL 60637
USA

J.E. Marsden
Department of Mathematics
University of California
Berkeley, CA 94720
USA

L. Sirovich
Division of
  Applied Mathematics
Brown University
Providence, RI 02912
USA

S. Wiggins
Applied Mechanics Department
Mail Code 104-44
California Institute of Technology
Pasadena, CA 91125
USA

Cover illustration: Period of the Porous Medium, from Figure 8.1.

Mathematics Subject Classification (1991): 35B27, 76S05, 73B27, 76T05

Library of Congress Cataloging-in-Publication Data
Hornung, Ulrich, 1941–
    Homogenization and porous media / Ulrich Hornung.
       p.   cm. – (Interdisciplinary applied mathematics ; 6)
    Includes bibliographical references (pp. 259–275) and index.
    ISBN 978-1-4612-7339-4   ISBN 978-1-4612-1920-0 (eBook)
    DOI 10.1007/978-1-4612-1920-0
    1. Fluid dynamics.   2. Homogenization (Differential equations)
    3. Porous materials – Mathematical models.   I. Title.   II. Series:
    Interdisciplinary applied mathematics ; v. 6.
    QA911.H667   1996
    535′.05 – dc20   96-22801

Printed on acid-free paper.

© 1997 Springer Science+Business Media New York
Originally published by Springer-Verlag New York in 1997
Softcover reprint of the hardcover 1st edition 1997

Production managed by Bill Imbornoni; manufacturing supervised by Jacqui Ashri.
Camera-copy produced from the author's LaTeX files.

9 8 7 6 5 4 3 2 1

ISBN 978-1-4612-7339-4

Preface

For several decades developments in porous media have taken place in almost independent areas. In civil engineering, many papers were published dealing with the foundations of flow and transport through porous media. The method used in most cases is called *averaging*, and the notion of a *representative elementary volume* (REV) plays an important role. In chemical engineering, papers on conceptual models were written on the *theory of mixtures*. In theoretical physics and stochastic analysis, percolation theory has emerged, providing probabilistic models for systems where the connectedness properties of some component dominate the behavior. In mathematics, a theory has been developed called *homogenization* which deals with partial differential equations having rapidly oscillating coefficients.

Early work in these and related areas was – among others – done by the following scientists: Maxwell [Max81] and Rayleigh [Ray92] studied the effective conductivity of media with small concentrations of randomly and periodically, respectively, arranged inclusions. Einstein [Ein06] investigated the effective viscosity of suspensions with hard spherical particles in compressible viscous fluids. Marchenko and Khrouslov [MK64] looked at the asymptotic nature of homogenization; they introduced a general approach of averaging based on asymptotic tools which can handle a variety of different physical problems.

Unfortunately, up to now, little effort has been made to bridge the gap between these different fields of research. Consequently, many results were and are discovered independently, and scientists are almost unable to understand each other because the respective languages have been developing in different directions.

The purpose of this volume is to present a survey on the method of *homogenization* applied to problems in *porous media*. The authors have tried to write their contributions so that the text is understood by engineers with only a basic knowledge of the theory of partial differential equations. On the other hand, the collection of descriptions of problems, models, mathematical results, and applications will also be of interest to mathematicians who want to learn about "real-life" problems.

Darcy published his observations and conclusions as early as in 1856 ([Dar56]), and one may wonder why it has taken more than one hundred and twenty years before a mathematical derivation of the law named after him was given (see [Kel80] and [Tar80]). One might argue that for practical applications mathematical convergence proofs are irrelevant, but this is not at all true. Not only does a mathematical treatment clarify on which assumptions the law is based, but it also allows development of techniques that can be applied to similar and related problems. This has, indeed, been the case. In the last fifteen years, rapid mathematical development in this area has taken place, and now there is a far better understanding of many of the basic phenomena. It was only recently that several nonstandard models were discovered or could be justified. Among these are Darcy's law with memory, nonlinear Darcy laws, and the correct transmission conditions between free and porous media flow (all studied in Chapter 3). Furthermore, the new nonlinear models describing non-Newtonian flow have to be emphasized (see Chapter 4). In addition, homogenization has allowed proper derivation of many macroscopic laws that are known in two-phase flow and transport of solutes and heat, and poroelastic porous media (Chapters 5, 6, 7, and 8). *Dual-porosity models* have been known since 1960. It was only in the nineties that these models were put on a sound mathematical basis, i.e., well-posedness was shown, they were derived from micromodels, and they could be used for numerical computations and applications (Chapters 9 and 10).

At this moment, we have not yet taken advantage of the full power of the mathematical tools from *homogenization* theory. Basic problems of modeling are still open. For example, the justification of two- or multiphase flow equations, such as those based on the concept of *relative permeability* and *capillary pressure* are not fully satifsfactory (cf. Chapter 5). It would be desirable to include dispersion and to better understand the effects of upscaling in this context, i.e., to explain the relation between micro- and macrodispersion. The *secondary flux model* (Chapter 9) and the *viscous dual-porosity model* (Chapter 10) have not yet been derived by homogenization. There seems to be a very good chance that at least some of these problems may be successfully solved in the near future.

On the other hand, many known theoretical results have not yet been used for numerical simulations of practical problems in research and industry. The editor of this volume would be more than glad if this text could contribute to a better understanding, interaction, and cooperation between engineers and mathematicians in all directions.

The general outline of the book is as follows:

Chapter 1 describes the formal approach of homogenization, namely, the method of *asymptotic expansions* using two scales. This tool is useful for many problems in standard form and yields the correct homogenization limit in most cases. Its main drawback is that it does not supply any rigorous proofs, but, in any case, it helps to understand the basic ideas, at least, from a formal point of view.

Chapter 2 deals with *percolation theory*. This theory is a well-established mathematical theory that has already proven its usefulness for a large number of applications. Especially for two- or three-phase flow, there are regimes for which a percolation theoretical approach seems to be more appropriate than the standard Darcy-type description.

The following chapters are devoted to different physical phenomena of flow and transport through porous media.

Most homogenization results have been obtained for *one-phase Newtonian flow*, i.e., for the standard single-phase flow problem, in other words, for flow in a saturated medium. Chapter 3 shows how to derive not only Darcy's law but also under which conditions Brinkman's law is obtained. In addition, the inclusion of inertial effects and time-dependent problems are studied. Only very recently, the question of how to set up interface conditions between free Stokes flow and flow in a porous medium was settled; section 3.5 is devoted to this problem.

Recent results for *one-phase, non-Newtonian flow*, such as Bingham flow, and for polymers, are dealt with in Chapter 4. Due to the nonlinearity of the underlying equations, advanced tools from nonlinear functional analysis have to be used when the homogenization process converges.

In practical applications, the problem of two-phase flow is of greatest importance. Soil science very often deals with water and air in the unsaturated zone. A standard problem in oil-reservoir simulation is the displacement of oil by water. There is a long history of papers treating this *immiscible displacement* problem. One of the successful applications of homogenization is the derivation of the well-known *dual-porosity* model. Chapter 5 studies homogenization of two-phase flow from the porescale to the mesoscale, but also from the mesoscale to the macroscale. In addition, flow in media with random heterogeneities is investigated.

Miscible displacement, the problem of transporting solutes in fluids, is dealt with in Chapter 6. This process is becoming more and more important in soil pollution and remediation techniques. It is closely related to absorption and adsorption, and to chemical reactions.

Coupled transport of fluids and *thermal energy* are treated in Chapter 7.

Poroelastic media are of great relevance, e.g., in the use of sound waves in geological formations. Only a good understanding of the physics of sound propagation in porous media allows the application of acoustic techniques for seismically exploring reservoirs. Homogenization is a promising method for deriving the basic equations in this area. This topic is studied in Chapter 8.

Chapter 9 gives a survey on the application of models that contain two scales that are called *microstructure* models. There is a well-developed mathematical theory independent of the derivation of these models by homogenization, but of course, the models themselves can only be well understood in the context of homogenization.

Numerical applications of models, derived by homogenization, to oil reservoir simulation problems are described in Chapter 10. Dual-porosity models are compared with *mesoscopic models*, i.e., with models that explicitly account for two-phase flow in fissures and blocks of a fractured medium. It is demonstrated that dual-porosity models show far better performance as far as computer time is concerned.

For the more mathematically inclined reader, the important methods, tools, and results of a general nature from *homogenization theory* are collected in Appendix

A. The symbols, notations, and definitions most often used in this volume are compiled in Appendix B for quick reference.

The editor of this volume would like to point out that a project like this would have been impossible a few years ago. There has never been a meeting of all the authors who have contributed to this book, and the editor met only a few of them during the time the text was written. The exchange of information took place and was possible only via "e-mail" and "ftp". One might say that this volume results from taking advantage of the so-called "information super-highway", namely, Internet.

Unfortunately, during the preparation of this book, Serguei Kozlov died under tragic circumstances. Serguei had planned to contribute a chapter on *Geometries of Porous Media*, a project that he could not finish. Because his death is a great loss to the community of reasearchers working in homogenization, the authors are dedicating this book to his memory.

Acknowledgement

Parts of this book were supported by the project "Homogenization and Variational Methods for the Mechanics of Composite Materials and Flexible Structures (EURHomogenization)", Grant SC1*-CT91-0732 (TSTS) of the European Program "Human Capital and Mobility (HCM)".

Contents

Dedicated to the memory of Serguei M. Kozlov

Contributors

Grégoire Allaire

Commissariat à l'Energie Atomique
DRN/DMT/SERMA, CEA Saclay
F-91191 Gif sur Yvette, France
& Laboratoire d'Analyse Numérique
Université Paris 6, France
e-mail: allaire@ann.jussieu.fr

Todd Arbogast

Department of Mathematics
The University of Texas
Austin, TX 78712, USA
e-mail: arbogast@math.utexas.edu

Jean-Louis Auriault

Laboratoire Sols Solides Structures
Domaine Universitaire, BP 53 X
F-38041 Grenoble Cedex, France
e-mail: auriault@img.fr

Alain Bourgeat

ELF Aquitaine et Equipe d'Analyse Numérique
Université Jean Monnet
23, rue du Dr. Paul Michelon
F-42023 Saint-Etienne-Cedex 02, France
e-mail: bourgeat@anumsun1.univ-st-etienne.fr

Horia Ene

Institute of Mathematics
Romanian Academy
P.O. Box 1-764
70700 Bucharest, Romania
e-mail: hene@imar.ro

Kenneth Golden

Department of Mathematics
University of Utah
Salt Lake City, UT 84112-1107, USA
e-mail: golden@math.utah.edu

Ulrich Hornung

Fakultät für Informatik, UniBwM
D-85577 Neubiberg, Germany
e-mail: ulrich@informatik.unibw-muenchen.de

Andro Mikelić

Laboratoire d'Analyse Numérique, Bât 101
Université Lyon 1
43, bd. du onze novembre
F-69622 Villeurbanne Cedex, France
e-mail: andro@lan1.univ-lyon1.fr

Ralph Showalter

Department of Mathematics
The University of Texas
Austin, TX 78712, USA
e-mail: show@math.utexas.edu

1

Introduction

Ulrich Hornung

1.1 Basic Idea

Roughly speaking, homogenization is a mathematical method that allows us to "upscale" differential equations. This method not only offers formulas for upscaling but also provides tools for producing rigorous mathematical convergence proofs.

In principle, there are two different approaches that one might consider, namely the *representative elementary volume* (REV) technique or *homogenization*. The idea of the REV approach is the following. Let u be a real valued function on a domain Ω in space which describes a certain physical quantity (such as the relative pore volume or the pressure of a fluid in a porous medium) with rapid spatial oscillations. To smoothen this function, one considers local averages of the form,

$$\langle u \rangle(x) = \int_{V(x)} u(y)\, dy,$$

where $V(x)$ is a small, but not too small, neighborhood of point x of the size of a REV (several hundreds or thousands of pores). If the oscillations of u reflect the behavior of the physical quantity in question on a "microscale" (on the pore scale), the averaged function $\langle u \rangle$ is supposed to describe its properties on a larger scale, i.e., on a "macroscale" (laboratory scale).

The basic idea of homogenization is somewhat different. Instead of working with only one function u, one considers a whole family of functions u^ε where $\varepsilon > 0$ is a spatial (length) scale parameter (the typical size of a pore). In other words, one does not work with only one situation for which certain mathematical manipulations are made, but the specific problem in question is imbedded in a family of problems parametrized by the scale parameter $\varepsilon > 0$. Now – loosely speaking – the "trick" consists in determining the limit

$$u = \lim_{\varepsilon \to 0} u^\varepsilon \tag{1.1}$$

and considering this limit as the result of the "upscaling" procedure (see Fig. 1.1). The work done by homogenization consists of finding differential equations that the limit u satisfies and proving that formula (1.1) holds.

It is well-known that any kind of mathematical "limit" has to be studied carefully. One fundamental mistake, unfortunately, made too frequently is to interchange

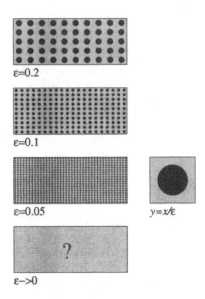

$\varepsilon=0.2$

$\varepsilon=0.1$

$\varepsilon=0.05$ $y=x/\varepsilon$

?

$\varepsilon\to0$

FIGURE 1.1. Homogenization limit.

limits with other mathematical operations, such as forming differential equations, calculating their coefficients, etc. It is by no means obvious in which sense formula (1.1) has to be understood. In most cases the limit cannot be taken "pointwise" but only in the sense of certain function spaces (see Appendices A and B).

The basic difference between the two methodologies may be looked at in the following way: the REV approach uses smoothing and spatial averaging formulas; homogenization does the upscaling by letting the microscale tend to zero.

1.2   First Examples

1.2.1   One-Dimensional Diffusion

As a first introduction, we consider a simple, one-dimensional diffusion problem. Let u be the variable in question defined on the unit interval $\Omega = [0, 1]$. We study the boundary-value problem

$$\begin{cases} \frac{d}{dx}(a(x)\frac{d}{dx}u(x)) = 0, & 0 < x < 1, \\ u(x) = 0, & x = 0, \\ u(x) = 1, & x = 1, \end{cases} \qquad (2.1)$$

with a smooth, strictly positive function a. Obviously, the differential equation in (2.1) requires that the flux $q(x) = -a(x)\frac{d}{dx}u(x)$ is constant in Ω. Let us denote this constant by q^*. Then, immediately, we get

$$\frac{d}{dx}u(x) = -\frac{q^*}{a(x)},$$

and, therefore, we have the solution formula,

$$u(x) = a^* \int_0^x \frac{d\tilde{x}}{a(\tilde{x})}, \tag{2.2}$$

with

$$a^* = -q^* = \frac{1}{\int_0^1 \frac{d\tilde{x}}{a(\tilde{x})}}. \tag{2.3}$$

As a consequence, we have determined the *effective* diffusion constant. Because we have used a "unit gradient" in the domain Ω, the constant value a^* of a in (2.1) would have given the solution $u^*(x) = x$ and, thus, the same flux $q = -a^*\frac{d}{dx}u^*(x) = q^*$.

In principle, this example has the essential features of *homogenization*: If we consider the function $a(.)$ as a description of a nonhomogeneous property (such as diffusivity, conductivity, permeability, etc.) of a medium located in the domain Ω, then, the number a^* is an *effective parameter* of this medium.

Let us now modify this example in the following way: We study a family of problems, indexed by the scale parameter $\varepsilon = \frac{1}{n}$, namely,

$$\begin{cases} \frac{d}{dx}(a^\varepsilon(x)\frac{d}{dx}u^\varepsilon(x)) = 0, & 0 < x < 1, \\ u^\varepsilon(x) = 0, & x = 0, \\ u^\varepsilon(x) = 1, & x = 1, \end{cases}$$

where we make the assumption that the functions $a^\varepsilon(.)$ are given in the form $a^\varepsilon(x) = a(\frac{x}{\varepsilon})$ with a fixed periodic function $a(.)$ having periodicity length 1. The meaning of this is that the functions $a^\varepsilon(.)$ are oscillatory with increasing frequencies for $\varepsilon \to 0$ (see the left part of Fig. 1.2). Applying formula (2.2) to this case,

$$u^\varepsilon(x) = \frac{\int_0^x \frac{d\tilde{x}}{a^\varepsilon(\tilde{x})}}{\int_0^1 \frac{d\tilde{x}}{a^\varepsilon(\tilde{x})}} = \frac{\int_0^{nx} \frac{dy}{a(y)}}{\int_0^n \frac{dy}{a(y)}} = x + \frac{\int_0^{nx} (\frac{1}{a(y)} - 1)dy}{\int_0^n \frac{dy}{a(y)}}.$$

Obviously, this may be rewritten in the form,

$$u^\varepsilon(x) = u(x) + \varepsilon u_1(\frac{x}{\varepsilon}), \tag{2.4}$$

where $u(x) = x$ and u_1 is a 1-periodic function with

$$u_1(y) = \frac{\int_0^y \frac{d\tilde{y}}{a(\tilde{y})}}{\int_0^1 \frac{d\tilde{y}}{a(\tilde{y})}} - y$$

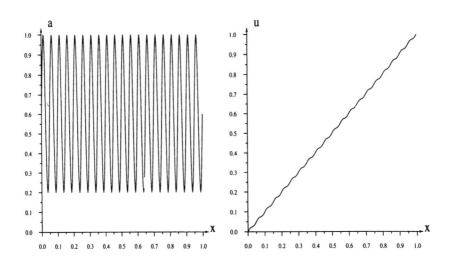

FIGURE 1.2. Oscillatory function and solution.

for $0 \leq y \leq 1$ (see the right part of Fig. 1.2). As a consequence, in this case, we get pointwise convergence in (1.1). Here, the function u describes the "global" behavior of the function u^ε, whereas the "corrector" u_1 describes the local oscillations of u^ε.

The previous examples will be generalized to a two-dimensional situation in section 1.2.3.

1.2.2   Resistor Networks

The boundary-value problem (2.1) can be looked at as the continuous analogon of a discrete series of finitely many electric resistors, where the resistance of each individual resistor corresponds to the reciprocal of the conductivity a. If $R_{i+1/2}$ is the resistance between the ith and $(i + 1)$th node and if u_i is the voltage at the ith node, then, according to Ohm's law, the current q_i through the ith resistor is given by $q_i = -\frac{u_{i+1}-u_i}{R_{i+1/2}}$. All these currents must have the same value q^*. If the first voltage is 0 and the last voltage is 1,

$$\sum_i u_{i+1} - u_i = -\sum_i R_{i+1/2}q^* = 1,$$

and, from this,

$$q^* = -\frac{1}{\sum_i R_{i+1/2}}.$$

In other words, the total resistance of this series of resistors is given by

$$R^* = \sum_i R_{i+1/2} \tag{2.5}$$

which corresponds to formula (2.3).

Of course, the previous argument is well-known in electricity. What we want to point out here is another aspect. One may also consider a very large number of resistors n and ask for the limit of the behavior of the resistor series for $n \to \infty$. Of course, this question is meaningful only if the difference between the voltages at the beginning and the end tends to infinity or if the resistances tend to zero which means that we have to "scale" the resistances in a certain way. Let us study the latter case. A natural assumption is $R_{i+1/2} = \frac{1}{n} R(\frac{i+1/2}{n})$ where R is a continuous function on the interval $[0, 1]$. Then, formula (2.5) gives the total resistance

$$\frac{1}{n} \sum_i R(\frac{i + 1/2}{n})$$

which is a riemannian sum for the integral

$$R^* = \int_0^1 R(x)dx.$$

This formula is exactly the same as formula (2.3) with $R(x) = \frac{1}{a(x)}$. The point to be emphasized here is that we have "homogenized" a series of resistors. Even though this is almost a trivial example, it shows another important feature of *homogenization*, namely, that upscaling may change the type of equations. Here the microstructure is described by discrete resistances, difference quotients, and sums, whereas the homogenized limit is expressed in terms of continuous functions, differentials, and integrals.

The next example is a resistor network, as shown in Fig. 1.3. As before, we assume that the voltage is 0 at the left side and 1 at the right side. We denote the voltages at the node (i, j) by $u_{i,j}$ with a horizontal index i and a vertical index j, the resistance between the nodes with indices (i, j) and $(i + 1, j)$ by $R_{i+1/2,j}$, and the resistance between the nodes with indices (i, j) and $(i, j+1)$ by $R_{i,j+1/2}$. Then, Ohm's law and Kirchhoff's rule require, at the node (i, j), the relation

$$\frac{u_{i+1,j} - u_{i,j}}{R_{i+1/2,j}} + \frac{u_{i,j} - u_{i-1,j}}{R_{i-1/2,j}} + \frac{u_{i,j+1} - u_{i,j}}{R_{i,j+1/2}} + \frac{u_{i,j} - u_{i,j-1}}{R_{i,j-1/2}} = 0. \tag{2.6}$$

This "five point star" is well-known in numerical analysis for elliptic differential equations. Formula (2.6) is used when solving the equation

$$\nabla \cdot (a\nabla u) = 0$$

with $a = \frac{1}{R}$ by means of finite differences. Here, we have assumed $R_{i+1/2,j} = \frac{1}{nm} R(\frac{i+1/2}{n}, \frac{j}{m})$ and $R_{i,j+1/2} = \frac{1}{nm} R(\frac{i}{n}, \frac{j+1/2}{m})$ with a smooth function R of two variables defined in the square $\Omega = [0, 1] \times [0, 1]$.

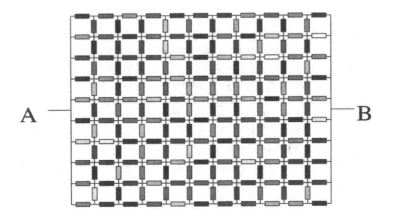

$$A \text{ ——————————— } B$$

FIGURE 1.3. Electric resistor network.

Any textbook on finite difference methods states that, for $n, m \to \infty$, the solutions of system (2.6) converge to the solution u of the boundary-value problem,

$$\begin{cases} \nabla \cdot (a(x_1, x_2)\nabla u(x_1, x_2)) = 0, & 0 < x_1, x_2 < 1, \\ u(x_1, x_2) = 0, & \text{for } x_1 = 0 \text{ and } 0 < x_2 < 1, \\ u(x_1, x_2) = 1, & \text{for } x_1 = 1 \text{ and } 0 < x_2 < 1, \\ \partial_{x_2} u(x_1, x_2) = 0, & \text{for } 0 < x_1 < 1 \text{ and } x_2 = 0, \\ \partial_{x_2} u(x_1, x_2) = 0, & \text{for } 0 < x_1 < 1 \text{ and } x_2 = 1, \end{cases} \quad (2.7)$$

provided that the function $a = \frac{1}{R}$ is sufficiently smooth. Here again, we have homogenized the resistor networks. The homogenized limit is the boundary-value problem (2.7).

1.2.3 Layered Media

In the unit square $\Omega = \{(x_1, x_2) : 0 < x_1, x_2 < 1\}$ in $\mathbb{R}^2$, we study a diffusion process governed by

$$\nabla \cdot (a^\varepsilon(x)\nabla u^\varepsilon(x)) = 0.$$

We assume that a^ε is a function only of x_2 and is periodic with periodicity length ε. In other words,

$$a^\varepsilon(x_1, x_2) = a\left(\frac{x_2}{\varepsilon}\right)$$

with a fixed 1-periodic function $a : \mathbb{R} \to \mathbb{R}$. This means that we have a layered medium (see Fig. 1.4) in which we want to calculate the effective properties.

FIGURE 1.4. Layered medium.

Naturally, we expect that the medium can be described in an averaged sense by a diffusion equation of the type,

$$\nabla \cdot (A \nabla u_0) = 0, \tag{2.8}$$

where A is a tensor of the form

$$A = \begin{pmatrix} \bar{a} & 0 \\ 0 & a^* \end{pmatrix}.$$

The easiest way to find the numbers $\bar{a}$ and a^* is to study the following two simple boundary-value problems:

Case 1:

$$\begin{cases} u = 0, & x_1 = 0, \\ u = 1, & x_1 = 1, \\ \partial_\nu u = 0, & x_2 = 0 \text{ or } = 1. \end{cases}$$

Here, $u^\varepsilon(x_1, x_2) = u_0(x_1, x_2)$ with $u_0(x_1, x_2) = x_1$ for all ε, and, hence, $a^\varepsilon \nabla u^\varepsilon = a^\varepsilon \vec{e}_1$ with $\vec{e}_1 = (1, 0)$. Therefore, the total flux through the square in the x_1 direction is given by

$$\int_0^1 a^\varepsilon \vec{e}_1 \cdot \nabla u^\varepsilon \, dx_2 = \int_0^1 a^\varepsilon(x_2) \, dx_2 = \int_0^1 a(\eta) \, d\eta.$$

This quantity equals $\bar{a}$, because Eq. (2.8) is independent of ε.

Case 2:

$$\begin{cases} u = 0, & x_2 = 0, \\ u = 1, & x_2 = 1, \\ \partial_\nu u = 0, & x_1 = 0 \text{ or } = 1. \end{cases}$$

Here, by a simple calculation of the type used in section 1.2.1,

$$u^\varepsilon(x_1, x_2) = u_0(x_1, x_2) + \varepsilon\sigma_2(\frac{x_2}{\varepsilon}),$$

where $u_0(x_1, x_2) = x_2$, and

$$\sigma_2(y_2) = a^* \int_0^{y_2} \frac{d\eta}{a(\eta)} - y_2$$

with

$$a^* = \frac{1}{\int_0^1 \frac{d\eta}{a(\eta)}}.$$

Immediately, one gets

$$a^\varepsilon \nabla u^\varepsilon = a^* \vec{e}_2.$$

Therefore, the total flux through the square is given by

$$\int_0^1 a^\varepsilon \vec{e}_2 \cdot \nabla u^\varepsilon \, dx_1 = \int_0^1 a^* dx_1 = a^*.$$

In this case, we see Eq. (2.8) again. In the next section, we are going to see how to unify the theory for these two cases.

1.3   Diffusion in Periodic Media

We start with the following problem. Let Ω be a bounded domain in $\mathbb{R}^N$ with smooth boundary $\partial\Omega$. Then we study the family of boundary-value problems

$$\begin{cases} \nabla \cdot (a^\varepsilon(x)\nabla u^\varepsilon(x)) + f(x) = 0, & x \in \Omega, \\ u(x) = u_D(x), & x \in \partial\Omega. \end{cases} \tag{3.1}$$

We assume that the coefficient a^ε is rapidly oscillating, i.e., it is of the form

$$a^\varepsilon(x) = a(\frac{x}{\varepsilon}), \tag{3.2}$$

for all $x \in \Omega$, where the function a is Y-periodic in $\mathbb{R}^N$ with periodicity cell $Y = \{y = (y_1, \ldots, y_N) : 0 < y_i < 1 \text{ for } i = 1, \ldots, N\}$ and ε a scale parameter. For smaller and smaller ε, the coefficient a^ε oscillates more and more rapidly, and it is natural to ask what the solution u^ε does in the limit as $\varepsilon \to 0$. Note that the weak form of problem (3.1) is given by

$$(a^\varepsilon \nabla u^\varepsilon, \nabla\varphi)_\Omega - (f, \varphi)_\Omega = 0 \tag{3.3}$$

for all test functions φ on Ω with $\varphi|\partial\Omega = 0$. Here, we have used the L^2-scalar product on Ω

$$(u, v)_\Omega = \int_\Omega u(x)v(x)\, dx.$$

1.3.1   Formal Asymptotic Expansion

To derive the limit problem in a formal way, one starts from the *ansatz* that the unknown function u^ε has an asymptotic expansion, with respect to ε, of the form,

$$u^\varepsilon(x) = u_0(x, y) + \varepsilon u_1(x, y) + \varepsilon^2 u_2(x, y) + \ldots, \tag{3.4}$$

where the coefficient functions $u_i(x, y)$ are Y-periodic with respect to the variable $y = \frac{x}{\varepsilon}$. The derivatives obey the law,

$$\nabla = \nabla_x + \frac{1}{\varepsilon}\nabla_y,$$

where the subscripts indicate the partial derivatives with respect to x and y, respectively. Therefore, from Eq. (3.1) one, immediately, gets the formula

$$\begin{aligned}
\varepsilon^{-2}&\nabla_y \cdot (a(y)\nabla_y u_0(x, y)) \\
&+\varepsilon^{-1}\big(\nabla_y \cdot (a(y)\nabla_y u_1(x, y)) + \nabla_y \cdot (a(y)\nabla_x u_0(x, y)) \\
&+a(y)\nabla_x \cdot \nabla_y u_0(x, y)\big) \\
&+\varepsilon^0\big(\nabla_y \cdot (a(y)\nabla_y u_2(x, y) + a(y)\nabla_x u_1(x, y)) \\
&+a(y)\nabla_x \cdot \nabla_y u_1(x, y) + a(y)\nabla_x \cdot \nabla_x u_0(x, y)\big) \\
&+\varepsilon^1(\ldots) + \ldots + f(x) = 0.
\end{aligned} \tag{3.5}$$

The next step consists of comparing the coefficients of the different ε powers in this equation. The term with ε^{-2} gives

$$\nabla_y \cdot (a(y)\nabla_y u_0(x, y)) = 0 \text{ for } y \in Y.$$

Because $u_0(x, y)$ is Y-periodic in the variable y, $u_0(x, y) = u_0(x)$ is a function of x alone, independent of y. Using this and the term with ε^{-1} in Eq. (3.5),

$$\nabla_y \cdot (a(y)\nabla_y u_1(x, y)) = -\nabla_y \cdot (a(y)\nabla_x u_0(x)) \text{ for } y \in Y.$$

At this point one expresses the function $u_1(x, y)$ in terms of the function $u_0(x)$. First, use the obvious identity,

$$\nabla_x u_0(x) = \sum_{j=1}^N \vec{e}_j \partial_{x_j} u_0(x).$$

Therefore, we can write

$$\nabla_y \cdot (a(y)\nabla_y u_1(x, y)) = -\sum_{j=1}^N \partial_{y_j} a(y)\partial_{x_j} u_0(x) \text{ for } y \in Y.$$

Now, for $j = 1, \ldots, N$, let $w_j(y)$ be a Y-periodic solution of the *cell problem*,

$$\boxed{\nabla_y \cdot (a(y)\nabla_y w_j(y)) = -\nabla_y \cdot (a(y)\vec{e}_j) \text{ for } y \in Y,}$$ (3.6)

the weak form of which is

$$(a(\nabla w_j + \vec{e}_j), \nabla\varphi)_Y = 0$$

for all Y-periodic test functions φ on Y.

Using these functions $w_j(y)$, we solve for $u_1(x, y)$, namely,

$$u_1(x, y) = \sum_{j=1}^{N} w_j(y)\partial_{x_j} u_0(x) + u_1(x),$$

where $u_1(x)$ is independent of y. Differentiating this, we, immediately, get

$$\nabla_y u_1(x, y) = \sum_{j=1}^{N} \nabla_y w_j(y)\partial_{x_j} u_0(x).$$ (3.7)

We proceed further and look at the term with ε^0 in Eq. (3.5). We get

$$\nabla_y \cdot \left(a(y)\nabla_y u_2(x, y) + a(y)\nabla_x u_1(x, y)\right)$$
$$+ a(y)\nabla_x \cdot \nabla_y u_1(x, y) + a(y)\nabla_x \cdot \nabla_x u_0(x) + f(x) = 0 \text{ for } y \in Y.$$

We integrate this identity over Y and obtain

$$\int_Y \nabla_y \cdot \left(a(y)\nabla_y u_2(x, y) + a(y)\nabla_x u_1(x, y)\right) dy$$
$$+ \int_Y a(y)\nabla_x \cdot \nabla_y u_1(x, y)dy$$
$$+ \int_Y a(y)dy\, \Delta_x u_0(x) + f(x) = 0,$$ (3.8)

because the volume of Y is 1. We integrate the first integral by parts and get

$$\int_Y \nabla_y \cdot \left(a(y)\nabla_y u_2(x, y) + a(y)\nabla_x u_1(x, y)\right) dy$$
$$= \int_{\partial Y} \vec{v} \cdot \left(a(y)\nabla_y u_2(x, y) + a(y)\nabla_x u_1(x, y)\right) d\Gamma(y),$$

where $\vec{v}$ is the normal vector on ∂Y. This boundary integral vanishes because of the Y-periodicity of the functions $u_2(x, y)$ and $u_1(x, y)$. For the second term in Eq. (3.8), we use Eq. (3.7) and get

$$\nabla_x \cdot \nabla_y u_1(x, y) = \sum_{i,j=1}^{N} \partial_{y_i} w_j(y)\partial_{x_i x_j} u_0(x).$$

In this way, from Eq. (3.8), we obtain

$$\sum_{i,j=1}^{N} \int_Y a(y)\partial_{y_i} w_j(y)\partial_{x_i x_j} u_0(x) + \int_Y a(y)dy \, \Delta_x u_0(x) + f(x) = 0.$$

It is convenient to introduce the abbreviation

$$\boxed{a_{ij} = \int_Y a(y)(\delta_{ij} + \partial_{y_i} w_j(y)) \, dy,}$$

with which we get the final result

$$\sum_{i,j=1}^{N} a_{ij}\partial_{x_i x_j} u_0(x) + f(x) = 0.$$

This elliptic differential equation is the *homogenized* limit of the equation in problem (3.1). For short, we simplify the notation and write

$$\sum_{i,j=1}^{N} a_{ij}\partial_{ij} u(x) + f(x) = 0.$$

Using the tensor $A = (a_{ij})_{i,j}$, we sum up and get the following statement.

Proposition 3.1 *The differential operator $\nabla \cdot (A\nabla u(x))$ is the homogenization of the operator family $\nabla \cdot (a^\varepsilon(x)\nabla u^\varepsilon(x))$, i.e., the homogenization of problem (3.1) is given by*

$$\begin{cases} \nabla \cdot (A\nabla u(x)) + f(x) = 0, & x \in \Omega, \\ u(x) = u_D(x), & x \in \partial\Omega. \end{cases} \tag{3.9}$$

Remark Writing the weak form of this problem, one gets

$$(A\nabla u, \nabla\varphi)_\Omega - (f, \varphi)_\Omega = 0$$

for all test functions φ on Ω with $\varphi|\partial\Omega = 0$. For a rigorous mathematical proof, see Appendix A, sections A.2.2 and A.3.3.

It has to be emphasized that the coefficients in the equation obtained by homogenization are, in general, not diagonal. In other words, we have derived an equation that describes an anisotropic medium. However, in general, the tensor A is symmetric and positive definite.

Proposition 3.2 *(a) The tensor A is symmetric.*
(b) If the coefficient a in equation (3.2) satisfies

$$a(y) \geq \alpha > 0 \text{ for all } y \in Y,$$

then, A is positive definite.

Proof (a) The coefficients a_{ij} can be written as

$$a_{ij} = \int_Y a(y)(\nabla w_j(y) + \vec{e}_j) \cdot \vec{e}_i \, dy.$$

The weak formulation of problem (3.6) is given by

$$\int_Y a(y)(\nabla w_j(y) + \vec{e}_j) \cdot \nabla \varphi(y) \, dy = 0 \text{ for all } Y\text{-periodic } \varphi.$$

Therefore,

$$\begin{aligned}
0 &= \int_Y a(y)(\nabla w_j(y) + \vec{e}_j) \cdot \nabla w_i(y) \, dy \\
&= \int_Y a(y)\nabla w_j(y) \cdot \nabla w_i(y) \, dy + \int_Y a(y)\vec{e}_j \cdot \nabla w_i(y) \, dy,
\end{aligned}$$

and, thus,

$$\begin{aligned}
\int_Y a(y)\vec{e}_j \cdot \nabla w_i(y) \, dy &= -\int_Y a(y)\nabla w_j(y) \cdot \nabla w_i(y) \, dy \\
&= -\int_Y a(y)\nabla w_i(y) \cdot \nabla w_j(y) \, dy \\
&= \int_Y a(y)\vec{e}_i \cdot \nabla w_j(y) \, dy.
\end{aligned}$$

From this, $a_{ij} = a_{ji}$.

(b) We know from the previous calculations

$$a_{ij} = \int_Y a(y)(\nabla w_j(y) + \vec{e}_j) \cdot \vec{e}_i \, dy = \int_Y a(y)(\nabla w_j(y) + \vec{e}_j) \cdot (\nabla w_i(y) + \vec{e}_i) \, dy.$$

Therefore, for real numbers α_i,

$$\alpha_i a_{ij} \alpha_j = \int_Y a(y)\nabla(\alpha_j(w_j(y) + y_j)) \cdot \nabla(\alpha_i(w_i(y) + y_i)) \, dy.$$

Whenever at least one of the α_i differs from zero, the quadratic form built by summing over these terms does not vanish.   Q.E.D.

Remark Though the functions w_j are uniquely defined only up to an additive constant, the numbers a_{ij} are unique.

In this section, we have assumed that the coefficient a is perfectly periodic. As a consequence, the homogenized tensor A is constant. Without any difficulty, one can generalize and consider coefficients $a : \Omega \times Y \to \mathbb{R}$ that depend on the *slow* variable $x \in \Omega$ and are periodic with respect to the *fast* variable $y \in Y$. In such a situation, the homogenized tensor is a function of the space variable $x \in \Omega$. For more details, see section A.3.

1.3.2   Application to Layered Media

Here, we consider a special case of problem (3.1) that was already encountered in section 1.2.3. We assume that the function a in Eq. (3.2) depends only on y_N and not on y_i for $i \neq n$. In this case, the functions w_j can be given explicitly.

Proposition 3.3 *(a) If $a(y_1, \ldots, y_N) = \tilde{a}(y_N)$, then, solutions of Eq. (3.6) are given by*

$$w_N(y_1, \ldots, y_N) = \frac{\int_0^{y_N} \frac{d\eta}{\tilde{a}(\eta)}}{\int_0^1 \frac{d\eta}{\tilde{a}(\eta)}} - y_N$$

and $w_j = 0$ for $j \neq n$.

(b) In this case, the coefficients a_{ij} are given by

$$a_{ij} = \begin{cases} a^*, & i = j = n, \\ \bar{a}\delta_{ij}, & else, \end{cases}$$

with $a^* = \int_0^1 \frac{d\eta}{\tilde{a}(\eta)}$ and $\bar{a} = \int_0^1 \tilde{a}(\eta)d\eta$.

Proof The formulas are easily verified by differentation and integration. Q.E.D.

Remark The physical meaning of part (b) of the proposition is that the effective *conductivity* or *diffusivity* of a layered medium is given by the *arithmetic* mean in directions parallel to the layers but by the *geometric* mean in the direction normal to the layers. This fact is analogous to the rules that apply to electric resistances which are connected in series or parallel; see, e.g., Lurie and Cherkaev [LC86] and Mûrat and Tartar [MT95a]; see also [BS89].

1.3.3 Estimates of the Effective Conductivity Tensor

An important question that arises immediately after deriving the formulas for the homogenized tensor A (Proposition 3.1) is which types of tensors can be obtained. This problem is closely related to estimating the homogenized tensor in terms of the data, i.e., in terms of the diffusion coefficient function a given. Not surprisingly, it turns out that the cases studied in the previous section are extreme in the sense that the arithmetic and the geometric mean give upper and lower bounds for the eigenvalues of A, called the Voigt-Reiss inequality. For this and related results, such as the Hashin-Shtrikman bounds, see [JKO94] section 1.6 and Chapter 6. Because estimates and attainability are of great theoretical and practical relevance, there is quite a large number of papers devoted to these problems, e.g., [KM86], [LC86], [Mil91], and [MT95a].

1.3.4 Media with Obstacles and Diffusion in Perforated Domains

In this section, we assume that the medium in which we study diffusion processes has a periodic arrangement of obstacles. The formal description of this geometry goes along the following lines.

We assume that, in the standard periodicity cell Y, there is a standard obstacle $\mathfrak{S} \subset\subset Y$ with a piecewise smooth boundary Γ. The remainder is denoted by $\mathfrak{P} = Y \setminus \mathfrak{S}$ (see Figs. 1.5 and 8.1). We assume that this standard geometry is repeated periodically all over $\mathbb{R}^N$. The geometric structure within the fixed domain

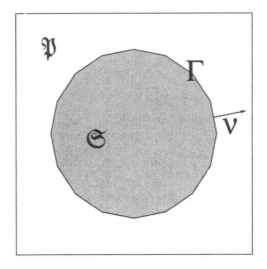

FIGURE 1.5. Standard microcell.

Ω is obtained by intersecting the ε-multiple of this periodic geometry with Ω. The result is denoted by

$$\mathfrak{P}^\varepsilon = \Omega \cap (\varepsilon \mathfrak{P}) \text{ and } \Gamma^\varepsilon = \Omega \cap (\varepsilon \Gamma).$$

To avoid too many technical difficulties, we assume that $\partial\Omega \cap \varepsilon\Gamma = \emptyset$ (see Fig. 1.6). In this situation, we want to study a diffusion process that takes place only in the domain left by the obstacles, namely, in $\mathfrak{P}^\varepsilon$. A simple model is the following boundary-value problem.

$$\begin{cases} \nabla \cdot (a^\varepsilon(x)\nabla u^\varepsilon(x)) + f(x) = 0, & x \in \mathfrak{P}^\varepsilon, \\ \vec{\nu} \cdot a^\varepsilon(x)\nabla u^\varepsilon(x) = 0, & x \in \Gamma^\varepsilon, \\ u(x) = u_D, & x \in \partial\Omega, \end{cases} \qquad (3.10)$$

where $\vec{\nu}$ is the normal vector on Γ^ε. For simplicity, we assume that

$$a^\varepsilon(x) = a(\frac{x}{\varepsilon})$$

with a Y-periodic function $a : \mathfrak{P} \to \mathbb{R}$. The zero flux condition on Γ^ε is used to model the obstacles as impermeable. Problem (3.10) can be looked at as a generalization of the situation from section 1.3. To do this, one has to introduce the characteristic function of $\mathfrak{P}$, namely,

$$\chi(x) = \begin{cases} 1, & x \in \mathfrak{P}, \\ 0, & x \in \mathfrak{S}, \end{cases}$$

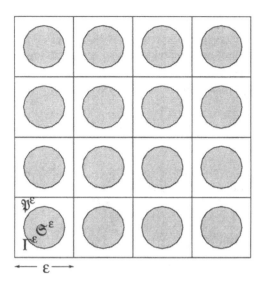

FIGURE 1.6. Periodic medium.

and to define $\chi^{\varepsilon}(x) = \chi(\frac{x}{\varepsilon})$. Then, as the weak form of (3.10), one gets the equation

$$(\chi^{\varepsilon} a^{\varepsilon} \nabla u^{\varepsilon}, \nabla \varphi) = (f, \varphi) \text{ for all } \varphi \in V_0, \qquad (3.11)$$

which has the same formal structure as we had in Eq. (3.3). The well-posedness of the ε-problem follows from standard arguments. Therefore, the same analysis applies as in section 1.3, the difference being that now the *cell problems* are written in strong form to determine Y-periodic functions $w_j(y)$ which satisfy

$$\begin{aligned} \nabla_y \cdot (a(y)\nabla w_j(y)) &= -\nabla_y \cdot (a(y)\vec{e}_j), & y \in \mathfrak{P}, \\ \vec{v} \cdot \nabla_y w_j(y) &= -\vec{v} \cdot \vec{e}_j, & y \in \Gamma, \end{aligned} \qquad (3.12)$$

where $\vec{v}$ is now the normal vector on Γ and, in weak form,

$$(\chi a \nabla (w_j + \vec{e}_j), \nabla \varphi)_Y = 0 \text{ for all } Y\text{-periodic } \varphi.$$

Remark In the case that a is constant on $\mathfrak{P}$, the differential equation for w_j simplifies to

$$\Delta w_j(y) = 0 \text{ for } y \in \mathfrak{P}.$$

In any case, the tensor $A = (a_{ij})_{i,j}$ is defined by

$$a_{ij} = (a(\nabla w_j + \vec{e}_j), \vec{e}_i)_{\mathfrak{P}} \qquad (3.13)$$

analogous to section 1.3. In this way, one gets the following result.

Proposition 3.4 *The homogenization of problem (3.10) is problem (3.9).*

For a rigorous mathematical proof, see [JKO94] Chapter 8.

What makes this problem relatively simple is the fact that we have imposed homogeneous Neumann boundary conditions on the surfaces Γ^ε of the obstacles. The nonhomogeneous case is treated in [CD88].

1.4   Formal Derivation of Darcy's Law

In this section we will give only a formal derivation of Darcy's law. Rigorous proofs and generalizations are given in Chapters 3 and 4.

We start from the following problem on the pore scale. The geometry is the same as in section 1.3.4:

$$
\begin{cases}
\varepsilon^2 \mu \Delta \vec{v}^\varepsilon(x) = \nabla p^\varepsilon(x), & x \in \mathfrak{P}^\varepsilon, \\
\nabla \cdot \vec{v}^\varepsilon(x) = 0, & x \in \mathfrak{P}^\varepsilon, \\
\vec{v}^\varepsilon(x) = 0, & x \in \Gamma^\varepsilon.
\end{cases}
\tag{4.1}
$$

The reason that ε^2 is chosen to enter into the problem is that we want to scale the velocity field $\vec{v}^\varepsilon$ so that it has a limit $\vec{v}$. Here, we present the formal asymptotics of this problem. We are also going to sketch the proof. For details, refer to section 3.1, the book [SP80], and the paper [Tar80]. As usual, we start from the assumption that there is an asymptotic expansion of the unknown functions in the form,

$$
\vec{v}^\varepsilon(x) = \vec{v}_0(x, y) + \varepsilon \vec{v}_1(x, y) + \varepsilon^2 \vec{v}_2(x, y) \ldots
$$

and

$$
p^\varepsilon(x) = p_0(x, y) + \varepsilon p_1(x, y) + \varepsilon^2 p_2(x, y) \ldots
$$

with Y-periodicity of the coefficient functions w.r.t. the variable y. The following equations result:

$$
\begin{aligned}
\varepsilon^0 \mu \Delta_y \vec{v}_0(x, y) + \varepsilon^1(\ldots) + \ldots & \\
= \varepsilon^{-1} \nabla_y p_0(x, y) + \varepsilon^0 & \left(\nabla_y p_1(x, y) + \nabla_x p_0(x, y) \right) \\
+ \varepsilon^1(\ldots) + \ldots, \; & y \in \mathfrak{P},
\end{aligned}
\tag{4.2}
$$

$$
\begin{aligned}
\varepsilon^{-1} \nabla_y \cdot \vec{v}_0(x, y) + \varepsilon^0 & \left(\nabla_y \cdot \vec{v}_1(x, y) + \nabla_x \cdot \vec{v}_0(x, y) \right) \\
+ \varepsilon^1(\ldots) + \ldots = 0, \; & y \in \mathfrak{P},
\end{aligned}
\tag{4.3}
$$

$$
\varepsilon^0 \vec{v}_0(x, y) + \varepsilon^1 \vec{v}_1(x, y) + \ldots = 0, \; y \in \Gamma.
\tag{4.4}
$$

Comparing the coefficients goes along the following lines. The ε^{-1} term in Eq. (4.2) yields

$$
\nabla_y p_0(x, y) = 0, \; y \in \mathfrak{P},
$$

hence, $p_0(x, y) = p_0(x)$ independently of y. The ε^0 term in Eq. (4.2) yields

$$
\mu \Delta_y \vec{v}_0(x, y) = \nabla_y p_1(x, y) + \nabla_x p_0(x), \; y \in \mathfrak{P},
$$

and the ε^{-1} term in Eq. (4.3) yields

$$\nabla \cdot \vec{v}_0(x, y) = 0, \quad y \in \mathfrak{P}.$$

In the usual way,

$$\nabla_x p_0(x) = \sum_j \vec{e}_j \partial_j p_0(x).$$

The *cell problems* now are to find Y-periodic vector fields $\vec{w}_j(y)$ with components $\vec{w}_{ji}(y)$ that solve the Stokes problems,

$$\boxed{\begin{array}{ll} \Delta_y \vec{w}_j(y) = \nabla_y \pi_j(y) - \vec{e}_j, & y \in \mathfrak{P}, \\ \nabla \cdot \vec{w}_j(y) = 0, & y \in \mathfrak{P}, \\ \vec{w}_j(y) = 0, & y \in \Gamma, \end{array}} \qquad (4.5)$$

where the functions $\pi_j(y)$ are the corresponding Y-periodic pressure fields. Using these cell functions, we easily get

$$\vec{v}_0(x, y) = -\frac{1}{\mu} \sum_j \vec{w}_j(y) \partial_j p_0(x).$$

The averaged vector field is defined by

$$\bar{u}(x) = \int_{\mathfrak{P}} \vec{v}_0(x, y) dy,$$

and its ith component is expressed by

$$\bar{u}_i(x) = -\frac{1}{\mu} \sum_j k_{ij} \partial_{x_j} p_0(x)$$

with

$$\boxed{k_{ij} = \int_{\mathfrak{P}} w_{ji}(y) dy.} \qquad (4.6)$$

If one introduces the tensor $K = (k_{ij})_{i,j}$, one gets the short expression

$$\bar{u}(x) = -\frac{1}{\mu} K \nabla p_0(x)$$

which is *Darcy's law*. It remains to be shown that the velocity field $\bar{u}$ is divergence-free. The term with ε^0 in Eq. (4.3) yields

$$\nabla_y \cdot \vec{v}_1(x, y) + \nabla_x \cdot \vec{v}_0(x, y) = 0, \quad y \in \mathfrak{P}.$$

Integrating this over $\mathfrak{P}$ (it is irrelevant in which direction we take the normal vector $\vec{v}$, because the integrals vanish),

$$\begin{aligned} \nabla_x \cdot \bar{u}(x) &= \int_{\mathfrak{P}} \nabla \cdot \vec{v}_0(x, y) dy = -\int_{\mathfrak{P}} \nabla_y \cdot \vec{v}_1(x, y) dy \\ &= -\int_{\partial \mathfrak{P}} \vec{v} \cdot \vec{v}_1(x, y) d\Gamma(y) \\ &= -\int_{\Gamma} \vec{v} \cdot \vec{v}_1(x, y) d\Gamma(y) - \int_{\partial Y} \vec{v} \cdot \vec{v}_1(x, y) d\Gamma(y). \end{aligned}$$

The boundary integral over Γ is zero due to the term with ε^1 in Eq. (4.4), and the boundary integral over ∂Y is zero due to the Y-periodicity of $\vec{v}_1(x, y)$ with respect to y. Therefore, we have shown the following:

Proposition 4.1 *Homogenization of the Stokes problem (4.1) is given by*

$$\bar{u}(x) = -\frac{1}{\mu} K \nabla p(x), \quad \nabla \cdot \bar{u} = 0. \tag{4.7}$$

Similarly to section 1.3.1, one gets the symmetry and positive definiteness of the tensor K.

Lemma 4.2 *The tensor K is symmetric and positive definite.*

Proof The weak form of the cell problems for the vector fields $\vec{w}_j$ is expressed by

$$((\vec{w}_j, \vec{\varphi}))_{\mathfrak{P}} = (\vec{e}_j, \vec{\varphi})_{\mathfrak{P}}$$

for all Y-periodic vector fields $\vec{\varphi}$ with $\nabla \cdot \vec{\varphi} = 0$ in $\mathfrak{P}$ and $\vec{\varphi} = 0$ on Γ, where we have used the bilinear form

$$((\vec{\varphi}, \vec{\psi}))_{\mathfrak{P}} = \sum_{l,k} (\partial_l \varphi_k, \partial_l \psi_k)_{\mathfrak{P}}.$$

Thus, with $\vec{w}_i$ as a test function,

$$((\vec{w}_j, \vec{w}_i))_{\mathfrak{P}} = (\vec{e}_j, \vec{w}_i)_{\mathfrak{P}} = k_{ji}.$$

With indices j and i interchanged,

$$((\vec{w}_i, \vec{w}_j))_{\mathfrak{P}} = (\vec{e}_i, \vec{w}_j)_{\mathfrak{P}} = k_{ij}.$$

Because the left-hand sides are equal, so are the right-hand sides. Therefore, K is symmetric. If real numbers $\alpha_1, \ldots, \alpha_N$ are given, then,

$$\alpha_i k_{ij} \alpha_j = \alpha_i (\vec{e}_i, \vec{w}_j)_{\mathfrak{P}} \alpha_j = ((\alpha_i \vec{w}_i, \alpha_j \vec{w}_j))_{\mathfrak{P}}.$$

The sum over these terms is positive, whenever one of the α_i differs from zero. Q.E.D.

1.5   Formal Derivation of a Distributed Microstructure Model

Here we use a configuration on the microscale which differs from that in section 1.3.4. We allow flow to take place in the interior of the aggregates. We assume that, in the standard periodicity cell Y, there is a standard aggregate $\mathfrak{B} \subset\subset Y$ with smooth boundary Γ the remainder of which is denoted by $\mathfrak{F} = Y \setminus \mathfrak{B}$. As before, we assume that this standard geometry is repeated periodically all over $\mathbb{R}^N$.

The geometric structure within the fixed domain Ω is obtained by intersecting the ε-multiple of this periodic geometry with Ω. The result is

$$\mathfrak{B}^\varepsilon = \Omega \cap (\varepsilon\mathfrak{B}),$$

$$\mathfrak{F}^\varepsilon = \Omega \cap (\varepsilon\mathfrak{F}),$$

and

$$\Gamma^\varepsilon = \Omega \cap (\varepsilon\Gamma).$$

First, we will study a diffusion process in which two scales for the diffusivity occur. We start from the following initial boundary-value problem:

$$\begin{cases}
\partial_t u^\varepsilon(t, x) = \nabla \cdot (a_{\mathfrak{F}}^\varepsilon(x)\nabla u^\varepsilon(t, x)), & t > 0,\ x \in \mathfrak{F}^\varepsilon, \\
\partial_t U^\varepsilon(t, x) = \varepsilon^2 \nabla \cdot (a_{\mathfrak{B}}^\varepsilon(x)\nabla U^\varepsilon(t, x)), & t > 0,\ x \in \mathfrak{B}^\varepsilon, \\
a_{\mathfrak{F}}^\varepsilon(x)\vec{\nu} \cdot \nabla u^\varepsilon(t, x) = \varepsilon^2 a_{\mathfrak{B}}^\varepsilon(x)\vec{\nu} \cdot \nabla U^\varepsilon(t, x), & t > 0,\ x \in \Gamma^\varepsilon, \\
\varepsilon d^\varepsilon(x)a_{\mathfrak{B}}^\varepsilon(x)\vec{\nu} \cdot \nabla U^\varepsilon(t, x) & \\
\quad = c_{\mathfrak{F}}^\varepsilon(x)u^\varepsilon(t, x) - c_{\mathfrak{B}}^\varepsilon(x)U^\varepsilon(t, x), & t > 0,\ x \in \Gamma^\varepsilon, \\
u^\varepsilon(t, x) = u_D(t, x), & t > 0,\ x \in \partial\Omega, \\
u^\varepsilon(t, x) = u_I(x), & t = 0,\ x \in \Omega, \\
U^\varepsilon(t, x) = U_I(x), & t = 0,\ x \in \Omega,
\end{cases} \qquad (5.1)$$

where we assume that the normal vector $\vec{\nu}$ is directed out of $\mathfrak{B}^\varepsilon$, i.e., into $\mathfrak{F}^\varepsilon$. We assume that the coefficients $a_{\mathfrak{F}}^\varepsilon$, $a_{\mathfrak{B}}^\varepsilon$, and d^ε are oscillating, i.e.,

$$a_{\mathfrak{F}}^\varepsilon(x) = a_{\mathfrak{F}}(\frac{x}{\varepsilon}),\ a_{\mathfrak{B}}^\varepsilon(x) = a_{\mathfrak{B}}(\frac{x}{\varepsilon}),\ c_{\mathfrak{F}}^\varepsilon(x) = c_{\mathfrak{F}}(\frac{x}{\varepsilon}),$$

$$c_{\mathfrak{B}}^\varepsilon(x) = c_{\mathfrak{B}}(\frac{x}{\varepsilon}),\ \text{and}\ d^\varepsilon(x) = d(\frac{x}{\varepsilon}),$$

where the functions $a_{\mathfrak{F}}, a_{\mathfrak{B}}, c_{\mathfrak{F}}, c_{\mathfrak{B}}$, and d are Y-periodic. The parameter function d^ε has been introduced to allow different types of models, namely, for $d \equiv 0$, a *Dirichlet* type transmission condition

$$c_{\mathfrak{F}}^\varepsilon(x)u^\varepsilon(x) = c_{\mathfrak{B}}^\varepsilon(x)U^\varepsilon(x)$$

on the interface Γ^ε, whereas, for $d > 0$, one has a boundary condition of the *Robin* type. The formal asymptotic is not exactly the same as in section 1.3.1, because we have to take care of the factors ε and ε^2 which appear in the model. The starting point is as usual the *ansatz* that the unknowns can be written as a power series in the form

$$u^\varepsilon(t, x) = u_0(t, x, y) + \varepsilon u_1(t, x, y) + \varepsilon^2 u_2(t, x, y)\ldots$$

and

$$U^\varepsilon(t, x) = U_0(t, x, y) + \varepsilon U_1(t, x, y) + \varepsilon^2 U_2(t, x, y)\ldots$$

with the usual assumption of Y-periodicity of the coefficient functions. For simplicity of notation, we drop the variable t in the following. The differential equation for u^ε in $\mathfrak{F}^\varepsilon$ yields

$$
\begin{aligned}
\varepsilon^0 \partial_t u_0(x, y) + \varepsilon^1 \ldots &= \varepsilon^{-2} \nabla_y \cdot (a_{\mathfrak{F}}(y) \nabla_y u_0(x, y)) \\
&+ \varepsilon^{-1} \big(\nabla_y \cdot (a_{\mathfrak{F}}(y) \nabla_y u_1(x, y)) + \nabla_y \cdot (a_{\mathfrak{F}}(y) \nabla_x u_0(x, y)) \\
&+ a_{\mathfrak{F}}(y) \nabla_x \cdot \nabla_y u_0(x, y) \big) \\
&+ \varepsilon^0 \big(\nabla_y \cdot (a_{\mathfrak{F}}(y) \nabla_y u_2(x, y) + a_{\mathfrak{F}}(y) \nabla_x u_1(x, y)) \\
&+ a_{\mathfrak{F}}(y) \nabla_x \cdot \nabla_y u_1(x, y) + a_{\mathfrak{F}}(y) \nabla_x \cdot \nabla_x u_0(x, y) \big) \\
&+ \varepsilon^1 (\ldots) + \ldots, \quad y \in \mathfrak{F}.
\end{aligned}
\tag{5.2}
$$

The differential equation for U^ε in $\mathfrak{B}^\varepsilon$ yields

$$
\varepsilon^0 \partial_t U_0(x, y) + \varepsilon^1 \ldots = \varepsilon^0 \nabla_y \cdot (a_{\mathfrak{B}}(y) \nabla_y U_0(x, y)) + \varepsilon^1 \ldots, \quad y \in \mathfrak{B}.
\tag{5.3}
$$

The two conditions on the interface Γ^ε yield

$$
\begin{aligned}
\varepsilon^{-1} a_{\mathfrak{F}}(y) \vec{v} &\cdot \nabla_y u_0(x) \\
&+ \varepsilon^0 a_{\mathfrak{F}}(y) \vec{v} \cdot \big(\nabla_y u_1(x) + \nabla_x u_0(x) \big) \\
&+ \varepsilon^1 \big(a_{\mathfrak{F}}(y) \vec{v} \cdot (\nabla_y u_2(x) + \nabla_x u_1(x)) \big) \\
&+ \varepsilon^2 \ldots = \varepsilon^1 a_{\mathfrak{B}}(y) \vec{v} \cdot \nabla_y U_0(x, y) + \varepsilon^2 \ldots, \quad y \in \Gamma,
\end{aligned}
\tag{5.4}
$$

and

$$
\begin{aligned}
\varepsilon^0 d(y) a_{\mathfrak{B}}(y) \vec{v} &\cdot \nabla_y U_0(x, y) + \varepsilon^1 \ldots \\
&= \varepsilon^0 \big(c_{\mathfrak{F}}(y) u_0(x, y) - c_{\mathfrak{B}}(y) U_0(x, y) \big) + \varepsilon^1 \ldots, \quad y \in \Gamma.
\end{aligned}
\tag{5.5}
$$

Now we have to compare the coefficients of the different ε powers in these equations. The term with ε^{-2} in Eq. (5.2) yields

$$
\nabla_y \cdot (a_{\mathfrak{F}}(y) \nabla_y u_0(x, y)) = 0 \text{ for } y \in \mathfrak{F}.
$$

Because $u_0(x, y)$ is Y-periodic in the variable y, $u_0(x, y) = u_0(x)$ is a function of x alone, independently of y. Using this and the term with ε^{-1} in Eq. (5.2),

$$
\nabla_y \cdot (a_{\mathfrak{F}}(y) \nabla_y u_1(x, y)) = -\nabla_y \cdot (a_{\mathfrak{F}}(y) \nabla_x u_0(x)) \text{ for } y \in \mathfrak{F}.
$$

The function $u_1(x, y)$ is expressed in terms of the function $u_0(x)$ as in section 1.3.1. Using the functions $w_j(y)$ from (3.12),

$$
u_1(x, y) = \sum_{j=1}^{N} w_j(y) \partial_{x_j} u_0(x) + u_1(x),
$$

where $u_1(x)$ is independent of y. We proceed further and look at the term with ε^0 in Eq. (5.2):

$$
\begin{aligned}
\partial_t u_0(x, y) &= \nabla_y \cdot \big(a_{\mathfrak{F}}(y) \nabla_y u_2(x, y) + a_{\mathfrak{F}}(y) \nabla_x u_1(x, y) \big) \\
&+ a_{\mathfrak{F}}(y) \nabla_x \cdot \nabla_y u_1(x, y) + a_{\mathfrak{F}}(y) \nabla_x \cdot \nabla_x u_0(x) \text{ for } y \in \mathfrak{F}.
\end{aligned}
$$

Integrating this identity over $\mathfrak{F}$,

$$|\mathfrak{F}|\partial_t u_0(x) = \int_{\mathfrak{F}} \nabla_y \cdot \left(a_{\mathfrak{F}}(y)\nabla_y u_2(x, y) + a_{\mathfrak{F}}(y)\nabla_x u_1(x, y)\right) dy$$

$$+ \int_{\mathfrak{F}} a_{\mathfrak{F}}(y)\nabla_x \cdot \nabla_y u_1(x, y)dy + \int_{\mathfrak{F}} a_{\mathfrak{F}}(y)dy \, \Delta_x u_0(x). \tag{5.6}$$

Integrating the first integral on the right side by parts,

$$\int_{\mathfrak{F}} \nabla_y \cdot \left(a_{\mathfrak{F}}(y)\nabla_y u_2(x, y) + a_{\mathfrak{F}}(y)\nabla_x u_1(x, y)\right) dy$$

$$= - \int_{\partial\mathfrak{F}} a_{\mathfrak{F}}(y)\vec{v} \cdot \left(\nabla_y u_2(x, y) + \nabla_x u_1(x, y)\right) d\Gamma(y).$$

We have $\partial\mathfrak{F} = \partial Y + \Gamma$, and the boundary integral over ∂Y vanishes because of the Y-periodicity of the functions $u_2(x, y)$ and $u_1(x, y)$. For the integral over Γ we look at the term ε^1 in (5.4) and get

$$\int_{\Gamma} a_{\mathfrak{F}}(y)\vec{v} \cdot \left(\nabla_y u_2(x, y) + \nabla_x u_1(x, y)\right) d\Gamma(y) = \int_{\Gamma} a_{\mathfrak{B}}(y)\vec{v} \cdot \nabla_y U_0(x, y) \, d\Gamma(y).$$

For the second term in Eq. (5.6), in the well-known way,

$$\nabla_x \cdot \nabla_y u_1(x, y) = \sum_{i,j=1}^{N} \partial_{y_i} w_j(y)\partial_{x_i x_j} u_0(x).$$

Therefore, from Eq. (5.6),

$$|\mathfrak{F}|\partial_t u_0(x, y) + \int_{\Gamma} a_{\mathfrak{B}}(y)\vec{v} \cdot \nabla_y U_0(x, y) \, d\Gamma(y)$$

$$= \sum_{i,j=1}^{N} \int_{\mathfrak{F}} a_{\mathfrak{F}}(y)\partial_{y_i} w_j(y)dy \, \partial_{x_i x_j} u_0(x) + \int_{\mathfrak{F}} a_{\mathfrak{F}}(y)dy \, \Delta_x u_0(x).$$

The term with ε^0 in Eq. (5.3) yields

$$\partial_t U_0(x, y) = \nabla_y \cdot (a_{\mathfrak{B}}(y)\nabla_y U_0(x, y)), \quad y \in \mathfrak{B},$$

and the term with ε^1 in Eq. (5.5) yields

$$d(y)a_{\mathfrak{B}}(y)\vec{v} \cdot \nabla_y U_0(x, y) = c_{\mathfrak{F}}(y)u_0(x) - c_{\mathfrak{B}}(y)U_0(x, y), \quad y \in \Gamma.$$

We have shown the following result. For clarification, we write the time variable t again.

Proposition 5.1 *The homogenization of the initial boundary value problem (5.1) is given by*

$$
\begin{cases}
|\mathfrak{F}|\partial_t u(t, x) + Q(t, x) = \nabla \cdot (A\nabla u(t, x)), & t > 0, \ x \in \Omega, \\
Q(t, x) = \int_\Gamma a_\mathfrak{B}(y)\vec{v} \cdot \nabla_y U(t, x, y) \, d\Gamma(y), & t > 0, \ x \in \Omega, \\
\partial_t U(t, x, y) = \nabla_y \cdot (a_\mathfrak{B}(y)\nabla_y U(t, x, y)), & t > 0, \ x \in \Omega, \ y \in \mathfrak{B}, \\
d(y)a_\mathfrak{B}(y)\vec{v} \cdot \nabla_y U(t, x, y) \\
\quad = c_\mathfrak{F}(y)u(t, x) - c_\mathfrak{B}(y)U(t, x, y), & t > 0, \ x \in \Omega, \ y \in \Gamma, \\
u(t, x) = u_D(t, x), & t > 0, \ x \in \partial\Omega, \\
u(t, x) = u_I(x), & t = 0, \ x \in \Omega, \\
U(t, x) = U_I(x), & t = 0, \ x \in \Omega.
\end{cases}
\tag{5.7}
$$

The reader should notice that the system of Proposition 5.1 is a coupled system of a *global equation* for the global, or fast, variable u and a *local equation* for the local, or slow, variable U. The coupling is in both directions: The global equation has the sink term Q calculated from the local variable U, whereas the local equation has the boundary condition in which the global variable appears. At each point $x \in \Omega$, an extra local problem has to be considered, namely, the problem for the local variable U in which the local geometry enters explicitly. Therefore, a system of this type is called *distributed microstructure model*. They appear under different circumstances in the following chapters. In section 3.4, single-phase flow in a fractured medium is considered, in section 6.2.3 chromatography is studied, and in 6.3.1, miscible displacement influenced by mobile and immobile water is dealt with. Distributed microstructure models are studied in more detail in Chapter 9. Another simple example leading to a distributed microstructure model is described in section 9.3. Computational results for such models are presented in Chapter 10.

1.6   Remarks on Networks of Resistors, Capillary Tubes, and Cracks

In subsection 1.2.2, it was pointed out that one may look at a diffusion equation as the homogenized limit of a system of difference equations modeling electric currents in a network of resistors. Formally, the same systems of equations arise when one studies networks of capillary tubes so narrow that Poiseuille flow can be assumed. In that case, the flow rate is proportional to the pressure drop in the tubes, and, for the junctions, a mass balance similar to Kirchhoff's rule is adequate.

A situation which has something in common with the two previous is a system of cracks in rocks. In narrow cracks, flow of a liquid may be modeled as Hele Shaw (i.e., flow between two parallel plates), and, if the rock is practically impervious, one arrives at modeling diffusion in the crack system. Mathematically, in such a case, an appropriate formalism starts from a system of 2-D surfaces in 3-D space on which surface-like diffusion takes place. The diffusion equation on the surface Γ reads

$$
\nabla^\Gamma \cdot (a^\Gamma \nabla^\Gamma u) = 0
$$

where $\nabla^\Gamma\cdot$ denotes the surface divergence and $\nabla^\Gamma u$ is the surface gradient of the function u. Where the surfaces intersect, one has to use a law which is the 2-D analog of the 1-D Kirchhoff law, namely, conservation of mass (or charge in the case of electric resistors). If a point on an intersection curve C is chosen, if $\vec{v}_i$ are vectors perpendicular to C and tangential to the surfaces in all possible directions, and if $\vec{q}_i = -a_i^\Gamma \nabla^\Gamma u_i$ are the corresponding flow rates, then

$$\sum_i \vec{v}_i \cdot \vec{q}_i = 0.$$

Homogenization can also deal with these cases. If large networks or crack systems are periodically arranged in space, the notion of two-scale convergence (see section A.3) is useful. Basically, in a first step, the approach consists of allowing the homogenized limit u of the potential u^ε to be a function of two space variables, namely, the slow variable x and the fast variable y. Now, $u(x, y)$ is defined only for $y \in \Gamma$ where Γ is the standard tube or surface $\Gamma \subset Y$ in the standard periodicity cell Y. In a second step, cell solutions are being used that allow integration over the fast variable y which results in a diffusion equation written only in terms of the slow variable x. The effective diffusion tensor – very similarly to the situations of section 1.3 – depends on the geometry and the diffusion coefficients in the standard cell Y, i.e., on Γ. The mathematical tools needed for this procedure are worked out in [ADH95], and applications can be found in [Hor95].

1.6.1 Monte-Carlo Simulations

Resistor networks have also been studied by stochastics. One approach is to assume that, in a network of the kind shown in Fig. 1.3, the values of the resistances are independently chosen either ∞ or 1 with probability p and $1 - p$, respectively. This leads to the problem of percolation studied in detail in Chapter 2. Another interesting problem is to choose finite resistances according to a prescribed probability distribution and to find the effective resistance of the full network. For stochastically independent and log-normally distributed resistances R, the results from Monte-Carlo computations are shown in Figs. 1.7 and 1.8. Underlying Fig. 1.7 is a network of 20×22 resistors. The horizontal axis corresponds to the standard deviation σ of the logarithm of the individual resistances R. The vertical axis corresponds to the ratio $\frac{R_{eff}}{R_0}$ of the average R_{eff} obtained from 10,000 Monte-Carlo simulations, and the network resistance is R_0 for $\sigma = 0$, i.e., for the case where all resistances are held at the same constant. The figure shows a strong dependence on the standard deviation σ. For very large σ, this problem approaches the percolation problem with a probability close to the percolation threshold (see section 2.1).

In Fig. 1.8 the dependence of the effective resistance on the size of the network is shown. The horizontal axis corresponds to the number of resistances in each space dimension, and the vertical axis is the same as before. The standard deviation $\sigma = 2$ is held fixed. The figure shows that the variations of the resistances become more and more insignificant for large networks. In other words, large and small

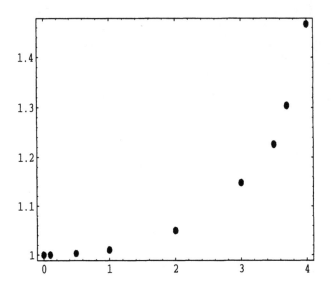

FIGURE 1.7. Random resistor networks: Dependence of the effective resistance on the standard deviation.

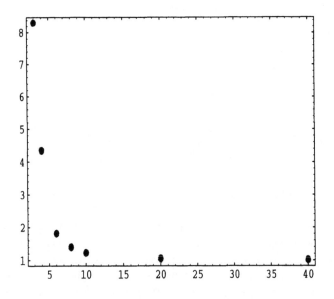

FIGURE 1.8. Random resistor networks: Dependence of the effective resistance on the size of the network.

values of R balance out. This observation agrees fully with theoretical results from the stochastic homogenization of diffusion in random media (see, e.g., [JKO94] section 7.3, and see section 5.2.5 for flow in random porous media).

2

Percolation Models for Porous Media

Kenneth M. Golden[1]

Recent progress in understanding the effective transport properties of percolation models for porous and conducting random media is reviewed. Both lattice and continuum models are studied. First, we consider the random flow network in $\mathbb{Z}^N$, where the pipes of the network are open with probability p and closed with probability $1 - p$. Near the percolation threshold p_c, the effective permeability $\kappa^*(p)$ exhibits the scaling behavior $\kappa^*(p) \sim (p - p_c)^e$, $p \to p_c^+$, where e is the permeability critical exponent. In the limit of low Reynolds number flow, this model is equivalent to a corresponding random resistor network. Here we discuss recent results for the resistor network problem which yield the inequalities $1 \le e \le 2$, $N = 2, 3$ and $2 \le e \le 3$, $N \ge 4$, assuming a hierarchical node-link-blob (NLB) structure for the backbone near p_c. The upper bound $t = 2$ in $N = 3$ virtually coincides with a number of recent numerical estimates. Secondly, we consider problems of transport in porous and conducting media with broad distribution in the local properties, which are often encountered. Here we discuss a continuum percolation model for such media, which is exactly solvable for the effective transport properties in the high disorder limit. The model represents such systems as fluid flowing through consolidated granular media and fractured rocks, as well as electrical conduction in matrix-particle composites near critical volume fractions. Moreover, the results for the model rigorously establish the widely used Ambegaokar, Halperin, and Langer critical path analysis [AHL71].

[1]**Acknowledgements:** K. Golden would like to specially thank S. Kozlov for many wonderful conversations on the transport properties of percolation models. He would also like to thank M. Aizenman, L. Berlyand, and P. Doyle for helpful conversations concerning the work reviewed here. L. Schwartz, M. Sahimi, VCH Publishers, and the Amercian Physical Society are kindly acknowledged for use of the figures that appear here. Finally, K. Golden gratefully acknowledges support from NSF Grant DMS-9307324 and ONR Grant N000149310141.

2.1   Fundamentals of Percolation Theory

Porous media represent an interesting class of materials which exhibit a wide range of microstructures. For a given microstructure, the central problem of single-phase flow through the medium is to find its effective permeability κ^* [Sah95, Dul92, Dag89, Mat67, BB93, Has59, Pra61, BM85, JKS86, AT91]. The microstructure can be characterized in many ways, the simplest being the pore volume fraction, or porosity ϕ. In general, however, the macroscopic transport properties of porous media depend in a complex way on the details of the pore structure, not simply its volume fraction. Due to the wide range of relevant microstructures, there have been many approaches to estimating the effective permeability κ^* of porous media. One widespread approach is to treat the porous medium as a network of open pores of varying size and "throats" of varying cross section which connect the pores [Kop82, Fat56, Sah95, Dul92, BB93]. In the limit of low Reynolds number flow, this "pipe" network becomes equivalent to a linear electrical network. In each pipe, the fluid flux is linearly related to the pressure drop across the pipe, which is a local version of Darcy's law. For electrical conduction, the current in Ohm's law is proportional to the potential drop across the resistor. In this case, the problem of finding the effective permeability κ^* of the network is mathematically equivalent to finding the effective conductivity σ^* of the corresponding resistor network.

The simplest network model which, nevertheless, exhibits complex macroscopic behavior is where the pipes form the standard bond lattice in $\mathbb{Z}^N$ and they are either open or closed with probabilities p or $1 - p$, respectively. This network is based on the classical percolation model of Broadbent and Hammersly [BH57], one of the original inspirations for which was the flow of gas through the porous substance inside a miner's gas mask. In this model, we assign a local permeability, or fluid conductance $\kappa(x)$, to the bonds, where $\kappa(x) = 1$ or $\varepsilon \geq 0$ with probabilitities p or $1 - p$. When $\varepsilon = 0$, the effective permeability $\kappa^*(p) = 0$ for $p \leq p_c$, yet $\kappa^*(p) > 0$ for $p > p_c$, where p_c is the percolation threshold, which equals $1/2$ for $N = 2$. Furthermore, it is generally believed (though not rigorously proven), that $\kappa^*(p)$ exhibits critical scaling behavior near the percolation threshold [FHS87, BB93, Sah95],

$$\kappa^*(p) \sim (p - p_c)^e, \qquad p \to p_c^+, \qquad (1.1)$$

where e is the permeablity critical exponent. Due to the above equivalence between the fluid and electrical conduction problems, we can just as well consider the random resistor network [Kir71, SA92, GK84, Gri89, Kes82, CC86], where the electrical conductivity $\sigma(x)$ of the bonds is defined similarly to $\kappa(x)$ above. Then, the effective conductivity $\sigma^*(p)$ is believed to behave as

$$\sigma^*(p) \sim (p - p_c)^t, \qquad p \to p_c^+, \qquad (1.2)$$

where t is the conductivity critical exponent. At the percolation threshold, the system undergoes a phase transition from an insulator to a conductor. Analogously, the fluid sytem undergoes a phase transition from a medium, through which no fluid can pass, to a porous or permeable medium. Furthermore, for such network models, the critical exponents for permeability and electrical conductivity are the same: [FHS87, BB93]

$$e = t \qquad (1.3)$$

(although this relation is not generally true for continuum systems [FHS87]). These network models have been used extensively to study problems in porous media [Sah95] and electrical transport [SA92].

In the first part of this chapter, we will review recent progress in understanding the behavior of $\kappa^*(p)$ or $\sigma^*(p)$ near the percolation threshold [Gol90, Gol92]. In particular, assuming a hierarchical, node-link-blob (NLB) structure [Sta77, Con82, BB93] for the conducting backbone of the percolation cluster near p_c (and certain technical assumptions), we have rigorously shown that the permeability exponent e or the conductivity exponent t satisfies the following inequalities:

$$1 \le e \le 2, \qquad N = 2, 3, \qquad (1.4)$$
$$2 \le e \le 3, \qquad N \ge 4.$$

Our approach was motivated by the simple observation that, in numerical simulations [Kir71, NvVE86, WL74], the graph of $\sigma^*(p)$ for bond or site models in $N \ge 2$ is always convex near p_c. We, then, analyze the asymptotic behavior of $\frac{d^2\sigma^*}{dp^2}$ near the percolation threshold p_c and investigate the consequences for the critical exponent t or e under the above mentioned assumptions. In particular, in the key assumption about backbone structure, our NLB model contains both singly and multiply connected bonds, has "loops" on arbitrarily many length scales in a self-similar fashion, and incorporates the few features rigorously known [Con82, CC86] about the backbone on a macroscopic scale.

Our results for $N = 3$ are particularly intriguing. First, the inequality $t \le 2$ excludes roughly one fourth of published numerical estimates of t for $N = 3$, which have ranged from 1.5 to 2.36. Furthermore, this inequality is based on an exact calculation of $t = 2$ for one particular member of our class of hierarchical NLB backbones, which provides an upper bound on t for the full class. In view of this result, it is quite striking that Gingold and Lobb [GL90] have obtained the estimate $t = 2.003 \pm 0.047$ for $N = 3$ from simulation on lattices up to $(80)^3$ and Adler et al. [AMA+90] have obtained $t = 2.02 \pm 0.05$ from a 13th order series expansion. Moreover, in [BB93] by Monte Carlo simulation, it is found that the permeability exponent e has a "universal" value of 2.0 under certain conditions of the distribution of local conductances. In addition, our inequality is compatible with the results of an $\varepsilon = 6 - N$ expansion [HKL84] and the general view that "roughly $t = 2$" [SA92]. The recent numerical results, in conjunction with our work, suggest the possibility that $t = e = 2$ is an *exact* result for $N = 3$. To our knowledge, the results in [Gol90, Gol92] are the only ones which relate t or e directly and naturally to the number 2, rather than to other (unknown) critical exponents of percolation theory.

Before discussing the second part of this chapter, we will mention other recent results in homogenization that may be of interest to researchers in porous media. In [Gol89] we began investigating the behavior of $\sigma^*(p)$ in the complex $p-$plane and were able to establish rigorously that, for $\varepsilon > 0$, $\sigma^*(p)$ is analytic in p in some open neighborhood containing $[0, 1]$. Investigating how this analyticity is lost and how singular behavior develops at criticality, as $\varepsilon \to 0$ or in the infinite volume limit, has led to a number of works, including [Gol91] and [BG90]. The investigation culminated in [Gol95b], where this question has been addressed in detail, by finding a direct correspondence between the random resistor network (or continuum analogs) and the Ising model for a ferromagnet, where there is a well-developed Lee–Yang theory [YL52, LY52, Gri72] for understanding the onset of critical behavior. The correspondence is established by introducing a partition function and free energy for conduction problems, which have equivalent representations to those for the Ising ferromagnet. Through this correspondence, the percolation threshold p_c for transport in the random flow or resistor network is characterized as an accumulation point of zeros of the partition function in the complex p-plane.

On a more practical note, we also mention that the author has been involved in studying the electromagnetic properties of sea ice, which is a composite of pure ice containing random brine and air inclusions. It is very interesting to note that sea ice exhibits a percolation threshold at a critical temperature $T_c \approx -5$ °C, above which the brine inclusions coalesce and the sea ice becomes porous, allowing transport of sea water and brine through the ice. The implications of this transition on homogenized electrical transport coefficients is discussed in [Gol95a] and [Gol94a]. Furthermore, we report briefly in [LG] on observations we made of transport and percolation processes in sea ice and their effect on microwave backscatter from sea ice in the Weddell Sea, Antarctica during July and August of 1994.

In the second part of this chapter, we will continue to review recent results in homogenization for percolation models which impact on the study of porous media. The work we now describe began in discussions with Serguei Kozlov in Moscow in 1991 and slowly evolved into [GK]. In many problems of technological importance, one meets systems which display wide distribution in the local properties characterizing the system. For example, in porous media, there is often a wide range of pore and neck sizes through which the fluid must flow [Sah95, Dul92, Fat56, Kop82, FHS87, Dag89, Mat67]. In associated network models, one must consider a broad distribution in the local fluid conductances in the bonds. As another example, in resistor network models of hoppping conduction in amorphous semiconductors, the bonds of the network are assigned a wide range of conductivities [AHL71, Sha77, Kir83]. A powerful idea which has been widely used [Sah95] to estimate the effective properties of such systems is the *critical path analysis*, first introduced by Ambegaokar, Halperin and Langer (AHL) [AHL71]. They proposed that transport in a medium with a broad range of local conductances g is dominated by a critical value g_c, which is the smallest conductance such that the set $\{g \mid g > g_c\}$ percolates or forms a connected cluster which spans the sample. This cluster is called the critical path. Then, the problem of estimating transport in a highly disordered medium with a wide range of local conductances is reduced to a percolation problem with threshold value g_c. The critical path analysis was further developed in the context of amorphous semiconductors in [Sha77] and [Kir83], and its accuracy was numerically confirmed for various conductance distributions in [BOJG86]. The AHL idea has also been used to model the permeability and electrical conductivity of porous media, such as sandstones, obtained through mercury injection [KT86, KT87, BJ87, Dou89], and fractured rocks with a broad distribution of fracture apertures [CGR87]. Recently, it has been applied to porous media saturated with a non-Newtonian fluid [Sah93].

Although the AHL idea has been used with substantial success, there has been little analysis of their fundamental observation that the critical conductance g_c dominates the effective behavior. In [GK], we introduce a continuum percolation model of conducting and porous random media which is exactly solvable in the high disorder limit, and the result we obtain for the effective conductivity or permeability rigorously establishes the AHL principle for this model. Furthermore, our model closely represents an important class of porous materials, including consolidated granular media and some fractured rocks, where the easiest flow paths or channels exhibit complex random topology, similar to a Voronoi network (see Fig. 2.2) [Sah95, Ker83, RS85, JHSD84, FHS87].

We obtain our model as a "long-range" generalization of the random checkerboard in $\mathbb{R}^2$ [Dyk71, SK82, Kel87, Mol91, Koz89, BG94, Gol94b], where the squares are assigned conductivities 1 with probability p and $\varepsilon > 0$ with probability $1 - p$. The random checkerboard has been used to model conducting materials which exhibit critical behavior too rich to be accurately handled by random resistor networks, such as graphite (conducting) particles embedded in a polymeric (insulating) matrix. For example, the presence of both corner and edge connections between squares produces two percolation thresholds with distinct asymptotic behavior of the effective conductivity as $\varepsilon \to 0$ (or $\varepsilon \to \infty$) in the different regimes of p separated by these thresholds [SK82, Mol91, Koz89, BG94]. In our model, we allow arbitrarily long-range connections between squares, which leads to infinitely many thresholds, and rather complex asymptotic behavior, which we can, nevertheless, obtain exactly. Our analysis is based on the variational formulation of effective properties, which allows us to obtain bounds required for the asymptotics by constructing trial fields which exploit relevant percolation structures [Koz89, BG94, Gol94b]. A key component of our percolation analysis is to connect our model to a Poisson distribution of discs in the plane.

Whereas the standard checkerboard model allows for only two types of connections between conducting particles (squares), our generalization allows for arbitrarily many. This leads us to expect that our model will serve as a good representation for high contrast matrix-particle composites, particularly in the regime where the conducting particles "percolate," yet are of low enough volume fraction so that the conducting phase is only "partially connected" [Gol94b] where there is a broad distribution of connection types between the conducting particles. In such regimes, the effective conductivity has been observed to vary over many orders of magnitude [MBN90] over small volume fraction ranges, as does our model. More precisely, in the high disorder limit, the paths of easiest current flow in our model form a (continuum) Voronoi network with a broad distribution of "bond" conductivities. This network can be identified with the one obtained from joining the centers of "touching" conducting particles. The quality of the connection between the particles determines the conductivity of the path joining them. Our model can be modified to handle such problems, and our variational analysis, which is quite general, can be suitably applied to this and other related problems.

2.2   Exponent Inequalities for Random Flow and Resistor Networks

The principal results of our investigation and the assumptions under which they are obtained are as follows. We will formulate the problem for the effective conductivity $\sigma^*(p)$, but, of course, everything carries over to the effective permeabitlity $\kappa^*(p)$. This connection will be examined in more detail below. First, the most serious assumption is that the conducting backbone near p_c has a hierarchical, self-similar node-link-blob (NLB) structure, as described in the Introduction. We further make some technical assumptions about $\sigma^*(p)$: it obeys the above scaling law (1.2) near p_c, has at least three derivatives for all $p > p_c$, and $\frac{d^2\sigma^*}{dp^2} + \frac{d\sigma^*}{dp} > 0$ at $p = 1$, which we have verified numerically. Under these assumptions, we prove exact asymptotics for $\frac{d^2\sigma^*}{dp^2}$ as $p \to p_c^+$. The proof employs a novel technique whereby $\frac{d^2\sigma^*}{dp^2}$, for the NLB model with $\varepsilon = 0$ and p near p_c, is computed using perturbation theory for $\sigma^*(p)$ (for two-and three-component resistor lattices) around $p = 1$, with a sequence of ε's converging to 1 as one goes deeper in the hierarchy. Our asymptotics yield not only convexity near p_c, which implies $t \geq 1$, but delineate in which dimensions $\frac{d^2\sigma^*}{dp^2} \to 0, +\infty$, or a positive constant as $p \to p_c^+$. Combining this information with the scaling law $\frac{d^2\sigma^*}{dp^2} \sim (p - p_c)^{t-2}$ yields the inequalities $1 \leq t \leq 2$ for $N = 2, 3$ and $2 \leq t \leq 3$ for $N \geq 4$. The inequality $t \leq 3$ for $N \geq 4$ is obtained by applying a similar analysis to $\frac{d^3\sigma^*}{dp^3}$ for the simpler node-link model, and can be viewed as a mean field bound, because it is believed that $t = 3$ for $N \geq 6$. We stress that the convexity and inequalities are not rigorous for the actual backbone near p_c for the original lattice, but *are* rigorous for the NLB model of the backbone, under the above technical assumptions.

Before we begin, we refer the reader to [Gol92]. In addition to containing the mathematical details of the results discussed here, we obtain, there, numerical and rigorous results concerning the regimes in ε and p of convexity of $\sigma^*(p)$ for bond and site models, the principal rigorous results being that, for the $N = 2$ bond problem, although $\sigma^*(p)$ cannot be convex for all p, when $\varepsilon = 0$, it is convex for every $\varepsilon > 0$ near $p_c = 1/2$.

We now formulate the bond conductivity problem for $\mathbb{Z}^N$, where, for simplicity, we begin with $N = 2$. Take an $L \times L$ sample G_L of the bond lattice with $M(\sim NL^N)$ bonds. Assigned to G_L are M independent random variables c_i, $1 \leq i \leq M$, the bond conductivities, which take the values 1 with probability p and $\varepsilon \geq 0$ with probability $1 - p$. We attach perfectly conducting bus bars to two opposite edges and let $\sigma_L(p)$ be the effective conductance of this network, averaged over realizations of the bond conductivities. This conductance is just the total current that flows through the network when there is a unit average potential difference between the bus bars. The potential at any lattice site in $\mathbb{Z}^N$ is determined by Kirchoff's laws. For $N \geq 1$, the bulk conductivity of the lattice is defined as

$$\sigma^*(p) = \lim_{L\to\infty} L^{2-N}\sigma_L(p). \tag{2.1}$$

For $\varepsilon > 0$, the infinite volume limit in (2.1) has been shown to exist [Koz78, PV82, GP83, Kün83], and for $\varepsilon = 0$ the existence of σ^* has been proven in the continuum [Zhi89].

At this point, it is useful to investigate the relationship between the conductivity and flow problems, particularly, in view of the fact [CC86] that, for effective permeability, we must replace (2.1) with

$$\kappa^*(p) = \lim_{L\to\infty} L^{1-N}\kappa_L(p), \tag{2.2}$$

where $\kappa_L(p)$ replaces $\sigma_L(p)$, and is the total fluid current in the network, which we now discuss. To elucidate the connection, we consider the following simple model [Dul92] in $N = 3$ of a cube of side L filled with n parallel cylindrical capillary tubes of length L and radius R, evenly spaced throughout the cube. Then in the Darcy regime, the total flow rate is given by

$$Q = n \frac{\pi R^4}{8\eta} \left(\frac{\Delta P}{L}\right), \tag{2.3}$$

where η is the fluid viscosity and ΔP is the pressure drop across the cube (or across each capilllary). If we identify $k = \frac{\pi R^4}{8}$ as the permeability of each capillary, then, because $n = L^2$, we can rewrite (2.3) as

$$\frac{Q}{L^2} = \frac{k}{\eta} \left(\frac{\Delta P}{L}\right), \tag{2.4}$$

where the left-hand side is the fluid current density. Now, we can associate this problem with the effective conductance of n parallel "networks" of L conductances in series, each with conductance

$$g = \frac{k}{\eta L}, \tag{2.5}$$

which is just the effective conductance of L bonds in series, each of conductance $\frac{k}{\eta}$. This relation accounts for the different scalings in (2.1) and (2.2).

Now, the calculation of $\frac{d^2\sigma^*}{dp^2}$ requires the following definition. For any graph B with bonds b_i of unit conductivity, define

$$\delta^2\sigma(B) = \sum_{b_i \neq b_j} \left[\sigma_{ij}(1, 1) + \sigma_{ij}(0, 0) - \sigma_{ij}(1, 0) - \sigma_{ij}(0, 1)\right], \tag{2.6}$$

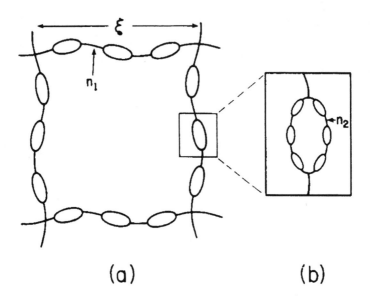

FIGURE 2.1. Node–link–blob model of the conducting backbone near p_c. In (a), the nodes are a correlation length ξ apart and are connected by necklaces of beads (blobs) and strings (links) with n_1 bonds connecting two beads. The beads have a self-similar structure, as shown in (b), with n_2 bonds connecting two beads.

where, in (2.6), $\sigma_{ij}(1, 1) = \sigma(B)$, the conductivity of B measured between the two bus bars, $\sigma_{ij}(0, 0)$ is the conductivity of B with b_i and b_j removed, and so on. This expression represents the discrete second derivative of σ with respect to p, as follows. Let G be the lattice in $N \geq 2$ with bond conductivities 1 and 0 and bulk conductivity function $\sigma^*(p)$. If $B = B(p)$ is a realization of occupied bonds of G at probability p, then [Gol92]

$$p^2 \frac{d^2\sigma^*}{dp^2} = \delta^2\sigma^*(B(p)), \tag{2.7}$$

where $\delta^2\sigma^*$ is the scaled infinite volume limit of (2.6), and the right-hand side in (2.7) is appropriately averaged. (We are assuming here that $\sigma^*(p)$ is twice differentiable for $p > p_c$ when $\varepsilon = 0$.) In (2.6), note that dangling bonds do not contribute, so that one may think of $B(p)$ as a realization of the backbone at bond fraction p. For clarity, note that, at $p = 1$, $B(p) = G$. We remark that analysis of simple graphs, typically, shows that positive contributions to (2.6) arise from series pairs, whereas negative contributions arise from pairs in parallel.

The idea now is to replace an actual backbone graph $B(p)$ for p near p_c by a node-link-blob (NLB) graph A, which is based on the work of Stanley [Sta77] and Coniglio [Con82]. This graph is a "super-lattice," constructed by replacing the bonds of the hypercubic lattice G in $N \geq 2$ by first-order necklaces composed of strings (links) and first-order beads (blobs), and separating the nodes of G by a correlation length ξ, as in Fig. 2.1a. The beads themselves have a hierarchical structure, as shown in Fig. 2.1b, consisting of two second-order necklaces in parallel, and so on, in a self-similar fashion to order J, for an arbitrary, large integer J. We assume that any kth order necklace has $\beta - 1$ beads on it, for an arbitrary, large integer β, and that each pair of beads is joined by a string of n_k bonds, so that there are a total of βn_k string bonds on each necklace. The βn_1 string bonds on any first-order necklace are called singly connected – because removal of one of them breaks the connection between nodes separated by ξ. All the rest of the bonds in the NLB graph are multiply connected, and among these it is useful to identify the βn_2 string bonds on a 2^{nN}-order necklace as doubly connected, because it is possible to remove two of them (in parallel) and break a connection between nodes. Based on a result of Coniglio's [Con82] implying, in our context, that the number of singly and doubly connected bonds between the nodes both diverge with exponent 1 as $p \to p_c^+$, we assume that $n_1 = 2\beta n_2$. Due to self-similarity, we assume that

$$n_{j-1} = 2\beta n_j \quad , \quad j = 2, \ldots, J. \tag{2.8}$$

Eq. (2.8) can be used to solve for the n_j, $j > 1$, in terms of n_1, with $n_2 = n_1/2\beta$, $n_3 = n_1/4\beta^2$, and so on, and we refer to the NLB graph as $A(n_1)$. In this model, the percolation limit $p \to p_c^+$ is characterized by the limits $n_1, \beta, J, \xi \to \infty$, so that the lengths of all orders of necklaces and the numbers and sizes of all orders of blobs, diverge as $p \to p_c^+$.

Before we give the asymptotics of $\delta^2\sigma^*(A(n_1))$, we must discuss the conditions under which they are proven. Consider $\sigma^*(q_1, q_2)$ for the bond lattice in $\mathbb{Z}^N$ with three conductivities 1, ε_1, and ε_2 in proportions p, q_1, and q_2, in addition to our standard two-component conductivity $\sigma^*(p)$. We require that $\sigma^*(q_1, q_2)$ has second-order partials at $q_1 = q_2 = 1$ for all $\varepsilon_1, \varepsilon_2 \geq 0$, and that $\sigma^*(p)$ has two derivatives at $p = 1$, for all $\varepsilon \geq 0$. For ε, ε_1, and $\varepsilon_2 > 0$, these conditions are satisfied by our general results [Gol92] that $\sigma^*(p)$ is analytic for all $p \in [0, 1]$ and $\sigma^*(q_1, q_2)$ is analytic for all $(q_1, q_2) \in [0, 1] \times [0, 1]$. The $\varepsilon = \varepsilon_1 = 0$ requirements will be assumed, although Kozlov [Koz89] has proven the existence of $\frac{d\sigma^*}{dp}\big|_{p=1}$ for a class of continuum analogs.

The second main condition is that, given the hypercubic base lattice G for $A(n_1)$,

$$K(G) = \frac{d\sigma^*}{dp}\bigg|_{p=1} + \frac{d^2\sigma^*}{dp^2}\bigg|_{p=1} > 0. \tag{2.9}$$

In any $N \geq 2$, $\frac{d\sigma^*}{dp}\big|_{p=1} = N/(N-1)$ [Kir71], whereas $\frac{d^2\sigma^*}{dp^2}\big|_{p=1}$, if negative, is quite small, e.g., ≈ -0.21 in $N = 2$ [NvVE86, Gol92], indicating that $\sigma^*(p)$ is quite straight near $p = 1$, so that (2.9) is satisfied. Condition (2.9) amounts to a consequence of the long-held view that effective medium theory (giving a straight-line solution) provides an accurate description of $\sigma^*(p)$ near $p = 1$, which also holds for general lattices. In fact, the asymptotics below can be proven for a variety of periodic base lattices G which satisfy (2.9) and, presumably, hold even for random lattices.

We may now state our principal result. Under the above assumptions, for fixed, large n_1, β, and J,

$$\delta^2\sigma(A(n_1)) = \alpha_J K(G)\beta n_1 + \sum_{i=0}^{\infty} \frac{a_i n_1 + b_i}{\beta^i}, \qquad (2.10)$$

where $(\alpha_J)^{-1} = \sum_{i=0}^{J}(\frac{1}{4})^i$ and the series in (2.10) converges, so that

$$\delta^2\sigma^*(A(n_1)) \sim \frac{\alpha_J K(G)\beta n_1}{\xi^{N-2}} > 0, \quad n_1, \beta, J, \xi \to \infty. \qquad (2.11)$$

The idea of the proof of (2.10) is first to write

$$\delta^2\sigma(A(n_1)) = \sum_{k \geq j} \delta_{jk}, \qquad (2.12)$$

where δ_{jk} is the sum of all contributions to $\delta^2\sigma(A(n_1))$ in (2.6) arising from pairs with one bond in a jth order string and the other in a kth order string, which is in either the same or a different first-order necklace. Now let z_k be the conductivity of a single first-order necklace with one bond removed from a kth order string, with $z_0 = \alpha_J/\beta n_1$ for no bond removed, $z_1 = 0$, and

$$z_k = z_0\left(1 + \frac{\gamma_k}{\beta^{k-1}}\right)^{-1}, \quad k \geq 2, \qquad (2.13)$$

where $\gamma_k \to 0$ as $k \to \infty$ geometrically fast. There are analogous formulas for the various forms of z_{jk} with two bonds removed, say, in series or in parallel. Then, through representations like (2.6) and (2.7), we obtain formulas for the δ_{jk} in terms of derivatives of $\sigma^*(p)$ and $\sigma^*(q_1, q_2)$ at $p = 1$, such as

$$\delta_{11} = z_0\left[\beta n_1(\beta n_1 - 1)\frac{d\sigma^*}{dp}(p = 1, h_1) + (\beta n_1)^2\frac{d^2\sigma^*}{dp^2}(p = 1, h_1)\right] \qquad (2.14)$$

and

$$\delta_{12} = z_0\left[(\beta n_1)^2\frac{d\sigma^*}{dp}(p = 1, h_1) + (\beta n_1)^2\frac{\partial^2\sigma^*}{\partial q_1 \partial q_2}(p = 1, h_1, h_2)\right], \qquad (2.15)$$

where $\frac{d\sigma^*}{dp}(p = 1, h_1)$, e.g., is for G with bond conductivities 1 and $h_1 = 0$, with $h_k = z_k/z_0$. As $k \to \infty$, $h_k \to 1$, and, as $\beta \to \infty$, $h_k \to 1$ for all $k \geq 2$, and similarly for $h_{jk} = z_{jk}/z_0$. The necessary control of the δ_{jk} is, then, obtained either from (2.9) or from perturbation theory around a homogeneous medium ($\varepsilon = 1$ or $\varepsilon_1 = \varepsilon_2 = 1$), which establishes (2.10).

We wish to make the following remarks concerning the above result. First, a result similar to (2.11) holds if we replace (2.8) by $n_{j-1} = \eta_j \beta_j n_j$, where the blobs of order $j - 1$ are made of η_j necklaces in parallel, with reasonable assumptions about η_j and β_j. Even if the blobs have a more complicated "super-lattice" structure themselves, an analog of (2.11), presumably, holds. Also, as noted above, (2.11) can be proven for a variety of base lattices G. Finally, although the principal assumption of the NLB graph replacing the actual backbone is quite serious, our proof of (2.10) shows that the dominant contribution to (2.11) comes from δ_{11}, which comes from macroscopic contributions in the NLB graph, where the model reflects the actual structure well. A similar result will hold for any reasonable assumption about microscopic backbone structure.

Now, we proceed to the implications of (2.11). First, its positivity establishes convexity of $\sigma^*(p)$ for the NLB model, which implies (under our assumptions, including scaling and the existence of three derivatives of $\sigma^*(p)$ for all $p > p_c$ when $\varepsilon = 0$) that $t \geq 1$, for any $N \geq 2$. Now, let $\lambda(n_1)$ be the length of a first-order necklace, so that $\lambda(n_1) \approx \beta n_1 + \beta^2 n_2 + \ldots + \beta^J n_J = \theta_J \beta n_1$, $\theta_J = \sum_{i=0}^{J} 2^{-i}$. By (2.11), then

$$\delta^2 \sigma^*(A(n_1)) \sim \frac{\rho_J \lambda(n_1)}{\xi^{N-2}} \quad , \quad n_1, \beta, J, \xi \to \infty, \tag{2.16}$$

where $\rho_J = \alpha_J K(G)/\theta_J$, so that $\rho_J \approx 2/3$ for large J in $N = 2$. Because all the parameters n_1, β, J and ξ are diverging as $p \to p_c^+$, we can define a whole class of NLB models by how fast $\lambda(n_1)$ scales to ∞ relative to ξ. By the structure of the model, clearly, $\lambda(n_1) \geq \xi$, and, typically, $\lambda/\xi \to \infty$. Thus, as a consequence of (2.16), in $N = 2$ and 3,

$$\delta^2 \sigma^*(A(n_1)) \to +\infty \quad , \quad n_1, \beta, J, \xi \to \infty, \tag{2.17}$$

except in $N = 3$ when $\lambda(n_1) = C\xi$, $C \geq 1$, in which case,

$$\delta^2 \sigma^*(A(n_1)) \to \rho C > 0, \tag{2.18}$$

where $\rho = \lim_{J \to \infty} \rho_J$. In $N \geq 4$, if λ and ξ are scaled so that $\lambda(n_1)/\xi^{N-2} \to 0^+$, then,

$$\delta^2 \sigma^*(A(n_1)) \to 0^+. \tag{2.19}$$

Under our assumptions, in particular, that $\frac{d^2\sigma^*}{dp^2} \sim (p - p_c)^{t-2}$, then, collecting our results,

$$1 \leq t \leq 2, \quad N = 2, 3 \tag{2.20}$$
$$2 \leq t \leq 3, \quad N \geq 4.$$

In (2.20), the last inequality $t \leq 3$ for $N \geq 4$ is obtained by a result that $\delta^3 \sigma^*(A'(n_1)) \sim C'\lambda^2(n_1)/\xi^{N-2}$ for a simpler node-link graph $A'(n_1)$, which is believed to be adequate in higher dimensions [Har83]. For models in $N = 4, 5$ which satisfy $\lambda^2(n_1)/\xi^{N-2} \to \infty$,

$$\delta^3 \sigma^*(A(n_1)) \to \infty,$$

so that $\frac{d^3\sigma^*}{dp^3} \sim (p - p_c)^{t-3} \to \infty$, which gives the inequality.

FIGURE 2.2. Two-dimensional Voronoi tessellation. The boundaries of the polygonal grains are formed from points which are equidistant from the dots in two neighboring grains. These boundaries form the channels of easiest flow in consolidated granular media and fractured rocks.

2.3   Critical Path Analysis in Highly Disordered Porous and Conducting Media

We now present a continuum percolation model which closely represents fluid flow in some fractured rocks and consolidated granular media, as well as electrical conduction in some matrix-particle composites [GK]. For simplicity, we first give the formulation for electrical conductivity in $N = 2$, but the model and result can be carried over to $N = 3$ and to fluid flow in a porous medium obeying Darcy's law. Consider the checkerboard of unit white squares in $\mathbb{R}^2$, with centers of the squares being the points of the lattice $\mathbb{Z}^2$. Randomly color the squares red with probability p, where the probability of coloring one square is independent from any other. Then for $x \in \mathbb{R}^2$, let $S(x)$ be the distance from x to the boundary of the nearest red square, with $S(x) = 0$ if x is inside a red square. Then define the local conductivity $\sigma(x)$ as

$$\sigma(x) = e^{\lambda S(x)}. \tag{3.1}$$

It is useful to think of the red squares as insulating particles, as we will be considering asymptotics as $\lambda \to +\infty$ (although we could just as easily consider $\lambda \to -\infty$). The medium defined by (3.1) can be thought of as being divided into grains associated with each red square, where each grain is the set of all points for which the distance to its red square is smaller than the distance to any other red square. The

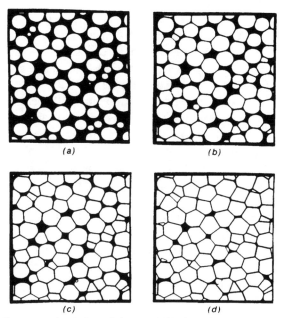

FIGURE 2.3. Computer simulation of the consolidation of spherical grains, showing decreasing porosity ϕ, with (a) $\phi = 0.364$, (b) $\phi = 0.200$, (c) $\phi = 0.100$, and (d) $\phi = 0.030$.

boundaries between these grains, where the distance to a red square is maximal, form the "channels" through which current (or fluid) passing through the medium will tend to flow. For small p, the set of boundaries forms a Voronoi network, as shown in Fig. 2.2 (from [Sah95]). The points in the figure represent the red squares.

Our goal is to find the $\lambda \to +\infty$ asymptotics of the effective conductivity $\sigma^*(p)$ of the medium in (3.1), which is defined as follows (e.g., [Koz78, GP83]). Let $E(x)$ and $J(x)$ be the stationary random electric and current fields in the medium satisfying $J(x) = \sigma(x)E(x)$, $\nabla \cdot J = 0$, $\nabla \times E = 0$, and $\langle E(x) \rangle = e_k$, where e_k is a unit vector in the kth direction, and $\langle \cdot \rangle$ denotes ensemble or infinite volume average. Then, the effective conductivity σ^* is defined via

$$\langle J \rangle = \sigma^* \langle E \rangle. \tag{3.2}$$

For fluid flow (with unit viscosity) in porous media [Sah95] obeying Darcy's law, $v = -\kappa(x)\nabla P$, where $\kappa(x)$ is the local permeability corresponding to (3.1), v is the fluid velocity satisfying $\nabla \cdot v = 0$, and P is the pressure (including gravity), one is interested in the effective permeability κ^*, defined analogously to (3.2),

$$\langle v \rangle = -\kappa^* \langle \nabla P \rangle \tag{3.3}$$

As briefly mentioned earlier, if $\kappa(x)$ has the form (3.1), then, for large λ, it is a close model for flow through consolidated granular media, where the grains themselves are permeable, with decreasing permeability as one approaches a hard core.

The degree of penetration into the grains decreases as λ increases. Fig. 2.3 (from [RS85]) shows a computer simulation of a grain consolidation process [RS85]. The sequence (a) - (d) shows increasing consolidation and, correspondingly, decreasing porosity. As $\lambda \to \infty$, the network of easiest flow paths in our model closely resembles the configuration in (d), which itself is similar to many types of sedimentary rocks, including Devonian sandstone [RS85].

To state the results for the asymptotics and to more fully describe our model, we first consider standard, nearest neighbor-site percolation on $\mathbb{Z}^2$ [SA92, Kes82, Gri89], which is equivalent to the percolation of nearest neighbor red squares (connected along an edge), with percolation threshold $p_c^s \approx 0.59$. Now, we relax the nearest neighbor restriction for connectedness and consider a generalized definition of percolation of the red squares. We say that two red squares $\hat{x}$ and $\hat{y}$, with centers x and y in $\mathbb{Z}^2$, are r-connected if there is a sequence $\hat{x}_0, \hat{x}_1, \cdots, \hat{x}_n, \hat{x}_0 = \hat{x}, \hat{x}_n = \hat{y}$ of red squares connecting them, so that dist $\{\hat{x}_i, \hat{x}_{i+1}\} \leq r$, where dist $\{\hat{x}_i, \hat{x}_{i+1}\}$ means the shortest distance between the boundaries of $\hat{x}_i$ and $\hat{x}_{i+1}$. With $\theta_r(p)$ the infinite cluster density of r-connected red squares, we define $p_c(r)$ as the percolation threshold for $\theta_r(p)$. We shall be concerned with a particular sequence r_j, $j \leq 1$, defined by squares which are increasingly distant from $\hat{0}$, the square centered at the origin, with $r_1 = 0$, $r_2 = 1$, $r_3 = \sqrt{2}$, $r_4 = 2$, $r_5 = \sqrt{5}$, $r_6 = \sqrt{8}, \cdots$. Note that it suffices to consider squares with centers $(m, n) \in \mathbb{Z}^2$ with $n \geq m \geq 0$, $n > 0$. For simplicity, we denote $p_c(r_j)$ as $p_c(j)$, and we also replace the term r_j-connected by j-connected. Note that the $j = 1$ ($r_1 = 0$) case includes both the nearest neighbor case above and the next nearest neighbor (diagonal) case (because the distance between connected squares is 0 in both cases), so that $p_c(1) = 1 - p_c^s \simeq 1 - 0.59 = 0.41$ [Koz89, BG94, Gol94b]. Furthermore $p_c(j+1) < p_c(j)$, which can be obtained from [AG91]. From analysis of a Poisson distribution of discs in the plane (see below), one can also find that

$$p_c(j) \sim \frac{\pi \mu_c}{j}, \qquad j \to \infty, \qquad (3.4)$$

where μ_c is the critical percolation intensity for unit discs [Gri89].

The last ingredient needed to state the results is the notion of critical values of $S(x)$, which are associated with the $p_c(j)$. These values $S_c(j)$ are defined by the observation that, for $p > p_c(j)$, the set $R_j = \{x \in \mathbb{R}^2 : S(x) \leq S_c(j)\}$ percolates in $\mathbb{R}^2$, where, for j associated with $(m, n) \in \mathbb{Z}^2$ as above,

$$S_c(j) = r_j/2 = \sqrt{(1 - \delta_{m0})(m - 1)^2 + (n - 1)^2} \Big/ 2, \qquad (3.5)$$

where δ_{m0} is the Kroneker delta. By percolating in $\mathbb{R}^2$, we mean that R_j contains an infinite polygonal line joining vertices of the red squares in R_j. Note that R_1 is just the set of red squares, which percolates in the above sense when $p > p_c(1)$. Now, in terms of the $S_c(j)$, define a step function $S_c(p)$ via

$$S_c(p) = S_c(j), \qquad p_c(j) < p < p_c(j - 1), \quad j \geq 1, \qquad (3.6)$$

where $p_c(0) = 1$. These critical distances $S_c(j)$ in our model correspond, via (3.1), to the critical conductances g_c in the AHL theory.

The principal results of our investigation are now stated as follows. For the effective conductivity $\sigma^*(p)$ of the medium $\sigma(x) = e^{\lambda S(x)}$, for $p \neq p_c(j)$,

$$\frac{1}{\lambda} \log \sigma^*(p) \sim S_c(p), \qquad \lambda \to \infty, \tag{3.7}$$

which establishes the validity of the AHL critical path analysis in the high disorder limit for our model. Furthermore, we have the following $p \to 0$ asymptotic for the exponents:

$$S_c(p) \sim \sqrt{\frac{\mu_c}{p}}, \qquad p \to 0, \tag{3.8}$$

where p is the density of red squares (i.e., with units of inverse square length). Via (3.7) and (3.8), we see that $\sigma^*(p)$ for our long range checkerboard model exhibits infinitely many thresholds $p_c(j) \to 0$, as $j \to \infty$ with an infinite set of asymptotics, as $\lambda \to \infty$. If, instead of $\lambda \to \infty$, we wish to set $\lambda = 1$, so that $\sigma(x) = e^{S(x)}$ and consider the asymptotics of $\sigma^*(p)$ as $p \to 0$, we also find that

$$\sqrt{p} \log \sigma^*(p) \sim \sqrt{\mu_c}, \qquad p \to 0. \tag{3.9}$$

Now, we give the analysis which leads to these results. The idea is to exploit the variational definition of σ^* equivalent to (3.2) and its dual to obtain upper and lower bounds on σ^*. Let $\Lambda_L = [0, L] \times [0, L] \subset \mathbb{R}^2$. Then the variational form of (3.2) is given by

$$\sigma^* = \lim_{L \to \infty} \frac{1}{L^2} \inf_{u \in \mathcal{P}} \int_{\Lambda_L} \sigma(x) |\nabla u|^2 dx, \tag{3.10}$$

where $\mathcal{P} = \{$continuous potentials u on $\Lambda_L : u(0, x_2) = 0, u(L, x_2) = L, \forall x_2 \in [0, L]\}$. We obtain bounds by inserting trial u in (3.10). To describe the construction, we recall certain properties of standard-site percolation. It has been shown [GK84, CC86] that, for $p > p_c^s$, the number α_L of disjoint crossings of Λ_L by nearest neighbor sites satisfies (roughly speaking) $\alpha_L = O(L)$ as $L \to \infty$. In our generalized model, recall that, for $p > p_c(j)$, $R_j = \{x \in \mathbb{R}^2 : S(x) \leq S_c(j)\}$ percolates in $\mathbb{R}^2$. In this case, the number $\alpha_L(j)$ of disjoint crossings of Λ_L by j-connected red squares also satisfies $\alpha_L(j) = O(L)$ as $L \to \infty$. We call the associated disjoint subsets of R_j that cross Λ_L pink "j-chains" (we say "pink" because such sets contain both red squares and parts of white ones). For our purposes, it will be necessary to consider only those chains which cross vertically. Now the trial u is constructed roughly as follows. In the regions between the j-chains, u is flat, so that $\nabla u = 0$. However, on the j-chains, u increases linearly across the chain, so that the total contribution to (3.10) of $|\nabla u|^2$ on the pink j-chains is $O(L^2)$. (Constructing u in the neighborhood of points where the j-chains have zero thickness is handled with asymptotic expansions, which are patched continuously to the rest of u, as in [Koz89] for the so-called Laplace–Dirichlet integral.) For such u, then, the integrand in (3.10) is zero off the j-chains, and, for x in the pink j-chains,

$$\sigma(x) \leq e^{\lambda S_c(j)}, \tag{3.11}$$

which leads to the inequality,

$$\sigma^*(p) \leq C_1 e^{\lambda S_c(j)}, \qquad p > p_c(j), \tag{3.12}$$

for some $C_1 > 0$ (depending on p), with $j \geq 1$.

To get the lower bound, we first note that the dual to (3.10) is obtained by replacing σ^* and $\sigma(x)$ by $(\sigma^*)^{-1}$ and $\sigma^{-1}(x)$, respectively. The key observation in the analysis, now, is that, for $p < p_c(j - 1)$, $j \geq 1$ (with $p_c(0) = 1$), $W_j = \{x \in \mathbb{R}^2 : S(x) \geq S_c(j)\}$ percolates in $\mathbb{R}^2$, which can be seen as follows. When $p < p_c(j - 1)$, R_{j-1} cannot percolate, $j \geq 2$. In this case, easy geometrical reasoning shows that infinite chains of white cells must exist, so that the minimal thickness r_j of these white chains (meaning that discs of radius $r_j/2$ percolate in these chains) is $2S_c(j)$. Then, $Wj = \{x \in \mathbb{R}^2 : S(x) \geq r_j/2 = S_c(j)\}$, which contains this set of white chains, percolates in $\mathbb{R}^2$. Now, constructing u similar to that above, one obtains

$$[\sigma^*(p)]^{-1} \leq C_2 e^{-\lambda S_c(j)}, \qquad p < p_c(j - 1), \tag{3.13}$$

for some $C_2 > 0$ (depending on p) , with $j \geq 1$. Combining (3.12) and (3.13) yields (3.7).

To obtain the asymptotic behavior of the thresholds in (3.4) and the exponents in (3.8), we connect our work to the analogous problem for a Poisson distribution of discs in $\mathbb{R}^2$. Let $\{x_k\}_{k=1}^{\infty}$ be a set of Poisson-distributed red points in the plane, with intensity μ. First, we define, analogously to (3.1), $S_\mu(x) = \text{dist} \{x, \text{ nearest } x_k\}$ and $\sigma_\mu(x) = e^{\lambda S_\mu(x)}$. Let S_μ^c be the smallest h for which $\{x \in \mathbb{R}^2 : S_\mu(x) \leq h\}$ percolates. Then, S_μ^c coincides with r_μ^c, the minimum radius so that the discs of radius r_μ^c, centered at the x_k, percolate. For the effective conductivity σ_μ^* in this case, the above arguments used for the long-range checkerboard yield

$$\frac{1}{\lambda} \log \sigma_\mu^* \sim S_\mu^c = r_\mu^c, \qquad \lambda \to \infty. \tag{3.14}$$

We remark that, via the scaling properties of the Poisson model, we can replace $\lambda S_\mu(x)$ by $\frac{\lambda}{\sqrt{\mu}} S_1(x)$, so that we may set $\lambda = 1$ and consider asymptotics, as $\mu \to 0$, with a result analogous to (3.14).

It is useful to note that the above Poisson model can be obtained by rescaling our checkerboard model, where the red squares of our model correspond to the x_k of the Poisson model, as follows. On the scaled lattice $h\mathbb{Z}^2, h > 0$, let the density of red squares be $p/h^2 = \mu$. As $p \to 0$, with $h = \sqrt{p/\mu} \to 0$ as well, $\hat{S}(x, p) = h S(x/h, p) \to S_\mu(x)$. Then, the critical values also converge, $\hat{S}_c(p) \to S_\mu^c$ as $p \to 0$. Then, with $\hat{S}_c = hS_c$ and $h = \sqrt{p/m}$, setting $\mu = 1$ yields $\sqrt{p}S_c(p) \to r_1^c$ as $p \to 0$, which is equivalent to (3.8), because $\mu_c = (r_1^c)^2$.

Furthermore, the critical density of red points on the rescaled lattice is $p_c(j)/h^2$, so that $\mu_c = \lim_{j \to \infty} \frac{p_c(j)}{h^2}$. Note that $p_c(j)$ is the critical p for which disks of radius r_j percolate. So, if we rescale the lattice with $h \sim 1/r_j$, as $j \to \infty$, then, unit discs percolate, so that

$$\mu_c = \lim_{j \to \infty} p_c(j) r_j^2 . \tag{3.15}$$

To relate r_j to j, we note that there are $O(j)$ integer points inside the disc of radius r_j, as $j \to \infty$, so that

$$j \sim \pi r_j^2 , \quad j \to \infty , \tag{3.16}$$

which, combined with (3.15), yields (3.4).

In closing, we wish to make a few remarks. Presumably, an effective medium approach as in [SK82] could account for the behavior of our model for small j. However, as j grows, the number of configurations of squares that must be considered grows extremely rapidly, and numerical calulations become intractable.

An interesting question is the transition between different exponents for large λ, as p crosses the threshold $p_c(j)$. We remark that, for example, the constant C_1 in estimate (3.12) diverges like $\xi(p)$, as $p \to p_c(j)^+$, where $\xi(p)$ is the correlation length for j−percolation.

3

One-Phase Newtonian Flow

Grégoire Allaire

3.1 Derivation of Darcy's Law

3.1.1 Presentation of the Results

This section is devoted to the derivation of Darcy's law for an incompressible viscous fluid flowing in a porous medium. Starting from the steady Stokes equations in a periodic porous medium, with a no-slip (Dirichlet) boundary condition on the solid pores, Darcy's law is rigorously obtained by periodic homogenization using the two-scale convergence method. The assumption of the periodicity of the porous medium is by no means realistic, but it allows casting this problem in a very simple framework and proving theorems without too much effort. We denote by ε the ratio of the period to the overall size of the porous medium. It is the small parameter of our asymptotic analysis because the pore size is usually much smaller than the characteristic length of the reservoir. The porous medium is contained in a domain Ω, and its fluid part is denoted by $\mathfrak{P}^\varepsilon$. From a mathematical point of view, $\mathfrak{P}^\varepsilon$ is a periodically perforated domain, i.e., it has many small holes of size ε which represent solid obstacles that the fluid cannot penetrate.

The motion of the fluid in $\mathfrak{P}^\varepsilon$ is governed by the steady Stokes equations, complemented with a Dirichlet boundary condition. We denote by $\vec{v}^\varepsilon$ and p^ε the velocity and pressure of the fluid and $\vec{f}$ the density of forces acting on the fluid ($\vec{v}^\varepsilon$ and $\vec{f}$ are vector-valued functions, whereas p^ε is scalar). The fluid viscosity is a positive constant μ that we scale by a factor ε^2 (where ε is the period). The Stokes equations are

$$\begin{cases} \nabla p^\varepsilon - \varepsilon^2 \mu \Delta \vec{v}^\varepsilon = \vec{f} & \text{in } \mathfrak{P}^\varepsilon, \\ \nabla \cdot \vec{v}^\varepsilon = 0 & \text{in } \mathfrak{P}^\varepsilon, \\ \vec{v}^\varepsilon = 0 & \text{on } \partial\mathfrak{P}^\varepsilon. \end{cases} \tag{1.1}$$

The above scaling for the viscosity is such that the velocity $\vec{v}^\varepsilon$ has a nontrivial limit as ε goes to zero. Physically speaking, the very small viscosity, of order ε^2, exactly balances the friction of the fluid on the solid pore boundaries due to the no-slip boundary condition. Note that this scaling is perfectly legitimate because, by linearity of the equations, one can always replace $\vec{v}^\varepsilon$ by $\varepsilon^2 \vec{v}^\varepsilon$. To obtain an existence and uniqueness result for (1.1), the forcing term is assumed to have the usual regularity: $\vec{f}(x) \in L^2(\Omega)^N$. Then, as is well-known (see e.g., [Tem79]), the

Stokes equations (1.1) admit a unique solution:

$$\vec{v}^{\varepsilon} \in H_0^1(\mathfrak{P}^{\varepsilon})^N, \quad p^{\varepsilon} \in L^2(\mathfrak{P}^{\varepsilon})/\mathbb{R}, \tag{1.2}$$

the pressure being uniquely defined up to an additive constant. The homogenization problem for (1.1) is to find the effective equation satisfied by the limits of $\vec{v}^{\varepsilon}$, p^{ε}. From the point of view of homogenization, the mathematical originality of system (1.1) is that the periodic oscillations are not in the coefficients of the operator but in the geometry of the porous medium $\mathfrak{P}^{\varepsilon}$.

Before stating the main result, let us describe more precisely the assumptions on the porous domain $\mathfrak{P}^{\varepsilon}$. As usual in periodic homogenization, a periodic structure is defined by a domain Ω and an associated microstructure, or periodic cell $Y = (0, 1)^N$, made of two complementary parts, the fluid part $\mathfrak{P}$ and the solid part $\mathfrak{S}$ ($\mathfrak{P} \cup \mathfrak{S} = Y$ and $\mathfrak{P} \cap \mathfrak{S} = \emptyset$). We assume that Ω is a smooth, bounded, connected set in $\mathbb{R}^N$ and that $\mathfrak{P}$ is a smooth, connected, open subset of Y, identified with the unit torus (i.e., $\mathfrak{P}$, repeated by Y-periodicity in $\mathbb{R}^N$, is a smooth, connected, open set of $\mathbb{R}^N$). The domain Ω is covered by a regular mesh of size ε. Each cell Y_i^{ε} is of the type $(0, \varepsilon)^N$ divided into a fluid part $\mathfrak{P}_i^{\varepsilon}$ and a solid part $\mathfrak{S}_i^{\varepsilon}$, is similar to the unit cell Y rescaled to size ε. The fluid part $\mathfrak{P}^{\varepsilon}$ of a porous medium is defined by

$$\mathfrak{P}^{\varepsilon} = \Omega \setminus \bigcup_{i=1}^{N(\varepsilon)} \mathfrak{S}_i^{\varepsilon} = \Omega \cap \bigcup_{i=1}^{N(\varepsilon)} \mathfrak{P}_i^{\varepsilon}, \tag{1.3}$$

where the number of cells $N(\varepsilon) = |\Omega|\varepsilon^{-N}(1 + o(1))$.

A final word of caution is in order : the sequence of solutions $(\vec{v}^{\varepsilon}, p^{\varepsilon})$ is not defined in a fixed domain *independent of ε* but rather in a varying set, $\mathfrak{P}^{\varepsilon}$. To state the homogenization theorem, convergences in fixed Sobolev spaces (defined on Ω) are used which require first that $(\vec{v}^{\varepsilon}, p^{\varepsilon})$ be extended to the whole domain Ω. Recall that, by definition, an extension $(\tilde{v}^{\varepsilon}, \tilde{p}^{\varepsilon})$ of $(\vec{v}^{\varepsilon}, p^{\varepsilon})$ is defined on Ω and coincides with $(\vec{v}^{\varepsilon}, p^{\varepsilon})$ on $\mathfrak{P}^{\varepsilon}$.

Theorem 1.1 *There exists an extension $(\tilde{v}^{\varepsilon}, \tilde{p}^{\varepsilon})$ of the solution $(\vec{v}^{\varepsilon}, p^{\varepsilon})$ of (1.1) such that the velocity $\tilde{v}^{\varepsilon}$ converges weakly in $L^2(\Omega)^N$ to $\vec{v}$ and the pressure $\tilde{p}^{\varepsilon}$ converges strongly in $L^2(\Omega)/\mathbb{R}$ to p, where $(\vec{v}, p)$ is the unique solution of the homogenized problem, a Darcy law,*

$$\begin{cases} \vec{v}(x) = \frac{1}{\mu} A \left(\vec{f}(x) - \nabla p(x) \right) & in \ \Omega, \\ \nabla \cdot \vec{v}(x) = 0 & in \ \Omega, \\ \vec{v}(x) \cdot \vec{v} = 0 & on \ \partial\Omega, \end{cases} \tag{1.4}$$

where A is a symmetric, positive definite tensor (the so-called permeability tensor) defined by its entries

$$A_{ij} = \int_{\mathfrak{P}} \nabla \vec{w}_i(y) \cdot \nabla \vec{w}_j(y) dy, \tag{1.5}$$

where, $(\vec{e}_i)_{1 \leq i \leq N}$ *being the canonical basis of* $\mathbb{R}^N$, $\vec{w}_i(y)$ *denotes the unique solution in* $H^1_\#(\mathfrak{P})^N$ *of the local, or unit cell, Stokes problem,*

$$\begin{cases} \nabla \pi_i - \Delta \vec{w}_i = \vec{e}_i & in \; \mathfrak{P} \\ \nabla \cdot \vec{w}_i = 0 & in \; \mathfrak{P} \\ \vec{w}_i = 0 & in \; \mathfrak{S} \\ y \to \pi_i, \vec{w}_i & Y\text{-periodic.} \end{cases} \tag{1.6}$$

The formulas correspond to those of section 1.4.

The next subsection is devoted to the proof of Theorem 1.1, and the last subsection is concerned with the delicate task of constructing the extension $(\tilde{v}^\varepsilon, \tilde{p}^\varepsilon)$. The weak convergence of the velocity can be further improved by the following *corrector* result (see [All91a] for a proof, or adapt the ideas used for establishing Eq. (3.24) in Appendix A).

Proposition 1.2 *With the same notations as in Theorem 1.1, the velocity satisfies*

$$\left(\tilde{v}^\varepsilon - \sum_{i=1}^{N} \vec{w}_i(\frac{x}{\varepsilon}) v_i(x) \right) \to 0 \; strongly \; in \; L^2(\Omega)^N, \tag{1.7}$$

where $(\vec{w}_i)_{1 \leq i \leq N}$ *are the local, unit-cell velocities and* $(v_i)_{1 \leq i \leq N}$ *are the components of the homogenized velocity* $\tilde{v}(x)$.

Remark The homogenized problem (1.4) is a Darcy law, i.e., the flow rate $\tilde{v}$ is proportional to the balance of forces including the pressure. The permeability tensor A depends only on the microstructure $\mathfrak{P}$ of the porous media (not on the exterior forces or on the physical properties of the fluid). Quite early, many papers have been devoted to the derivation of Darcy's law by homogenization, using formal asymptotic expansions (see, for example, [Kel80], [Lio81], and [SP80]). The first rigorous proof (including the difficult construction of a pressure extension) appeared in [Tar80]. Further extensions are to be found in [All89], [LA90], and [Mik91]. A good reference for physical aspects of this problem, as well as mathematical ones, is the book [EP87]. Of course, more complicated models than the incompressible Stokes equations can be homogenized to derive various variants of Darcy's law. The next sections investigate such more general microscopic models (we also refer to [Aga93], [Aur92], [Con85], [Con87a], [Con88], [ESP75], [LSP75], [MA87]). Of course there are other methods, apart from periodic homogenization, which permit deriving Darcy's law. It can also be established by stochastic homogenization, representative volume averaging, and so on.

3.1.2   Proof of the Homogenization Theorem

This subsection is devoted to proving Theorem 1.1 by the method of two-scale convergence (see Appendix A). For the moment, we assume the existence of bounded extensions of the velocity and pressure of the fluid in the porous medium.

Lemma 1.3 *There exists an extension $(\tilde{v}^\varepsilon, \tilde{p}^\varepsilon)$ of the solution $(\vec{v}^\varepsilon, p^\varepsilon)$ satisfying the a priori estimates,*

$$\|\tilde{v}^\varepsilon\|_{L^2(\Omega)^N} + \varepsilon\|\nabla\tilde{v}^\varepsilon\|_{L^2(\Omega)^{N\times N}} \leq C \tag{1.8}$$

and

$$\|\tilde{p}^\varepsilon\|_{L^2(\Omega)/\mathbb{R}} \leq C, \tag{1.9}$$

where the constant C does not depend on ε.

The proof of Lemma 1.3 is postponed until the next subsection. By applying the two-scale convergence method, we, first, prove

Theorem 1.4 *The extension $(\tilde{v}^\varepsilon, \tilde{p}^\varepsilon)$ of the solution of (1.1) two-scale converges to the unique solution $(\vec{v}_0(x, y), p(x))$ of the two-scale homogenized problem*

$$\begin{cases} \nabla_y p_1(x, y) + \nabla_x p(x) - \mu\Delta_{yy}\vec{v}_0(x, y) = \vec{f}(x) & \text{in } \Omega \times \mathfrak{P}, \\ \nabla_y \cdot \vec{v}_0(x, y) = 0 & \text{in } \Omega \times \mathfrak{P}, \\ \nabla_x \cdot \left(\int_Y \vec{v}_0(x, y)dy\right) = 0 & \text{in } \Omega, \\ \vec{v}_0(x, y) = 0 & \text{in } \Omega \times \mathfrak{S}, \\ \left(\int_Y \vec{v}_0(x, y)dy\right) \cdot \vec{v} = 0 & \text{on } \partial\Omega, \\ y \to \vec{v}_0(x, y), \, p_1(x, y) & Y\text{-periodic.} \end{cases} \tag{1.10}$$

Remark The two-scale homogenized problem is also called a two-pressure Stokes system. It is a combination of the usual homogenized and cell problems (see Chapter 1). By eliminating the y variable, the homogenized Darcy's law will be recovered in the end.

Proof of Theorem 1.4. Applying Theorem 3.6 of Appendix A, there exists a two-scale limit $\vec{v}_0(x, y) \in L^2\left(\Omega; H^1_\#(Y)^N\right)$ such that, up to a subsequence, the sequences $\tilde{v}^\varepsilon$ and $\varepsilon\nabla\tilde{v}^\varepsilon$ two-scale converge to $\vec{v}_0$ and $\nabla_y\vec{v}_0$, respectively. Furthermore, $\vec{v}_0$ satisfies

$$\begin{cases} \nabla_y \cdot \vec{v}_0(x, y) = 0 & \text{in } \Omega \times Y, \\ \nabla_x \cdot \left(\int_Y \vec{v}_0(x, y)dy\right) = 0 & \text{in } \Omega, \\ \vec{v}_0(x, y) = 0 & \text{in } \Omega \times \mathfrak{S}, \\ \left(\int_Y \vec{v}_0(x, y)dy\right) \cdot \vec{v} = 0 & \text{on } \partial\Omega. \end{cases} \tag{1.11}$$

By Theorem 3.2 of Appendix A, there exists a two-scale limit $p_0(x, y) \in L^2(\Omega\times Y)$ such that, up to a subsequence, $\tilde{p}^\varepsilon$ two-scale converges to p_0. Multiplying the momentum equation in (1.1) by $\varepsilon\psi(x, \frac{x}{\varepsilon})$, where $\psi(x, y)$ is a smooth, vector-valued, Y-periodic function, and integrating by parts, leads to

$$\lim_{\varepsilon\to 0} \int_\Omega \tilde{p}^\varepsilon \nabla_y \cdot \vec{\psi}(x, \frac{x}{\varepsilon})dx = \int_\Omega \int_Y p_0(x, y)\nabla_y \cdot \vec{\psi}(x, y)dxdy = 0. \tag{1.12}$$

Another integration by parts in (1.12) shows that $\nabla_y p_0(x, y)$ is 0. Thus, there exists $p(x) \in L^2(\Omega)/\mathbb{R}$ such that $p_0(x, y) = p(x)$.

The next step in the two-scale convergence method is to multiply system (1.1) by a test function, having the form of the two-scale limit $\vec{v}_0$, and to read off a variational formulation for the limit. Therefore, we choose a test function $\vec{\psi}(x, y) \in \mathcal{D}\left(\Omega; C_\#^\infty(Y)^N\right)$ with $\vec{\psi}(x, y) = 0$ in $\Omega \times \mathfrak{S}$ (thus, $\vec{\psi}(x, \frac{x}{\varepsilon}) \in H_0^1(\mathfrak{P}^\varepsilon)^N$). Furthermore, we assume that $\vec{\psi}$ satisfies the incompressibility conditions (1.11), i.e., $\nabla_y \cdot \vec{\psi}(x, y) = 0$ and $\nabla_x \cdot \left(\int_Y \vec{\psi}(x, y)dy\right) = 0$. Multiplying Eq. (1.1) by $\vec{\psi}(x, \frac{x}{\varepsilon})$ and integrating by parts yields

$$-\int_{\mathfrak{P}^\varepsilon} p^\varepsilon(x)\nabla_x \cdot \vec{\psi}(x, \frac{x}{\varepsilon})dx \tag{1.13}$$

$$+\mu \int_{\mathfrak{P}^\varepsilon} \varepsilon \nabla \vec{v}^\varepsilon(x) \cdot \nabla_y \vec{\psi}(x, \frac{x}{\varepsilon})dx = \int_{\mathfrak{P}^\varepsilon} \vec{f}(x) \cdot \vec{\psi}(x, \frac{x}{\varepsilon})dx + O(\varepsilon),$$

where $O(\varepsilon)$ stands for the the remaining terms of order ε. In (1.13) the domain of integration $\mathfrak{P}^\varepsilon$ can be replaced by Ω because the test function is zero in $\Omega \setminus \mathfrak{P}^\varepsilon$. Then, passing to the two-scale limit, the first term in (1.13) contributes nothing, because the two-scale limit of $\tilde{p}^\varepsilon$ does not depend on y and $\vec{\psi}$ satisfies $\nabla_x \cdot \left(\int_Y \vec{\psi}(x, y)dy\right) = 0$, whereas the other terms yield

$$\mu \int_\Omega \int_\mathfrak{P} \nabla_y \vec{v}_0(x, y) \cdot \nabla_y \vec{\psi}(x, y)dxdy = \int_\Omega \int_\mathfrak{P} \vec{f}(x) \cdot \vec{\psi}(x, y)dxdy. \tag{1.14}$$

By density, (1.14) holds for any function $\vec{\psi}$ in the Hilbert space V defined by

$$V = \left\{ \vec{\psi}(x, y) \in L^2\left(\Omega; H_\#^1(Y)^N\right) \text{ such that } \begin{array}{l} \nabla_y \cdot \vec{\psi}(x, y) = 0 \text{ in } \Omega \times Y \\ \nabla_x \cdot \left(\int_Y \vec{\psi}(x, y)dy\right) = 0 \text{ in } \Omega \end{array} \right.$$

$$\left. \text{and} \quad \begin{array}{l} \vec{\psi}(x, y) = 0 \text{ in } \Omega \times \mathfrak{S} \\ \left(\int_Y \vec{\psi}(x, y)dy\right) \cdot \vec{v} = 0 \text{ on } \partial\Omega \end{array} \right\}. \tag{1.15}$$

It is not difficult to check that the hypothesis of the Lax–Milgram lemma holds for the variational formulation (1.14) in the Hilbert space V, which, by consequence, admits a unique solution $\vec{v}_0$ in V. Furthermore, by Lemma 1.5 below, the orthogonal of V, a subset of $L^2\left(\Omega; H_\#^{-1}(Y)^N\right)$, is made of gradients of the form $\nabla_x q(x) + \nabla_y q_1(x, y)$ with $q(x) \in H^1(\Omega)/\mathbb{R}$ and $q_1(x, y) \in L^2\left(\Omega; L_\#^2(\mathfrak{P})/\mathbb{R}\right)$. Thus, integrating by parts, the variational formulation (1.14) is equivalent to the two-scale homogenized system (1.10). (There is a subtle point here; one must check that the pressure $p(x)$ arising as a Lagrange multiplier of the incompressibility constraint $\nabla_x \cdot \left(\int_Y \vec{v}_0(x, y)dy\right) = 0$ is the same as the two-scale limit of the pressure $\tilde{p}^\varepsilon$. This can easily be done by multiplying Eq. (1.1) by a test function $\vec{\psi}$, which is divergence-free only in y, and by identifying limits.) Because (1.10) admits a unique solution, then, the entire sequence $(\tilde{v}^\varepsilon, \tilde{p}^\varepsilon)$ converges to its unique solution $(\vec{v}_0(x, y), p(x))$. This completes the proof of Theorem 1.4.   Q.E.D.

We now arrive at the last step of the two-scale convergence method, eliminating, if possible, the microscopic variable y in the homogenized system. This allows deducing Theorem 1.1 from Theorem 1.4.

Proof of Theorem 1.1. The derivation of the homogenized Darcy law (1.4) from the two-scale homogenized problem (1.10) is just a matter of algebra. From the first equation of (1.10), the velocity $\vec{v}_0(x, y)$ is computed in terms of the macroscopic forces and the local velocities

$$\vec{v}_0(x, y) = \sum_{i=1}^{N} \left(f_i(x) - \frac{\partial p}{\partial x_i}(x) \right) \vec{w}_i(y). \qquad (1.16)$$

Averaging (1.16) on Y and denoting by $\vec{v}$ the average of $\vec{v}_0$, i.e., $\vec{v}(x) = \int_{\mathfrak{P}} \vec{v}_0(x, y) dy$, yields the Darcy relationship

$$\vec{v}(x) = \frac{1}{\mu} A \left(\vec{f}(x) - \nabla p(x) \right), \qquad (1.17)$$

because the matrix A satisfies

$$A_{ij} = \int_{\mathfrak{P}} \nabla \vec{w}_i(y) \cdot \nabla \vec{w}_j(y) dy = \int_{\mathfrak{P}} \vec{w}_i(y) \cdot \vec{e}_j dy. \qquad (1.18)$$

Equation (1.18) is obtained by multiplying the ith local problem (1.6) by $\vec{w}_j$ and integrating by parts (the boundary integrals cancel out, due to the periodic boundary condition). Combining (1.17) with the divergence-free condition on $\vec{v}$ yields the homogenized Darcy's law. Note that it is a well-posed problem because it is just a second-order elliptic equation for the pressure p, complemented by a Neumann boundary condition. To complete the proof of Theorem 1.1, it remains to be shown that the convergence of the pressure $\tilde{p}^\varepsilon$ to p is not only weak, but also strong, in $L^2(\Omega)/\mathbb{R}$: this will be done in the next subsection.   Q.E.D.

Lemma 1.5 *Let V be the subspace of $L^2(\Omega; H^1_\#(Y)^N)$ defined by (1.15). Its orthogonal (with respect to the usual scalar product in $L^2(\Omega \times Y)$) $V^\perp$, a subspace of the dual space $L^2(\Omega; H^{-1}_\#(Y)^N)$, has the following characterization*

$$V^\perp = \left\{ \nabla_x \varphi(x) + \nabla_y \varphi_1(x, y) \text{ with } \varphi \in H^1(\Omega) \text{ and } \varphi_1 \in L^2 \left(\Omega; L^2_\#(\mathfrak{P})/\mathbb{R} \right) \right\}. \qquad (1.19)$$

Proof Note that $V = V_1 \cap V_2$ with

$$V_1 = \left\{ \vec{v}(x, y) \in L^2(\Omega; H^1_\#(Y)^N) \text{ such that } \nabla_y \cdot \vec{v} = 0 \text{ in } \Omega \times Y, \vec{v} = 0 \text{ in } \Omega \times \mathfrak{S} \right\}$$

and

$$V_2 = \left\{ \vec{v}(x, y) \in L^2(\Omega; H^1_\#(Y)^N) \right.$$
$$\left. \text{such that } \nabla_x \cdot \left(\int_{\mathfrak{P}} \vec{v} dy \right) = 0 \text{ in } \Omega, \left(\int_{\mathfrak{P}} \vec{v} dy \right) \cdot \vec{v}_x = 0 \text{ on } \partial\Omega \right\}.$$

It is a well-known result (see, e.g., [Tem79]) that

$$V_1^\perp = \left\{ \nabla_y \varphi_1(x, y) \text{ with } \varphi_1 \in L^2\left(\Omega; L_\#^2(\mathfrak{P})/\mathbb{R}\right) \right\}$$

and

$$V_2^\perp = \left\{ \nabla_x \varphi(x) \text{ with } \varphi \in H^1(\Omega) \right\}.$$

Because V_1 and V_2 are two closed subspaces, it is equivalent to say $(V_1 \cap V_2)^\perp = V_1^\perp + V_2^\perp$ or $V_1 + V_2 = \overline{V_1 + V_2}$. Indeed, we are going to prove that $V_1 + V_2$ is equal to $L^2(\Omega; H_\#^1(\mathfrak{P})^N)$, which establishes that $V_1 + V_2$ is closed, thus, completing the proof of this lemma.   Q.E.D.

Introducing the divergence-free solutions $(\vec{w}_i(y))_{1 \leq i \leq N}$ of the local Stokes problem (1.6) defined below, for any given $\vec{v}(x, y) \in L^2(\Omega; H_\#^1(\mathfrak{P})^N)$, we define a unique solution $q(x)$ in $H^1(\Omega)/\mathbb{R}$ of the Neuman problem,

$$\begin{cases} -\nabla_x \cdot \left(A\nabla q(x) - \int_{\mathfrak{P}} \vec{v}(x, y) dy \right) = 0 & \text{in } \Omega, \\ \left(A\nabla q(x) - \int_{\mathfrak{P}} \vec{v}(x, y) dy \right) \cdot \vec{v} = 0 & \text{on } \partial\Omega, \end{cases} \tag{1.20}$$

where the positive definite matrix A, defined in (1.5), satisfies

$$A\vec{e}_i = \int_{\mathfrak{P}} \vec{w}_i(y) dy$$

$((\vec{e}_i)_{1 \leq i \leq N}$ being the orthonormal basis). Then, decomposing $\vec{v}$ as

$$\vec{v}(x, y) = \sum_{i=1}^N \vec{w}_i(y) \frac{\partial q}{\partial x_i}(x) + \left(\vec{v}(x, y) - \sum_{i=1}^N \vec{w}_i(y) \frac{\partial q}{\partial x_i}(x) \right), \tag{1.21}$$

it is easily seen that the first term of this decomposition belongs to V_1, whereas the second belongs to V_2.

3.1.3   A Priori Estimate of the Pressure in a Porous Medium

This subsection is devoted to proving Lemma 1.3 which constructs extensions and establishes uniform estimates for the velocity and pressure of Stokes flow in a porous medium. This proof is rather technical and readers who are willing to accept it can safely skip this section, which is provided for completeness. Basically, we reproduce the original proof of Tartar [Tar80] which has been further generalized in [All89], [LA90], and [Mik91]. To simplify the details, we assume that the solid part $\mathfrak{S}$ is strictly included and isolated in the unit cell Y (the solid part of the porous medium is, thus, a periodic array of particles). We emphasize that Lemma 1.3 also holds true in more realistic situations where the solid part of the porous medium is a connected body.

As already pointed out, a technical difficulty is to extend the solution $(\vec{v}^\varepsilon, p^\varepsilon)$ of Stokes equations (1.1), defined only in the fluid domain $\mathfrak{P}^\varepsilon$, to the entire porous medium Ω. Let us explain briefly why. The usual strategy in homogenization is to obtain a priori estimates for $(\vec{v}^{\varepsilon}, p^{\varepsilon})$, which are *independent* of ε. It allows extracting weakly convergent subsequences, which, in turn, permits finding the homogenized problem. However, this weak compactness property (i.e., extracting weakly convergent subsequences) holds true only in a fixed Sobolev space. The problem is that $(\vec{v}^{\varepsilon}, p^{\varepsilon})$ is defined in $H_0^1(\mathfrak{P}^{\varepsilon})^N \times L^2(\mathfrak{P}^{\varepsilon})$ which is a space varying with ε. Therefore, $(\vec{v}^{\varepsilon}, p^{\varepsilon})$ needs to be extended to the whole porous medium Ω.

It is easy to extend the velocity $\vec{v}^{\varepsilon}$ by zero in $\Omega \setminus \mathfrak{P}^{\varepsilon}$ (this is compatible with its Dirichlet boundary condition on $\partial \mathfrak{P}^{\varepsilon}$) to obtain a function $\tilde{v}^{\varepsilon}$ which belongs to $H_0^1(\Omega)^N$:

$$\begin{cases} \tilde{v}^{\varepsilon} = \vec{v}^{\varepsilon} & \text{in } \mathfrak{P}^{\varepsilon}, \\ \tilde{v}^{\varepsilon} = 0 & \text{in } \Omega \setminus \mathfrak{P}^{\varepsilon}. \end{cases} \tag{1.22}$$

The definition of the proposed extension $\tilde{p}^{\varepsilon}$ of the pressure is slightly more complicated:

$$\begin{cases} \tilde{p}^{\varepsilon} = p^{\varepsilon} & \text{in } \mathfrak{P}^{\varepsilon}, \\ \tilde{p}^{\varepsilon} = \frac{1}{|\mathfrak{P}_i^{\varepsilon}|} \int_{\mathfrak{P}_i^{\varepsilon}} p^{\varepsilon} & \text{in each } \mathfrak{S}_i^{\varepsilon}. \end{cases} \tag{1.23}$$

Definition (1.23) turns out to be convenient for obtaining an a priori estimate for the pressure.

The proof of Lemma 1.3 requires a few technical lemmas that follow. We begin with a precise estimate of the constant appearing in the Poincaré inequality in $\mathfrak{P}^{\varepsilon}$.

Lemma 1.6 *There exists a constant C independent of ε, such that any function $v \in H_0^1(\mathfrak{P}^{\varepsilon})$ satisfies*

$$\|v\|_{L^2(\mathfrak{P}^{\varepsilon})} \leq C\varepsilon \|\nabla v\|_{L^2(\mathfrak{P}^{\varepsilon})}. \tag{1.24}$$

Proof For any function $w(y) \in H^1(\mathfrak{P})$ such that $w = 0$ on $\partial \mathfrak{S}$, the Poincaré inequality in $\mathfrak{P}$ states that

$$\|w\|_{L^2(\mathfrak{P})}^2 \leq C\|\nabla w\|_{L^2(\mathfrak{P})}^2, \tag{1.25}$$

where the constant C depends only on $\mathfrak{P}$. By a change of variable $x = \varepsilon y$, we rescale (1.25) from $\mathfrak{P}$ to $\mathfrak{P}_i^{\varepsilon}$. Then, for any function $w(x) \in H^1(\mathfrak{P}_i^{\varepsilon})$, such that $w = 0$ on $\partial \mathfrak{S}_i^{\varepsilon}$,

$$\|w\|_{L^2(\mathfrak{P}_i^{\varepsilon})}^2 \leq C\varepsilon^2 \|\nabla w\|_{L^2(\mathfrak{P}_i^{\varepsilon})}^2, \tag{1.26}$$

with the same constant C as in (1.25). Summing the inequality (1.26) arising from all the fluid cells $\mathfrak{P}_i^{\varepsilon}$, which cover $\mathfrak{P}^{\varepsilon}$, gives the desired result (1.24).   Q.E.D.

Now, we introduce a restriction operator from $H_0^1(\Omega)^N$ into $H_0^1(\mathfrak{P}^{\varepsilon})^N$ preserving divergence-free vectors (the assumption is used that $\mathfrak{S}$ is strictly included and isolated in Y).

Lemma 1.7 *There exists a linear continuous operator R_{ε} acting from $H_0^1(\Omega)^N$ into $H_0^1(\mathfrak{P}^{\varepsilon})^N$, such that*

$$R_{\varepsilon}\vec{v} = \vec{v} \text{ in } \mathfrak{P}^{\varepsilon} \text{ if } \vec{v} \in H_0^1(\mathfrak{P}^{\varepsilon})^N, \tag{1.27}$$

$$\nabla \cdot (R_{\varepsilon}\vec{v}) = 0 \text{ in } \mathfrak{P}^{\varepsilon}, \text{ if } \nabla \cdot \vec{v} = 0 \text{ in } \Omega, \tag{1.28}$$

and

$$\|R_\varepsilon \vec{v}\|_{L^2(\mathfrak{P}^\varepsilon)} + \varepsilon \|\nabla(R_\varepsilon \vec{v})\|_{L^2(\mathfrak{P}^\varepsilon)} \leq C \left(\|\vec{v}\|_{L^2(\Omega)} + \varepsilon \|\nabla \vec{v}\|_{L^2(\Omega)} \right), \qquad (1.29)$$

for any $\vec{v} \in H_0^1(\Omega)^N$ *(the constant C is independent of* $\vec{v}$ *and* ε*).*

Proof As in Lemma 1.6, we proceed by rescaling a similar operator R acting from $H^1(Y)^N$ into $H^1(\mathfrak{P})^N$. For any function $\vec{v} \in H^1(Y)^N$, there exists a unique solution in $H^1(\mathfrak{P})^N$, denoted $R\vec{v}$, of the following Stokes problem:

$$\begin{cases} \nabla q - \Delta R\vec{v} = -\Delta \vec{v} & \text{in } \mathfrak{P}, \\ \nabla \cdot R\vec{v} = \nabla \cdot \vec{v} + \frac{1}{|\mathfrak{P}|} \int_\mathfrak{G} \nabla \cdot \vec{v} & \text{in } \mathfrak{P}, \\ R\vec{v} = 0 & \text{on } \partial\mathfrak{G}, \\ R\vec{v} = \vec{v} & \text{on } \partial Y. \end{cases} \qquad (1.30)$$

Note that, because $\mathfrak{G}$ is strictly included in Y, the boundary of $\mathfrak{P}$ is made of two disjoint parts, $\partial\mathfrak{G}$ and ∂Y. Note also that the compatibility condition for (1.30) is satisfied, namely, that the identity,

$$\int_\mathfrak{P} \nabla \cdot R\vec{v} = \int_{\partial\mathfrak{P}} (R\vec{v}) \cdot \vec{v}, \qquad (1.31)$$

is compatible with the right-hand side of (1.30). Furthermore, standard estimates for a nonhomogeneous Stokes system yields

$$\|R\vec{v}\|_{H^1(\mathfrak{P})^N} \leq C \|\vec{v}\|_{H^1(Y)^N}, \qquad (1.32)$$

where the constant C depends only on $\mathfrak{P}$. Thus, R is a linear continuous operator. Now, rescaling R from Y to any cell Y_i^ε, we obtain an operator R_ε acting from $H_0^1(\Omega)^N$ into $H_0^1(\mathfrak{P}^\varepsilon)^N$ and defined in each cell Y_i^ε by

$$\begin{cases} \nabla q_\varepsilon - \Delta R_\varepsilon \vec{v} = -\Delta \vec{v} & \text{in } \mathfrak{P}_i^\varepsilon, \\ \nabla \cdot R_\varepsilon \vec{v} = \nabla \cdot \vec{v} + \frac{1}{|\mathfrak{P}_i^\varepsilon|} \int_{\mathfrak{G}_i^\varepsilon} \nabla \cdot \vec{v} & \text{in } \mathfrak{P}_i^\varepsilon, \\ R_\varepsilon \vec{v} = 0 & \text{on } \partial\mathfrak{G}_i^\varepsilon, \\ R_\varepsilon \vec{v} = \vec{v} & \text{on } \partial Y_i^\varepsilon, \end{cases} \qquad (1.33)$$

and, by summation over i, satisfying the rescaled estimate

$$\|R_\varepsilon \vec{v}\|_{L^2(\mathfrak{P}^\varepsilon)}^2 + \varepsilon^2 \|\nabla(R_\varepsilon \vec{v})\|_{L^2(\mathfrak{P}^\varepsilon)}^2 \leq C \left(\|\vec{v}\|_{L^2(\Omega)}^2 + \varepsilon^2 \|\nabla \vec{v}\|_{L^2(\Omega)}^2 \right). \qquad (1.34)$$

Finally, the reader can easily check properties (1.27) and (1.28) for this operator R_ε.   Q.E.D.

We now have the main tools to complete the

Proof of Lemma 1.3. We begin with the estimate of the velocity. Multiplying Eq. (1.1) by $\vec{v}^\varepsilon$ and integrating by parts gives

$$\varepsilon^2 \mu \int_{\mathfrak{P}^\varepsilon} |\nabla \vec{v}^\varepsilon|^2 = \int_{\mathfrak{P}^\varepsilon} \vec{f} \cdot \vec{v}^\varepsilon. \qquad (1.35)$$

Using the Poincaré inequality (Lemma 1.6) in (1.35) leads to

$$\varepsilon^2 \|\nabla \vec{v}^\varepsilon\|^2_{L^2(\mathfrak{P}^\varepsilon)} \leq C\varepsilon \|\vec{f}\|_{L^2(\Omega)} \|\nabla \vec{v}^\varepsilon\|_{L^2(\mathfrak{P}^\varepsilon)}.$$

Thus,

$$\varepsilon \|\nabla \vec{v}^\varepsilon\|_{L^2(\mathfrak{P}^\varepsilon)} \leq C \|\vec{f}\|_{L^2(\Omega)}, \tag{1.36}$$

and, again, using the Poincaré inequality,

$$\|\vec{v}^\varepsilon\|_{L^2(\mathfrak{P}^\varepsilon)} \leq C \|\vec{f}\|_{L^2(\Omega)}. \tag{1.37}$$

We turn to the case of the pressure. Let us explain briefly why things are more delicate in this case. From Eq. (1.1), we easily find that ∇p^ε is uniformly bounded in $H^{-1}(\mathfrak{P}^\varepsilon)^N$. Then, a well-known theorem of functional analysis (see, e.g., Proposition 1.2, Chapter 1, [Tem79]) states that p^ε belongs to $L^2(\mathfrak{P}^\varepsilon)$ with the following estimate:

$$\|p^\varepsilon\|_{L^2(\mathfrak{P}^\varepsilon)/\mathbb{R}} \leq C(\mathfrak{P}^\varepsilon) \|\nabla p^\varepsilon\|_{H^{-1}(\mathfrak{P}^\varepsilon)^N}. \tag{1.38}$$

Unfortunately, (1.38) is useless because the constant depends on the domain $\mathfrak{P}^\varepsilon$ and, thus, may be not uniformly bounded when ε goes to zero. Consequently, a more subtle argument is required, as follows.

By duality, we define an extension of ∇p^ε, denoted by F_ε,

$$\langle F_\varepsilon, \vec{v} \rangle_{H^{-1}, H^1_0(\Omega)} = \langle \nabla p^\varepsilon, R_\varepsilon \vec{v} \rangle_{H^{-1}, H^1_0(\mathfrak{P}^\varepsilon)} \text{ for any } \vec{v} \in H^1_0(\Omega)^N. \tag{1.39}$$

Because R_ε is a linear operator from $H^1_0(\Omega)^N$ into $H^1_0(\mathfrak{P}^\varepsilon)^N$, F_ε belongs to $H^{-1}(\Omega)^N$. Eliminating ∇p^ε by using (1.1) and integrating by parts yields

$$\langle F_\varepsilon, \vec{v} \rangle_{H^{-1}, H^1_0(\Omega)} = \int_{\mathfrak{P}^\varepsilon} \vec{f} \cdot R_\varepsilon \vec{v} \, dx + \varepsilon^2 \int_{\mathfrak{P}^\varepsilon} \nabla \vec{v}^\varepsilon \cdot \nabla(R_\varepsilon \vec{v}) \, dx. \tag{1.40}$$

Using the estimates on $\vec{v}^\varepsilon$ and R_ε, (1.40) leads to

$$\left| \langle F_\varepsilon, \vec{v} \rangle_{H^{-1}, H^1_0(\Omega)} \right| \leq C \left(\|\vec{v}\|_{L^2(\Omega)} + \varepsilon \|\nabla \vec{v}\|_{L^2(\Omega)} \right), \tag{1.41}$$

which shows that F_ε is uniformly (i.e., independently of ε) bounded in $H^{-1}(\Omega)^N$. By property (1.28), if $\vec{v}$ is divergence-free, so is $R_\varepsilon \vec{v}$, and an integration by parts in (1.39) gives

$$\langle F_\varepsilon, \vec{v} \rangle_{H^{-1}, H^1_0(\Omega)} = 0, \text{ if } \nabla \cdot \vec{v} = 0.$$

Thus, F_ε, being orthogonal to divergence-free functions, is the gradient of some function P_ε which is bounded in $L^2(\Omega)$ (see e.g., Proposition 1.1, Chapter 1, [Tem79]). Furthermore, the weakly convergent subsequences of P_ε are actually strongly convergent in $L^2(\Omega)$. Replacing $\vec{v}$ in (1.41) by a sequence $\vec{v}_\varepsilon$, which converges weakly to 0 in $H^1_0(\Omega)$, and using Rellich theorem, which states that $\vec{v}_\varepsilon$ converges strongly to 0 in $L^2(\Omega)$,

$$\lim_{\varepsilon \to 0} \langle \nabla P_\varepsilon, \vec{v}_\varepsilon \rangle_{H^{-1}, H^1_0(\Omega)} = 0,$$

which proves the strong convergence of ∇P_ε in $H^{-1}(\Omega)^N$. This is equivalent to the strong convergence of P_ε in $L^2(\Omega)/\mathbb{R}$.

Property (1.27) implies that

$$\langle \nabla P_\varepsilon, \vec{v} \rangle_{H^{-1}, H_0^1(\Omega)} = \langle \nabla p^\varepsilon, \vec{v} \rangle_{H^{-1}, H_0^1(\mathfrak{P}^\varepsilon)} \text{ for any } \vec{v} \in H_0^1(\mathfrak{P}^\varepsilon)^N,$$

and, by virtue of inequality (1.38), that P_ε and p^ε are equal in $\mathfrak{P}^\varepsilon$ up to a constant (its value does not matter because a pressure is always defined up to a constant).

It remains to be proven that P_ε is identical to the extension $\tilde{p}^\varepsilon$ introduced in (1.23), i.e., that

$$P_\varepsilon = \frac{1}{|\mathfrak{P}_i^\varepsilon|} \int_{\mathfrak{P}_i^\varepsilon} p^\varepsilon \text{ in each } \mathfrak{S}_i^\varepsilon. \tag{1.42}$$

This is done in two steps. First, in definition (1.39), we introduce a smooth function $\vec{v}_\varepsilon$, with compact support in one of the solid parts $\mathfrak{S}_i^\varepsilon$. For such a function, $R_\varepsilon \vec{v}_\varepsilon$ is zero in $\mathfrak{P}_i^\varepsilon$, and, thus,

$$\langle \nabla P_\varepsilon, \vec{v} \rangle_{H^{-1}, H_0^1(\mathfrak{S}_i^\varepsilon)} = 0, \tag{1.43}$$

which implies that P_ε is constant in $\mathfrak{S}_i^\varepsilon$. In a second step, we choose a test function $\vec{v}_\varepsilon$, with compact support in the entire cell Y_i^ε. Integrating by parts in (1.39) leads to

$$\int_{Y_i^\varepsilon} P_\varepsilon \nabla \cdot \vec{v}_\varepsilon = \int_{\mathfrak{P}_i^\varepsilon} p^\varepsilon \nabla \cdot (R_\varepsilon \vec{v}_\varepsilon). \tag{1.44}$$

Using definition (1.33) of $\nabla \cdot (R_\varepsilon \vec{v}_\varepsilon)$ and the properties of P_ε (constant in $\mathfrak{S}_i^\varepsilon$, equal to p^ε in $\mathfrak{P}_i^\varepsilon$), (1.44) becomes

$$\int_{\mathfrak{P}_i^\varepsilon} p^\varepsilon \nabla \cdot \vec{v}_\varepsilon + P_{\varepsilon|\mathfrak{S}_i^\varepsilon} \int_{\mathfrak{S}_i^\varepsilon} \nabla \cdot \vec{v}_\varepsilon = \int_{\mathfrak{P}_i^\varepsilon} p^\varepsilon \nabla \cdot \vec{v}_\varepsilon + \frac{1}{|\mathfrak{P}_i^\varepsilon|} \left(\int_{\mathfrak{S}_i^\varepsilon} \nabla \cdot \vec{v}_\varepsilon \right) \left(\int_{\mathfrak{P}_i^\varepsilon} p^\varepsilon \right), \tag{1.45}$$

which gives the desired value

$$P_{\varepsilon|\mathfrak{S}_i^\varepsilon} = \frac{1}{|\mathfrak{P}_i^\varepsilon|} \left(\int_{\mathfrak{P}_i^\varepsilon} p^\varepsilon \right). \tag{1.46}$$

Q.E.D.

3.2   Inertia Effects

This section is devoted to a generalization of the previous model in which inertial effects are added to the Stokes equations, then, becoming the Navier–Stokes equations. To simplify the exposition we shall consider, successively and separately, the different inertial terms arising in the equations. A first subsection is concerned with the linear, evolutionary Stokes equations. A second focuses on nonlinear, steady Navier–Stokes equations. Of course, these two cases could be combined together with no special difficulties, but at the price of unnecessary and lengthy technical details.

The geometrical situation is the same as that of section 3.1, namely, a periodic porous domain Ω and its fluid part $\mathfrak{P}^\varepsilon$ are considered, with period ε and unit cell Y. For a precise description of $\mathfrak{P}^\varepsilon$, the reader is referred to definition (1.3) above.

3.2.1   Darcy's Law with Memory

We consider the unsteady Stokes equations in the fluid domain $\mathfrak{P}^\varepsilon$ with a no-slip (Dirichlet) boundary condition. We denote, by $\vec{v}^\varepsilon$ and p^ε, the velocity and pressure of the fluid, $\vec{f}$ the density of forces acting on the fluid, and $\vec{v}_\varepsilon^0$ an initial condition for the velocity. We assume that the density of the fluid is equal to 1, whereas its viscosity is very small, and, indeed, is exactly $\mu\varepsilon^2$ (where ε is the pore size). The system of equations is

$$\begin{cases} \frac{\partial \vec{v}^\varepsilon}{\partial t} + \nabla p^\varepsilon - \varepsilon^2 \mu \Delta \vec{v}^\varepsilon = \vec{f} & \text{in } (0, T) \times \mathfrak{P}^\varepsilon, \\ \nabla \cdot \vec{v}^\varepsilon = 0 & \text{in } (0, T) \times \mathfrak{P}^\varepsilon, \\ \vec{v}^\varepsilon = 0 & \text{on } (0, T) \times \partial \mathfrak{P}^\varepsilon, \\ \vec{v}^\varepsilon(t = 0, x) = \vec{v}_\varepsilon^0(x) & \text{in } \mathfrak{P}^\varepsilon \text{ at time } t = 0. \end{cases} \tag{2.1}$$

The scaling ε^2 of the viscosity is the same as in section 3.1. However, here it is not a simple change of variable because the density in front of the inertial term has been scaled to 1. The scalings in system (2.1) are precisely those which give a nonzero limit for the velocity $\vec{v}^\varepsilon$ and a limit problem depending on time. In particular, (2.1) is not equivalent to the following system (which gives rise to a different homogenized system with no inertial term in the limit):

$$\begin{cases} \frac{\partial \vec{v}^\varepsilon}{\partial t} + \nabla p^\varepsilon - \mu \Delta \vec{v}^\varepsilon = \vec{f} & \text{in } (0, T) \times \mathfrak{P}^\varepsilon, \\ \nabla \cdot \vec{v}^\varepsilon = 0 & \text{in } (0, T) \times \mathfrak{P}^\varepsilon, \\ \vec{v}^\varepsilon = 0 & \text{on } (0, T) \times \partial \mathfrak{P}^\varepsilon, \\ \vec{v}^\varepsilon(t = 0, x) = \vec{v}_\varepsilon^0(x) & \text{in } \mathfrak{P}^\varepsilon \text{ at time } t = 0. \end{cases} \tag{2.2}$$

In some sense, the scaling of the viscosity in (2.1) can also be interpreted as a choice of the time scale of the order of the pore size squared. System (2.1) has been first studied by J.–L. Lions [Lio81], using formal asymptotic expansions. Rigorous homogenization results were proven later in [All92b]. A study of the different system (2.2) may be found in [Mik91].

To obtain an existence result and convenient a priori estimates for the solution of (2.1), the force $\vec{f}(t, x)$ is assumed to belong to $L^2((0, T) \times \Omega)^N$, and the initial condition $\vec{v}_\varepsilon^0(x)$ to $H_0^1(\mathfrak{P}^\varepsilon)^N$. Furthermore, denoting by $\tilde{v}_\varepsilon^0$ the extension by zero in the solid part $\Omega \setminus \mathfrak{P}^\varepsilon$ of the initial condition, we assume that it satisfies

$$\begin{cases} \|\tilde{v}_\varepsilon^0\|_{L^2(\Omega)} + \varepsilon \|\nabla \tilde{v}_\varepsilon^0\|_{L^2(\Omega)} \leq C, \\ \nabla \cdot \tilde{v}_\varepsilon^0 = 0 \text{ in } \Omega, \\ \tilde{v}_\varepsilon^0(x) \text{ two-scale converges to a unique limit } \vec{v}^0(x, y). \end{cases} \tag{2.3}$$

Then, standard theory (see, e.g., [Tem79] and [LM68]) yields the following:

Proposition 2.1 *Under assumption (2.3) on the initial condition, the Stokes equations (2.1) admit a unique solution $\vec{v}^\varepsilon \in L^2\left((0, T); H_0^1(\mathfrak{P}^\varepsilon)\right)^N$ and $p^\varepsilon \in L^2\left((0, T); L^2(\mathfrak{P}^\varepsilon)/\mathbb{R}\right)$. Furthermore, the extension by zero in the solid part $\Omega \setminus \mathfrak{P}^\varepsilon$ of the velocity $\tilde{v}^\varepsilon$ satisfies the a priori estimates,*

$$\|\tilde{v}^\varepsilon\|_{L^\infty((0,T);L^2(\Omega))} + \varepsilon\|\nabla\tilde{v}^\varepsilon\|_{L^\infty((0,T);L^2(\Omega))} \leq C \text{ and } \|\frac{\partial\tilde{v}^\varepsilon}{\partial t}\|_{L^2((0,T)\times\Omega)} \leq C, \quad (2.4)$$

where the constant C does not depend on ε.

The following homogenization theorem states that the limit problem is a Darcy law with memory (due to the convolution in time) which generalizes the usual Darcy law.

Theorem 2.2 *There exists an extension $(\tilde{v}^\varepsilon, \tilde{p}^\varepsilon)$ of the solution $(\vec{v}^\varepsilon, p^\varepsilon)$ of (2.1) which converges weakly in*

$$L^2\left((0, T); L^2(\Omega)^N\right) \times L^2\left((0, T); L^2(\Omega)/\mathbb{R}\right)$$

to the unique solution $(\vec{v}, p)$ of the homogenized problem:

$$\begin{cases} \vec{v}(t, x) = \vec{v}_{init}(x), \\ \quad + \frac{1}{\mu}\int_0^t A(t - s)\left(\vec{f} - \nabla p\right)(s, x)ds & in\ (0, T) \times \Omega, \\ \nabla \cdot \vec{v}(t, x) = 0 & in\ (0, T) \times \Omega, \\ \vec{v}(t, x) \cdot \vec{\nu} = 0 & on\ (0, T) \times \partial\Omega, \end{cases} \quad (2.5)$$

where $\vec{v}_{init}(x)$ is an initial condition, which depends only on the sequence $\vec{v}_\varepsilon^0$ and on the microstructure $\mathfrak{P}$, and $A(t)$ is a symmetric, positive definite, time-dependent, permeability tensor which depends only on the microstructure $\mathfrak{P}$ (their precise form is to be found in the proof of the present theorem).

The complicated form of the homogenized problem (2.5), which is not a parabolic p.d.e. but an integro-differential equation, is due to the elimination of a *hidden* microscopic variable. Actually, to prove Theorem 2.2, we first prove a result of the corresponding two-scale homogenized system which includes this hidden variable and has a much nicer form (for an introduction to two-scale convergence, see Appendix A).

Theorem 2.3 *Under assumption (2.3) on the initial condition, there exists an extension $(\tilde{v}^\varepsilon, \tilde{p}^\varepsilon)$ of the solution of (2.1) which two-scale converges to the unique* *solution $(\vec{v}_0(x, y), p(x))$ of the two-scale homogenized problem:*

$$
\begin{cases}
\frac{\partial \vec{v}_0}{\partial t}(x, y), \\
\quad + \nabla_y p_1(x, y) + \nabla_x p(x) - \mu \Delta_{yy} \vec{v}_0(x, y) = \vec{f}(x) & \text{in } (0, T) \times \Omega \times \mathfrak{P}, \\
\nabla_y \cdot \vec{v}_0(x, y) = 0 & \text{in } (0, T) \times \Omega \times \mathfrak{P}, \\
\nabla_x \cdot \left(\int_Y \vec{v}_0(x, y) dy \right) = 0 & \text{in } (0, T) \times \Omega, \\
\vec{v}_0(x, y) = 0 & \text{in } (0, T) \times \Omega \times \mathfrak{S}, \\
\left(\int_Y \vec{v}_0(x, y) dy \right) \cdot \vec{v} = 0 & \text{on } (0, T) \times \partial\Omega, \\
y \rightarrow \vec{v}_0, p_1 & Y\text{-periodic}, \\
\vec{v}_0(0, x, y) = \vec{v}^0(x, y) & \text{at time } t = 0.
\end{cases}
$$

$$(2.6)$$

Remark The two-scale homogenized problem (2.6) is also called a two-pressure Stokes system (see [Lio81]). Eliminating y in (2.6) yields Darcy's law with memory (2.5). It is not difficult to check that both $\vec{v}(t, x)$ and $A(t, x)$ decay exponentially in time. Thus, the permeability keeps track, mainly, of the recent history. If the force $\vec{f}$ is steady (i.e., does not depend on t), asymptotically, for large time t, the usual steady Darcy's law for $\vec{v}$ and p is recovered. As we shall see, the two-scale homogenized problem (2.6) is equivalent to (2.5), complemented with the cell problems (2.7)–(2.8).

The proof of Theorem 2.3 is completely parallel to that of Theorem 1.4 in section 3.1. In particular, the extension of the pressure is built in the same way. Therefore, we skip this proof which may be found in [All92b], if necessary. Note that the notion of two-scale convergence (which is heavily used here) is easily adapted to time-dependent functions.

Proof of Theorem 2.2. The only thing to prove is that eliminating the microscopic variable y in (2.6) leads to Darcy's law with memory (2.5). The solution $\vec{v}_0$ is decomposed into two parts $\vec{v}_1 + \vec{v}_2$, where $\vec{v}_1$ is just the evolution (without any forcing term) of the initial condition $\vec{v}^0$. Thus, at each point $x \in \Omega$, $\vec{v}_1$ is the unique solution of an equation posed solely in $\mathfrak{P}$:

$$
\begin{cases}
\frac{\partial \vec{v}_1}{\partial t}(x, y) + \nabla_y q(x, y) - \mu \Delta_{yy} \vec{v}_1(x, y) = 0 & \text{in } (0, T) \times \mathfrak{P}, \\
\nabla_y \cdot \vec{v}_1(x, y) = 0 & \text{in } (0, T) \times \mathfrak{P}, \\
\vec{v}_1(x, y) = 0 & \text{in } (0, T) \times \mathfrak{S}, \\
y \rightarrow \vec{v}_1, q & Y\text{-periodic}, \\
\vec{v}_1(0, x, y) = \vec{v}^0(x, y) & \text{at time } t = 0.
\end{cases}
$$

$$(2.7)$$

The average of $\vec{v}_1$ in y is just $\vec{v}(t, x)$ (the initial condition in the homogenized system (2.5)). On the other hand, $\vec{v}_2$ is given by

$$
\vec{v}_2(t, x, y) = \int_0^t \sum_{i=1}^N \left(f_i - \frac{\partial p}{\partial x_i} \right)(s, x) \frac{\partial \vec{w}_i}{\partial t}(t - s, y) ds,
$$

where, for $1 \leq i \leq N$, $\vec{w}_i$ is the unique solution of the cell problem, which does not depend on the macroscopic variable x. The cell problem is defined by

$$
\begin{cases}
\frac{\partial \vec{w}_i}{\partial t}(y) + \nabla_y \pi_i(y) - \mu \Delta_{yy} \vec{w}_i(y) = \vec{e}_i & \text{in } (0, T) \times \mathfrak{P}, \\
\nabla_y \cdot \vec{w}_i(y) = 0 & \text{in } (0, T) \times \mathfrak{P}, \\
\vec{w}_i(y) = 0 & \text{in } (0, T) \times \mathfrak{S}, \\
y \rightarrow \vec{w}_i, \pi_i & Y\text{-periodic}, \\
\vec{w}_i(0, y) = 0 & \text{at time } t = 0.
\end{cases}
\tag{2.8}
$$

Introducing the matrix A defined by

$$
A_{ij}(t) = \mu \int_{\mathfrak{P}} \frac{\partial \vec{w}_i}{\partial t}(t, y) \vec{e}_j \, dy,
\tag{2.9}
$$

Darcy's law with memory is easily deduced from the two-scale homogenized problem by averaging $\vec{v}_1$ and $\vec{v}_2$ with respect to y. Eventually, using semigroup theory and integrating by parts in the cell problem (2.9), one can prove that A is symmetric, positive definite, and decays exponentially with time.   Q.E.D.

3.2.2 Nonlinear Darcy's Law

We consider the steady Navier–Stokes equations

$$
\begin{cases}
\varepsilon^\gamma \vec{v}^\varepsilon \cdot \nabla \vec{v}^\varepsilon + \nabla p^\varepsilon - \varepsilon^2 \mu \Delta \vec{v}^\varepsilon = \vec{f} & \text{in } \mathfrak{P}^\varepsilon, \\
\nabla \cdot \vec{v}^\varepsilon = 0 & \text{in } \mathfrak{P}^\varepsilon, \\
\vec{v}^\varepsilon = 0 & \text{on } \partial \mathfrak{P}^\varepsilon.
\end{cases}
\tag{2.10}
$$

As before, the fluid viscosity μ has been scaled by a factor ε^2, implying precisely that the velocity $\vec{v}^\varepsilon$ has a nonzero limit. The nonlinear convective term has also been scaled by a factor ε^γ, where γ is a positive constant such that $\gamma \geq 1$. The limit case $\gamma = 1$ corresponds exactly to the scaling which yields a nonlinear homogenized problem. The case $\gamma = 4$ allows replacing $\varepsilon^2 \vec{v}^\varepsilon$ by a new velocity $\vec{v}_\varepsilon$ which satisfies the usual *unscaled* Navier–Stokes equations. Intermediate values of γ are analyzed below. For larger values, the convective terms are much smaller than the viscous ones, and the Navier–Stokes equations are just a small perturbation of the Stokes equations. For values smaller than 1, the opposite situation arises, and convective terms dominate viscous ones. Unfortunately, in this last case, the homogenized limit is unclear.

We begin with a result of Mikelić [Mik91] which states that, for $\gamma > 1$, the convective term of the Navier–Stokes equations disappears in the limit, and the homogenized system is the usual Darcy law (as in section 3.1). The only price to pay is a weaker convergence of the pressure: the closer γ to 1, the weaker the estimate on the pressure.

Theorem 2.4 *Let $\gamma > 1$. Define a constant $\beta > 1$ by*

$$
\beta = \min \left(2, \frac{N}{N-2}, \frac{N}{N+2-2\gamma} \right).
\tag{2.11}
$$

There exists an extension $(\tilde{v}^\varepsilon, \tilde{p}^\varepsilon)$ of the solution $(\vec{v}^\varepsilon, p^\varepsilon)$ of (2.10) such that the velocity $\tilde{v}^\varepsilon$ converges weakly in $L^2(\Omega)^N$ to $\vec{v}$, and the pressure $\tilde{p}^\varepsilon$ converges strongly in $L^{q'}(\Omega)/\mathbb{R}$ to p, for any $1 < q' < \beta$, where $(\vec{v}, p)$ is the unique solution of the homogenized problem, a linear Darcy law,

$$\begin{cases} \vec{v}(x) = \frac{1}{\mu} A\left(\vec{f}(x) - \nabla p(x)\right) & \text{in } \Omega, \\ \nabla \cdot \vec{v}(x) = 0 & \text{in } \Omega, \\ \vec{v}(x) \cdot \vec{v} = 0 & \text{on } \partial\Omega. \end{cases} \tag{2.12}$$

In (2.12), the permeability tensor A is the usual homogenized matrix for Darcy's law, defined by (1.5) in Theorem 1.1.

The proof of Theorem 2.4 is briefly sketched below. Before that, we consider the limit case $\gamma = 1$ which yields a nonlinear Darcy-type law (or a Dupuit–Forchheimer–Ergun law; see, e.g., [JNP82]). The nonlinear convective term does not disappear in the homogenized problem indicating quadratic behavior of Darcy's law.

Theorem 2.5 Let $\gamma = 1$, and let the space dimension be $N = 2$. Let $f \in \mathcal{D}(\Omega)$ be a smooth, compactly supported function such that its norm in $H^3(\Omega)$ is sufficiently small. Then, there exists an extension $(\tilde{v}^\varepsilon, \tilde{p}^\varepsilon)$ of the solution $(\vec{v}^\varepsilon, p^\varepsilon)$ of (2.10) and a unique solution $(\vec{v}_0, p_1, p)$ of the homogenized system,

$$\begin{cases} \nabla_y p_1 + \vec{v}_0 \cdot \nabla_y \vec{v}_0 - \mu \Delta_{yy} \vec{v}_0 = \vec{f}(x) - \nabla p(x) & \text{in } \mathfrak{P} \times \Omega, \\ \nabla_y \cdot \vec{v}_0 = 0 & \text{in } \mathfrak{P} \times \Omega, \\ \nabla_x \cdot \left(\int_Y \vec{v}_0 dy\right) = 0 & \text{in } \Omega, \\ \vec{v}_0 = 0 & \text{in } \mathfrak{S} \times \Omega, \\ \left(\int_Y \vec{v}_0 dy\right) \cdot \vec{v} = 0 & \text{on } \partial\Omega, \\ y \to (\vec{v}_0, p_1) & Y\text{-periodic}, \end{cases} \tag{2.13}$$

such that $\tilde{p}^\varepsilon$ converges strongly in $L^q(\Omega)/\mathbb{R}$ to p, for any $1 < q < 2$, and

$$\|\tilde{v}^\varepsilon(x) - \vec{v}_0(x, \frac{x}{\varepsilon})\|_{L^2(\Omega)^2} \le C\varepsilon. \tag{2.14}$$

The homogenized system (2.13), called a two-pressure Navier–Stokes system, is very similar to the two-scale homogenized system (1.10). It is not possible to eliminate the y variable to obtain an explicit macroscopic effective law. Therefore, the nonlinear Darcy law is not a local, explicit, partial differential equation. Such a homogenized problem has been derived, formally, in [SP80] and [Lio81]. A rigorous proof of convergence has recently been given in the two-dimensional case by Mikelić [Mik95]. It is conjectured that the result also holds true in 3-D. The proof of Theorem 2.5 is very technical and is not reproduced here (the key argument is to prove an existence and uniqueness result for the homogenized system (2.13) by using an abstract fixed-point argument).

Proof of Theorem 2.4. The proof is identical to that of Theorem 1.1 except for the part on estimating the pressure. Recall that a pressure extension $\tilde{p}^\varepsilon$ was defined by duality,

$$\langle \nabla \tilde{p}^\varepsilon, \vec{v} \rangle_{H^{-1}, H_0^1(\Omega)} = \langle \nabla p^\varepsilon, R_\varepsilon \vec{v} \rangle_{H^{-1}, H_0^1(\mathfrak{P}^\varepsilon)} \text{ for any } \vec{v} \in H_0^1(\Omega)^N, \quad (2.15)$$

where R_ε is a restriction operator from $H_0^1(\Omega)^N$ into $H_0^1(\mathfrak{P}^\varepsilon)^N$, defined in Lemma 1.7. It is not difficult to check, by a standard regularity result for Stokes equations (see e.g., [Tem79]), that R_ε is also a restriction operator from $W_0^{1,q}(\Omega)^N$ into $W_0^{1,q}(\mathfrak{P}^\varepsilon)^N$, for any $1 < q < \infty$, satisfying

$$\|R_\varepsilon \vec{v}\|_{L^q(\mathfrak{P}^\varepsilon)} + \varepsilon \|\nabla(R_\varepsilon \vec{v})\|_{L^q(\mathfrak{P}^\varepsilon)} \le C \left(\|\vec{v}\|_{L^q(\Omega)} + \varepsilon \|\nabla \vec{v}\|_{L^q(\Omega)} \right), \quad (2.16)$$

for any $v \in W_0^{1,q}(\Omega)^N$.

To obtain an estimate on $\nabla \tilde{p}^\varepsilon$ in $W^{-1,q'}(\Omega)^N$ (the dual of $W_0^{1,q}(\Omega)^N$ with $1/q + 1/q' = 1$), ∇p^ε is eliminated in (2.15) by using the Stokes equation. This yields

$$\langle \nabla \tilde{p}^\varepsilon, \vec{v} \rangle_{W^{-1,q'}, W_0^{1,q}(\Omega)} = \int_{\mathfrak{P}^\varepsilon} \vec{f} \cdot R_\varepsilon \vec{v} dx + \varepsilon^2 \int_{\mathfrak{P}^\varepsilon} \nabla \vec{v}^\varepsilon \cdot \nabla(R_\varepsilon \vec{v}) dx$$
$$+ \varepsilon^\gamma \int_{\mathfrak{P}^\varepsilon} \vec{v}^\varepsilon \cdot (\vec{v}^\varepsilon \cdot \nabla(R_\varepsilon \vec{v})) dx. \quad (2.17)$$

The two first terms in the right-hand side of (2.17) are bounded as before (recall that $W_0^{1,q}(\Omega) \subset H_0^1(\Omega)$ for $q \ge 2$). The third term is bounded by

$$\varepsilon^\gamma \|\tilde{v}^\varepsilon\|_{L^{q'}(\Omega)}^2 \|\nabla(R_\varepsilon \vec{v})\|_{L^q(\Omega)} \le \varepsilon^{\gamma - 1} \|\tilde{v}^\varepsilon\|_{L^{2q'}(\Omega)}^2 \|\vec{v}\|_{W_0^{1,q}(\Omega)}.$$

A combination of Hölder and Sobolev inequalities yields

$$\|\tilde{v}^\varepsilon\|_{L^{2q'}(\Omega)} \le C \|\tilde{v}^\varepsilon\|_{L^2(\Omega)}^{1 - N/2q} \|\nabla \tilde{v}^\varepsilon\|_{L^2(\Omega)}^{N/2q},$$

if $2q' \le \frac{2N}{N-2}$ (recall that $H^1(\Omega) \subset L^p(\Omega)$ for any $1 \le p \le \frac{2N}{N-2}$, excluding $p = \infty$ when $N = 2$). Therefore,

$$\left| \varepsilon^\gamma \int_{\mathfrak{P}^\varepsilon} \vec{v}^\varepsilon \cdot \vec{v}^\varepsilon \cdot \nabla(R_\varepsilon \vec{v}) dx \right| \le C \varepsilon^{\gamma - 1 - \frac{N}{2q}} \|\vec{v}\|_{W_0^{1,q}(\Omega)},$$

which goes to zero for q chosen sufficiently large. This allows obtaining the strong convergence of the pressure $\tilde{p}^\varepsilon$ as stated in Theorem 2.4.   Q.E.D.

3.3 Derivation of Brinkman's Law

3.3.1 Setting of the Problem

This section is devoted to the derivation of Brinkman's law for an incompressible viscous fluid flowing in a porous medium. As in the previous sections, we start from the steady Stokes equations in a periodic porous medium, with a no-slip (Dirichlet) boundary condition on the solid pores. We assume that the solid part of the porous medium is a collection of periodically distributed obstacles. We denote by ε the period, or the interobstacle distance. The major difference from the previous sections is that the solid obstacles are assumed to be much smaller than the period ε. Their size is denoted by $a_\varepsilon \ll \varepsilon$. There are now two small parameters in our asymptotic analysis which means that two-scale asymptotic expansions or two-scale convergence cannot be used in the sequel. Therefore, the homogenization process relies on the energy method (see Appendix A). The assumption on the periodicity of the porous medium allows simplifying the results greatly, although it is not strictly necessary (and not very realistic). As before, the porous medium is denoted by Ω, and its fluid part by $\mathfrak{P}^\varepsilon$.

The motion of the fluid in $\mathfrak{P}^\varepsilon$ is governed by the steady Stokes equations, complemented with a Dirichlet boundary condition. We denote by $\vec{v}^\varepsilon$ and p^ε the velocity and pressure of the fluid, μ its viscosity (a positive constant), and $\vec{f}$ the density of forces acting on the fluid ($\vec{v}^\varepsilon$ and $\vec{f}$ are vector-valued functions, whereas p^ε is scalar). The microsocopic model is

$$
\begin{cases}
\nabla p^\varepsilon - \mu \Delta \vec{v}^\varepsilon = \vec{f} & \text{in } \mathfrak{P}^\varepsilon, \\
\nabla \cdot \vec{v}^\varepsilon = 0 & \text{in } \mathfrak{P}^\varepsilon, \\
\vec{v}^\varepsilon = 0 & \text{on } \partial \mathfrak{P}^\varepsilon,
\end{cases}
\tag{3.1}
$$

which admits a unique solution $(\vec{v}^\varepsilon, p^\varepsilon)$ in $H_0^1(\mathfrak{P}^\varepsilon)^N \times L^2(\mathfrak{P}^\varepsilon)/\mathbb{R}$ if $\vec{f}(x) \in L^2(\Omega)^N$ (see, e.g., [Tem79]).

Let us describe the assumptions on the porous domain more precisely. It is contained in a bounded domain $\Omega \subset \mathbb{R}^N$, and its fluid part is denoted by $\mathfrak{P}^\varepsilon$. The set Ω is covered by a regular periodic mesh of period ε. At the center of each cell Y_i^ε (equal to $(0, \varepsilon)^N$, up to a translation), a solid obstacle $\mathcal{O}_i^\varepsilon$ is placed which is obtained by rescaling a unit obstacle $\mathcal{O}$ to the size a_ε (i.e., $\mathcal{O}_i^\varepsilon = a_\varepsilon \mathcal{O}$, up to a translation). The unit obstacle $\mathcal{O}$ is a nonempty, smooth, closed set included in the unit cell such that $Y \setminus \mathcal{O}$ is a smooth, connected, open set. The fluid part $\mathfrak{P}^\varepsilon$ of the porous medium is defined by

$$
\mathfrak{P}^\varepsilon = \Omega \setminus \bigcup_{i=1}^{N(\varepsilon)} \mathcal{O}_i^\varepsilon,
\tag{3.2}
$$

where the number of obstacles $N(\varepsilon) = |\Omega| \varepsilon^{-N} (1 + o(1))$. A fundamental assumption is that the obstacles are *much smaller* than the period,

$$
\lim_{\varepsilon \to 0} \frac{a_\varepsilon}{\varepsilon} = 0.
\tag{3.3}
$$

3.3.2 Principal Results

According to the various scalings of the obstacle size a_ε in terms of the inter-obstacle distance ε, different limit problems arise. To sort these different regimes, we introduce a ratio σ_ε defined by

$$
\sigma_\varepsilon =
\begin{cases}
\left(\dfrac{\varepsilon^N}{a_\varepsilon^{N-2}} \right)^{1/2} & \text{for } N \geq 3, \\[2ex]
\varepsilon \left| \log \left(\dfrac{a_\varepsilon}{\varepsilon} \right) \right|^{1/2} & \text{for } N = 2.
\end{cases}
\tag{3.4}
$$

As usual, in perforated domains like $\mathfrak{P}^{\varepsilon}$, the sequence of solutions $(\vec{v}^{\varepsilon}, p^{\varepsilon})$, undefined in a fixed Sobolev space and *independent of* ε, needs to be extended to the whole domain Ω. We denote by $(\tilde{v}^{\varepsilon}, \tilde{p}^{\varepsilon})$ such an extension, which coincides, by definition, with $(\vec{v}^{\varepsilon}, p^{\varepsilon})$ on $\mathfrak{P}^{\varepsilon}$.

Theorem 3.1 *According to the scaling of the obstacle size, there are three different flow regimes.*

1. *If the obstacles are too small, i.e.,* $\lim_{\varepsilon \to 0} \sigma_{\varepsilon} = +\infty$, *then, the extended solution* $(\tilde{v}^{\varepsilon}, \tilde{p}^{\varepsilon})$ *of (3.1) converges strongly in* $H_0^1(\Omega)^N \times L^2(\Omega)/\mathbb{R}$ *to* $(\vec{v}, p)$, *the unique solution of the homogenized Stokes equations,*

$$\begin{cases} \nabla p - \mu \Delta \vec{v} = \vec{f} & \text{in } \Omega, \\ \nabla \cdot \vec{v} = 0 & \text{in } \Omega, \\ \vec{v} = 0 & \text{on } \partial\Omega. \end{cases} \tag{3.5}$$

2. *If the obstacles have a critical size, i.e.,* $\lim_{\varepsilon \to 0} \sigma_{\varepsilon} = \sigma > 0$, *then, the extended solution* $(\tilde{v}^{\varepsilon}, \tilde{p}^{\varepsilon})$ *of (3.1) converges weakly in* $H_0^1(\Omega)^N \times L^2(\Omega)/\mathbb{R}$ *to* $(\vec{v}, p)$, *the unique solution of the Brinkman law,*

$$\begin{cases} \nabla p - \mu \Delta \vec{v} + \frac{\mu}{\sigma^2} M \vec{v} = \vec{f} & \text{in } \Omega, \\ \nabla \cdot \vec{v} = 0 & \text{in } \Omega, \\ \vec{v} = 0 & \text{on } \partial\Omega. \end{cases} \tag{3.6}$$

3. *If the obstacles are too big, i.e.,* $\lim_{\varepsilon \to 0} \sigma_{\varepsilon} = 0$, *then, the rescaled solution* $(\frac{\tilde{v}^{\varepsilon}}{\sigma_{\varepsilon}^2}, \tilde{p}^{\varepsilon})$ *of (3.1) converges strongly in* $L^2(\Omega)^N \times L^2(\Omega)/\mathbb{R}$ *to* $(\vec{v}, p)$, *the unique solution of Darcy's law,*

$$\begin{cases} \vec{v} = \frac{1}{\mu} M^{-1} \left(\vec{f} - \nabla p \right) & \text{in } \Omega, \\ \nabla \cdot \vec{v} = 0 & \text{in } \Omega, \\ \vec{v} \cdot \vec{v} = 0 & \text{on } \partial\Omega. \end{cases} \tag{3.7}$$

In all regimes, M is the same $N \times N$ symmetric matrix, which depends only on the model obstacle $\mathcal{O}$ (its inverse, M^{-1}, plays the role of a permeability tensor).

The following proposition gives the precise definition of M in terms of *local* problems around the unit obstacle $\mathcal{O}$. Mathematically speaking, M can be interpreted in terms of the *hydrodynamic capacity* of the set $\mathcal{O}$. From a physical point of view, the ith column of M is the drag force of the local Stokes flow around $\mathcal{O}$ in the ith direction. Thus, M may be interpreted as the slowing effect of the obstacles in the homogenized limit.

Proposition 3.2 *According to the space dimension N, the local Stokes problem and the matrix M are defined as follows.*

1. *If $N \geq 3$, for $1 \leq i \leq N$, the cell problems are*

$$\begin{cases} \nabla \pi_i - \Delta \vec{w}_i = 0 & in \; \mathbb{R}^N \setminus \mathcal{O}, \\ \nabla \cdot \vec{w}_i = 0 & in \; \mathbb{R}^N \setminus \mathcal{O}, \\ \vec{w}_i = 0 & in \; \mathcal{O}, \\ \vec{w}_i \to \vec{e}_i & at \; \infty. \end{cases} \tag{3.8}$$

The matrix M is defined by its entries

$$M_{ij} = \int_{\mathbb{R}^N \setminus \mathcal{O}} \nabla \vec{w}_i \cdot \nabla \vec{w}_j dx, \tag{3.9}$$

or, equivalently, by its columns, equal to the drag forces applied on $\mathcal{O}$ by the local Stokes flows

$$M\vec{e}_i = \int_{\partial \mathcal{O}} \left(\frac{\partial \vec{w}_i}{\partial n} - \pi_i \vec{v} \right). \tag{3.10}$$

2. *If $N = 2$, for $1 \leq i \leq 2$, the cell problems are*

$$\begin{cases} \nabla \pi_i - \Delta \vec{w}_i = 0 & in \; \mathbb{R}^2 \setminus \mathcal{O}, \\ \nabla \cdot \vec{w}_i = 0 & in \; \mathbb{R}^2 \setminus \mathcal{O}, \\ \vec{w}_i = 0 & in \; \mathcal{O}, \\ \vec{w}_i(x) \sim \vec{e}_i \log(|x|) & as \; |x| \to \infty. \end{cases} \tag{3.11}$$

The matrix M is defined by its columns, equal to the drag forces applied on $\mathcal{O}$ by the local Stokes flows

$$M\vec{e}_i = \int_{\partial \mathcal{O}} \left(\frac{\partial \vec{w}_i}{\partial n} - \pi_i \vec{v} \right). \tag{3.12}$$

Furthermore, whatever the shape or the size of the obstacle $\mathcal{O}$, in two space dimensions, the matrix M is always the same,

$$M = 4\pi \, Id. \tag{3.13}$$

In the supercritical case, the homogenized problem is a Darcy law with a permeability tensor M^{-1} (see (3.7)). This tensor has nothing to do with that, denoted by A, obtained in a two-scale periodic setting (see Theorem 1.1 in section 3.1). Recall that, here, the obstacles are much smaller than the period (see assumption (3.3)). The two matrices are not computed with the same local problems which are posed in a single periodicity cell in section 3.1 and in the whole space $\mathbb{R}^N$, here. However, it was proven in [All91a] that M^{-1} is the rescaled limit of A, when the obstacle size goes to zero in the unit periodic cell Y.

Remark The surprising result in two dimensions, that M is always equal to $4\pi \, Id$, is actually a consequence of the well-known *Stokes paradox*. This paradox asserts that there exists no solution of the local problem (3.8) in two dimensions (it explains why the growth condition at infinity in (3.11) is different from the higher dimensional cases). Notice also that the critical size yielding Brinkman's law changes drastically from 2-D, $a_\varepsilon = e^{-\sigma^2/\varepsilon^2}$, to 3-D, $a_\varepsilon = \sigma^2\varepsilon^3$.

The proof of Theorem 3.1 relies on the energy method of Tartar as adapted to the present framework by Cioranescu and Murat [CM82] and [CM95] (see Appendix A for a brief introduction to the energy method). Let us sketch the main idea of this method. It consists in multiplying the Stokes equation (3.1) by a test function, integrating by parts, and passing to the limit, as $\varepsilon \to 0$, to obtain the variational formulation of the homogenized problem. The key difficulty, here, is that the test function must belong to $H_0^1(\mathfrak{P}^\varepsilon)^N$, i.e., it has to vanish on the obstacles for any value of ε. Of course it is not the case for a nonzero, fixed, test function φ. Therefore, *boundary layers* $\left(\vec{w}_i^\varepsilon\right)_{1 \le i \le N}$ have to be constructed such that, $\vec{\varphi}$ being a smooth vector field, $\sum_{i=1}^N \varphi_i \vec{w}_i^\varepsilon$ belongs to $H_0^1(\mathfrak{P}^\varepsilon)^N$ and converges to $\vec{\varphi}$, as ε goes to zero. These boundary layers $\left(\vec{w}_i^\varepsilon\right)_{1 \le i \le N}$ are constructed with the help of the solutions $(\vec{w}_i)_{1 \le i \le N}$ of the local Stokes problems from Proposition 3.2 by rescaling them to the size a_ε around each obstacle and pasting each contribution at the cell boundary ∂Y_i^ε. Loosely speaking, $\vec{w}_i^\varepsilon$ is a divergence-free vector field which vanishes on the obstacles and is almost equal to the unit vector $\vec{e}_i$ far from the obstacles. Using this test function, $\sum_{i=1}^N \varphi_i \vec{w}_i^\varepsilon$, the energy method gives the desired result (see [All91c] and [All91d] for details). Another interest of the boundary layers $\left(\vec{w}_i^\varepsilon\right)_{1 \le i \le N}$ is that they permit to obtain *corrector results*. For example, in the critical case and under a mild smoothness assumption on the Brinkman velocity $\vec{v}$, the weak convergence in $H_0^1(\Omega)^N$ of $\tilde{v}^\varepsilon$ can be improved in

$$\left(\tilde{v}^\varepsilon - \sum_{i=1}^N v_i \vec{w}_i^\varepsilon\right) \to 0 \text{ strongly in } H_0^1(\Omega)^N.$$

Finally, as in section 3.1 on the derivation of Darcy's law, a technical difficulty is constructing a bounded extension of the pressure p^ε. Basically, the same technique described in section 3.1.3, works in the present situation at the price of many additional technicalities (see [All91c] for details). As usual, extending the velocity is easier: it suffices to take it equal to 0 inside the obstacles,

$$\begin{cases} \tilde{v}^\varepsilon = \vec{v}^\varepsilon & \text{in } \mathfrak{P}^\varepsilon, \\ \tilde{v}^\varepsilon = 0 & \text{in } \Omega \setminus \mathfrak{P}^\varepsilon. \end{cases}$$

Remark In space dimension $N = 2, 3$, Theorem 3.1 can easily be generalized to the nonlinear Navier–Stokes equations (see Remark 1.1.10 in [All91c]). The microscopic equations in the porous medium are

$$\begin{cases} \nabla p^\varepsilon + \vec{v}^\varepsilon \cdot \nabla \vec{v}^\varepsilon - \mu \Delta \vec{v}^\varepsilon = \vec{f} & \text{in } \mathfrak{P}^\varepsilon, \\ \nabla \cdot \vec{v}^\varepsilon = 0 & \text{in } \mathfrak{P}^\varepsilon, \\ \vec{v}^\varepsilon = 0 & \text{on } \partial\mathfrak{P}^\varepsilon. \end{cases} \tag{3.14}$$

Then, there are still three limit flow regimes, corresponding to the same obstacle sizes, and the definitions of the local problems and of the matrix M are still given by Proposition 3.2. In the critical case, the homogenized problem is a nonlinear Brinkman's law

$$\begin{cases} \nabla p + \vec{v} \cdot \nabla \vec{v} - \mu \Delta \vec{v} + \frac{\mu}{\sigma^2} M \vec{v} = \vec{f} & \text{in } \Omega, \\ \nabla \cdot \vec{v} = 0 & \text{in } \Omega, \\ \vec{v} = 0 & \text{on } \partial \Omega, \end{cases} \quad (3.15)$$

whereas in the supercritical case, it is still the same linear Darcy's law (3.7).

The rigorous derivation of Brinkman's law by homogenization of Stokes equations in a periodic porous medium was first established by Marchenko and Khrouslov [MK74]. The descripton of all limit regimes and the two-dimensional paradoxical result of Proposition 3.2 are due to Allaire [All91c]. Related results are obtained by Brillard [Bri86] in the framework of épiconvergence (or Γ-convergence). Brinkman's law has also been obtained, formally, using *three-scale* asymptotic expansion, by Lévy [Lev83] and Sanchez-Palencia [SP82]. Beside these works, which are concerned with periodic homogenization, Rubinstein [Rub86] used probabilistic methods to derive Brinkman's law from a 3-D Stokes flow past a random array of spheres. Finally, all the above results have been generalized in [All91b] to a different microscopic model: the no-slip boundary condition on the obstacles in (3.1) (i.e., $\vec{v}^\varepsilon = 0$ on $\partial \mathfrak{P}^\varepsilon$) can be replaced by a slip condition (i.e., $\vec{v}^\varepsilon \cdot \vec{v} = 0$ on $\partial \mathfrak{P}^\varepsilon$) and an additional condition on the tangential component of the normal stress.

3.4   Double Permeability

This section is devoted to deriving the so-called *double-porosity* model for describing single-phase flows in fractured porous media. This model is also often called *dual-porosity* or *double permeability*. It is well-known in the engineering literature [Arb88], [BZK60], [Kaz69], [Swa76], and [WR63]. Recently, it has been rigorously derived by homogenization techniques [ADH90], [ADH91]. A fractured porous medium possesses two porous structures, one associated with the system of cracks or fractures and the other with the matrix of porous rocks. In each of these structures, the fluid flow is assumed to be governed by Darcy's law. Contrary to the previous sections where the starting model was a *microscopic* model (Stokes equations at the pore level), here the original model is already an averaged model (Darcy's law in both the matrix and the fractures). Therefore, we shall obtain a *macroscopic* model starting from a *mesoscopic* one. More precisely, we shall prove, under suitable assumptions, that the homogenization of Darcy's law in a periodic, fractured, porous medium yields a double-porosity model.

As before, we denote by Ω the periodic porous medium with its period ε. The rescaled unit cell is $Y = (0, 1)^N$, which is made of two complementary parts, the matrix block $\mathfrak{B}$ and the fracture set $\mathfrak{F}$ ($\mathfrak{B} \cup \mathfrak{F} = Y$ and $\mathfrak{F} \cap \mathfrak{B} = \emptyset$). The matrix block $\mathfrak{B}$ is assumed to be completely surrounded by the fracture set $\mathfrak{F}$, i.e., $\mathfrak{B}$ is strictly included in Y. We define the matrix and fracture parts of Ω by

$$\mathfrak{B}^\varepsilon = \Omega \cap \bigcup_{i=1}^{N(\varepsilon)} \mathfrak{B}_i^\varepsilon \tag{4.1}$$

and

$$\mathfrak{F}^\varepsilon = \Omega \cap \bigcup_{i=1}^{N(\varepsilon)} \mathfrak{F}_i^\varepsilon, \tag{4.2}$$

where $\mathfrak{B}_i^\varepsilon$ and $\mathfrak{F}_i^\varepsilon$ are the ε-size copies of $\mathfrak{B}$ and $\mathfrak{F}$ covering Ω.

The reservoir Ω is periodic because its porosity ϕ_ε and permeability k_ε are periodic functions defined by

$$\begin{cases} k_\varepsilon(x) = \varepsilon^2 k_\mathfrak{B}, & \phi_\varepsilon(x) = \phi_\mathfrak{B} & \text{in } \mathfrak{B}^\varepsilon, \\ k_\varepsilon(x) = k_\mathfrak{F}, & \phi_\varepsilon(x) = \phi_\mathfrak{F} & \text{in } \mathfrak{F}^\varepsilon, \end{cases} \tag{4.3}$$

where $\phi_\mathfrak{B}$, $\phi_\mathfrak{F}$ are positive constants and $k_\mathfrak{B}$, $k_\mathfrak{F}$ are positive definite tensors (they could also depend on x and y). The fluid is assumed to be compressible, and the state law giving the relationship between its density ρ^ε and its pressure p^ε is assumed to be exponential:

$$\rho(p) = \rho_0\, e^{c(p - p_0)}.$$

Combining Darcy's law and the conservation of mass and neglecting gravitational effects yields the following equations:

$$\begin{cases} \phi_\varepsilon \frac{\partial \rho^\varepsilon}{\partial t} - \nabla \cdot \left(\frac{k_\varepsilon}{\mu c} \nabla \rho^\varepsilon \right) = f & \text{in } (0, T) \times \Omega, \\ \left(\frac{k_\varepsilon}{\mu c} \nabla \rho^\varepsilon \right) \cdot \vec{v} = 0 & \text{on } (0, T) \times \partial\Omega, \\ \rho^\varepsilon(0, x) = \rho^{init}(x) & \text{at time } t = 0. \end{cases} \tag{4.4}$$

Remark We emphasize the particular scaling of the permeability defined in (4.3): the matrix is much less permeable than the fractures. Equivalently, the time scale of filtration inside the matrix is much smaller than that inside the fractures. On the other hand, the porosities have the same order of magnitude in both regions. Arbogast, Douglas, and Hornung in [ADH90] have found that such scalings yield a homogenized, double-porosity model. If the permeabilities of both phases (matrix and fractures) were of the same order, the homogenized system would easily be seen to be a single Darcy law with effective coeficients computed by the usual rules of homogenization.

To obtain an existence result and convenient a priori estimates for the solution of (3.1), the source term $\vec{f}(t, x)$ is assumed to belong to

$$L^2((0, T) \times \Omega),$$

and the initial condition $\rho^{init}(x)$ to $H^1(\Omega)$ (the initial condition could vary with ε as soon as it converges sufficiently smoothly). Then, standard theory yields the following:

Proposition 4.1 *There exists a unique density $\rho^\varepsilon \in L^2\left((0, T); H^1(\Omega)\right)$ solution of system (2.1). Furthermore, it satisfies the a priori estimates,*

$$\|\rho^\varepsilon\|_{L^\infty((0,T);L^2(\Omega))} + \|\nabla\rho^\varepsilon\|_{L^\infty((0,T);L^2(\mathfrak{F}^\varepsilon))} + \varepsilon\|\nabla\rho^\varepsilon\|_{L^\infty((0,T);L^2(\mathfrak{B}^\varepsilon))} \leq C,$$
$$\left\|\tfrac{\partial\rho^\varepsilon}{\partial t}\right\|_{L^2((0,T)\times\Omega)} \leq C,$$

(4.5)

where the constant C does not depend on ε.

The following homogenization theorem states that the homogenized problem is a double-permeability model.

Theorem 4.2 *Under assumptions (4.3), the density ρ^ε solution of (4.4), two-scale converges to $\rho_0(x, y) \in L^2((0, T) \times \Omega \times Y)$, such that*

$$\begin{cases} \rho_0(x, y) = \rho_{\mathfrak{F}}(x) & \text{if } (x, y) \in \Omega \times \mathfrak{F}, \\ \rho_0(x, y) = \rho_{\mathfrak{B}}(x, y) & \text{if } (x, y) \in \Omega \times \mathfrak{B}, \end{cases}$$

where $\left(\rho_{\mathfrak{F}}(x), \rho_{\mathfrak{B}}(x, y)\right) \in L^2\left((0, T); H^1(\Omega)\right) \times L^2\left((0, T) \times \Omega; H^1(\mathfrak{B})\right)$ is the unique solution of the coupled homogenized problem,

$$\begin{cases} \theta\phi_{\mathfrak{F}}\frac{\partial\rho_{\mathfrak{F}}}{\partial t} - \nabla_x \cdot \left(\frac{k^*}{\mu c}\nabla_x\rho_{\mathfrak{F}}\right) = f - \phi_{\mathfrak{B}}\int_{\mathfrak{B}}\frac{\partial\rho_{\mathfrak{B}}}{\partial t}(x, y)dy & \text{in } (0, T) \times \Omega, \\ \left(\frac{k^*}{\mu c}\nabla\rho_{\mathfrak{F}}\right) \cdot \vec{v} = 0 & \text{on } (0, T) \times \partial\Omega, \\ \rho_{\mathfrak{F}}(0, x) = \rho^{init}(x) & \text{at time } t = 0, \\ \frac{\phi_{\mathfrak{B}}}{|Y|}\frac{\partial\rho_{\mathfrak{B}}}{\partial t} - \nabla_y \cdot \left(\frac{k_{\mathfrak{B}}}{\mu c}\nabla_y\rho_{\mathfrak{B}}\right) = f(t, x) & \text{in } (0, T) \times \mathfrak{B}, \\ \rho_{\mathfrak{B}}(x, y) = \rho_{\mathfrak{F}}(x) & \text{on } (0, T) \times \partial\mathfrak{B}, \\ \rho_{\mathfrak{B}}(0, x, y) = \rho^{init}(x) & \text{at time } t = 0, \end{cases}$$

(4.6)

where $\theta = \frac{|\mathfrak{F}|}{|Y|}$ is the volume fraction of fractures and k^* is the homogenized permeability tensor defined by its entries

$$k_{ij}^* = \frac{1}{|Y|}\int_{\mathfrak{F}} k_{\mathfrak{F}}\left(\vec{e}_i + \nabla_y w_i\right) \cdot \left(\vec{e}_j + \nabla_y w_j\right) dy,$$

where $w_i(y)$ are the unique solutions in $H^1_\#(\mathfrak{F})/\mathbb{R}$ of the cell problems

$$\begin{cases} -\nabla_y \cdot k_{\mathfrak{F}}\left(\vec{e}_i + \nabla_y w_i(y)\right) = 0 & \text{in } \mathfrak{F}, \\ k_{\mathfrak{F}}\left(\vec{e}_i + \nabla_y w_i(y)\right) \cdot \vec{v} = 0 & \text{on } \partial\mathfrak{B}, \\ y \to w_i(y) & Y\text{-periodic}, \end{cases}$$

with $(\vec{e}_i)_{1 \leq i \leq N}$ the canonical basis of $\mathbb{R}^N$.

The original proof of Theorem 4.2 may be found in [ADH90]. It turns out to be a simple application of the two-scale convergence method (see Appendix A).

3.5   On the Transmission Conditions at the Contact Interface between a Porous Medium and a Free Fluid

Andro Mikelić

3.5.1   Statement of the Problem and Existing Results from Physics

This section is devoted to the study of an incompressible viscous fluid flow by-passing a porous solid.

Supposing a creeping flow, we find out immediately that the effective flow in a porous solid is described by Darcy's law, i.e., the first two equations from (1.4) hold true. In the free fluid, we obviously keep the Stokes system. Hence, we have two completely different systems of partial differential equations (Eqs. (1.4) and (1.2), respectively) and it is not clear what type of conditions one should impose at the interface between the free fluid and porous solid. Clearly, due to the incompressibility, we have continuity of the normal mass flux. Other conditions classically used are the continuity of pressure and, for a free fluid, vanishing of the tangential velocity at the interface.

As we have already seen in section 3.1, derivation of the Darcy law involves weak convergence in $H(\Omega, \text{div})$ (respectively, two-scale convergence) of velocities, and only continuity of the normal velocities is preserved. Other continuity conditions at the interface are lost, in general.

Vanishing of the tangential velocity was found to be an unsatisfactory approximation and, in the article [BJ67], a new condition was proposed:

$$\nabla \vec{v}_F \vec{v} \cdot \vec{\tau} = \alpha A^{-1/2}(\vec{v}_F - \vec{v}_D) \cdot \vec{\tau}, \qquad (5.1)$$

where $\vec{v}_F$ is the effective velocity of the free fluid, $\vec{v}_D$ is the mean filtration velocity given by the Darcy law $\vec{v}_D = -\dfrac{1}{\mu} A \nabla p$, A is the permeability of the porous medium, $\vec{\tau}$ is a tangent vector to the interface, and $\vec{v}$ is the normal into the fluid. The scalar α is a function of the geometry of the porous medium. In [BJ67], the law (5.1) was justified experimentally. A theoretical attempt to derive (5.1) was undertaken in [Saf71] and, using a statistical approach, a Brinkman type approximation was derived. An argument in which terms are matched results in the formula

$$\vec{v}_F \cdot \vec{\tau} = \frac{1}{\alpha} A^{1/2} \nabla \vec{v}_F \vec{v} \cdot \vec{\tau} + O(A).$$

Interested readers can also consult the lecture notes [CR84].

A different consideration, presented in [ESP75] and [LSP75], distinguished two types of flow: a) The pressure gradient on the side of the porous solid of the interface is normal to it. Consequently, we have a balanced flow on both sides of the interface. Then using an asymptotic point of view, the following laws were obtained in [LSP75]:

$$\begin{cases} \vec{v}_F \cdot \vec{v} = \vec{v}_D \cdot \vec{v} \\ p = \text{constant}, \end{cases}$$

at the interface.

The second case, b) when the pressure gradient on the side of porous solid at the interface is not normal, was considered in the fundamental paper [ESP75]. After discussing the orders of magnitude of the unknowns and the intermediate layer around the interface, it was found that, at the interface, *the velocity of the free fluid is zero and the pressures are continuous.*

All results explained above do not have full mathematical rigor. Furthermore, different approaches have different results and two natural questions arise immediately:

(Q1) Can the correct matching conditions (i.e., conditions at the interface) between those two flow equations be found?

(Q2) Can the results obtained be verified with mathematical rigor?

The nonrigorous results above underline the importance of a mathematically rigorous proof. The correct answer, given in the paper [JM95], is explained in the next subsection.

3.5.2   Statement of the Mathematical Results and Comparison with the Literature

We follow [JM95] and start with the precise setting of the simple case.

Consider a slow, viscous, two-dimensional incompressible flow in a domain $\mathfrak{P}^\varepsilon$ consisting of the porous medium $\Omega_2 =]0, L[\times\mathbb{R}_-$, the free fluid domain $\Omega_1 =]0, L[\times\mathbb{R}_+$ and the interface $\Sigma =]0, L[\times\{0\}$ between them. We assume that the structure of the porous medium is periodic, generated by translations of a cell $Y^\varepsilon = \varepsilon Y$, where Y is the standard cell, $Y =]0, 1[^2$, consisting of an open set $\mathfrak{S}$, $\partial\mathfrak{S} \in C^\infty$, strictly included in Y. Let $\mathfrak{P} = Y \setminus \overline{\mathfrak{S}}$ be connected, and let χ be the characteristic function of $\mathfrak{P}$ extended by periodicity to $\mathbb{R}^2$. We set $\chi^\varepsilon(x) = \chi(\frac{x}{\varepsilon})$, $x \in \mathbb{R}^2$ and define P^ε by $P^\varepsilon = \{x : x \in \Omega_2, \chi^\varepsilon(x) = 1\}$. Furthermore, $\mathfrak{P}^\varepsilon = \Omega_1 \cup \Sigma \cup P^\varepsilon$. Assume that $L/\varepsilon \in \mathbb{N}$ (see Fig. 3.1).

Therefore, our porous medium is assumed to consist of a large number of periodically distributed channels of length ε, small compared with the length of the macroscopic domain.

Our objective is the systematic study of the effective behavior of the velocities $\vec{v}^\varepsilon$ and pressures p^ε as $\varepsilon \to 0$, i.e., when the size of the pores tends to zero. For a fixed $\varepsilon > 0$, $(\vec{v}^\varepsilon, p^\varepsilon)$ are defined through the equations of motion and mass conservation,

$$-\Delta\vec{v}^\varepsilon + \nabla p^\varepsilon = f^\varepsilon \qquad \text{in } \mathfrak{P}^\varepsilon, \tag{5.2}$$

$$\nabla \cdot \vec{v}^\varepsilon = 0 \qquad \text{in } \mathfrak{P}^\varepsilon, \tag{5.3}$$

where

$$f^\varepsilon = \begin{cases} \varepsilon^\gamma f & \text{in } \Omega_1 \text{ (the free fluid domain)}, \\ f & \text{in } \Omega_2 \text{ (the porous medium)}, \end{cases} \tag{5.4}$$

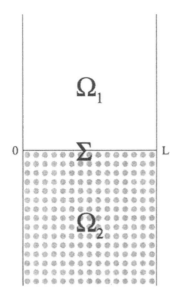

FIGURE 3.1. Fluid and porous domain.

with $f \in C_0^\infty(\Omega_1 \cup \Sigma \cup \Omega_2)^2$, $f \not\equiv 0$ on Σ. The motivation for different scalings in Ω_1 and P^ε comes from different values of the Reynolds' and Froude's numbers. We assume that the Reynolds' number is small in both domains to justify using the Stokes system (5.2)–(5.3). Then f^ε corresponds to the ratio between the Reynolds' and Froude's numbers and it is proportional to L^2/U. Rescaling gives (5.4).

Furthermore we assume that the velocity vanishes along the boundaries of the solid part of the porous medium and that $(\vec{v}^\varepsilon, p^\varepsilon)$ satisfy periodic boundary conditions on $(\{0\} \cup \{L\}) \times \mathbb{R}$, i.e.,

$$\vec{v}^\varepsilon = 0 \text{ on } \partial\mathfrak{P}^\varepsilon \setminus \partial\Omega, \qquad \vec{v}^\varepsilon \text{ is L-periodic in } x_1. \tag{5.5}$$

Despite the huge number of papers on the homogenization of flow in porous media, the articles addressing boundary effects are rare. Completely contrary to the situation with the Laplace operator and linear elasticity, where monographs such as [Lio81] and [OSY92] exist, here, we have few references. The paper [MA88] concerns the homogenization of fluid injected into the periodic porous medium, but gives only weak convergence of the velocity. Also, there are Conca's papers [Con87a], [Con87b] on the homogenization of flow through a sieve, but that problem has its own special structure. Let us also mention the paper [BMPM95] on effective equations for viscous, incompressible flow through a porous medium containing a thin fissure.

Here, the main difficulty comes from the appearance of the boundary layers in the neighborhoods of the contact surfaces, with the gradient of a solution greatly differing from its behavior inside the interiors of the domains.

Furthermore, the particular problem of the contact between a porous medium and a nonperforated domain under Dirichlet's conditions on the boundaries of the solid is the influence of the boundary layers on the effective behavior of solution. The corresponding problem for the Laplace's operator is solved in [JM93], and the boundary layer effects for the Stokes system in a porous medium are considered in [JM94]. Here, we address the contact problem for the Stokes system (5.2)–(5.5) and present some of the results from [JM95].

In view of the classical homogenization results of the Stokes system in porous media, it is clear that we expect Darcy's law to apply in Ω_2. In Ω_1 the flow should be governed by the Stokes system. Those two flows are to be coupled at the interface, and the main goal of this paper is to find the effective behavior of $(\vec{v}^\varepsilon, p^\varepsilon)$ at the interface Σ in the limit $\varepsilon \to 0$. We also point out that the Stokes system and the second-order equation for pressure are incompatible, posing additional complications.

We rigorously derive the laws governing flow at the interface by constructing the corresponding boundary layers. Furthermore, we compare our results with the well-known results from physics explained in the previous subsection, and find partial agreement, depending on the choice of γ. For example, the Beavers–Joseph's slip condition is not obtained in the first step, but it is an additional property of the solution for the homogenized problem.

We start by precisely defining the auxiliary problems.

The auxiliary problem determining permeability:

Let $(\vec{w}_j, \pi_j)$ be determined by (1.6). We define $\vec{w}_j^\varepsilon(x) = \vec{w}_j(\frac{x}{\varepsilon})$, $x \in P^\varepsilon$, and extend $\vec{w}_j^\varepsilon$ by zero to $\Omega_2 \setminus P^\varepsilon$. Furthermore, the permeability tensor A is given by (1.5), and

$$A_{2j} = \int_0^1 w_{j2}(y_1, y_2)\, dy_1, \qquad \forall y_2 \in]0, 1[.$$

The boundary layer created by the extension of $\vec{w}_j^\varepsilon$:

Let $Z^- = \cup_{k=1}^\infty \{\mathfrak{P} - (0, k)\}$, $X =]0, 1[\times\{0\}$, $Z^+ =]0, 1[\times]0, +\infty[$ and $Z_{BL} = Z^- \cup X \cup Z^+$ (see Fig. 3.2). Consider the following problem:

$$\begin{cases}
-\Delta_y \vec{w}^{j,bl} + \nabla_y \pi^{j,bl} = 0 & \text{in } Z^+ \cup Z^-, \\
\nabla_y \cdot \vec{w}^{j,bl} = 0 & \text{in } Z^+ \cup Z^-, \\
[\vec{w}^{j,bl}]_X(\cdot, 0) = \vec{w}_j(\cdot, 0) & \text{on } X, \\
[(\nabla \vec{w}^{j,bl} - \pi^{j,bl} I)\vec{e}_2]_X(\cdot, 0) & \\
\quad = (\nabla \vec{w}_j - q_j I)(\cdot, 0)\vec{e}_2 & \text{on } X, \\
\vec{w}^{j,bl} = 0 & \text{on } \cup_{k=1}^\infty \{\partial\mathfrak{S} - (0, k)\}, \\
(\vec{w}^{j,bl}, \pi^{j,bl}) & \text{is } y_1 - \text{periodic.}
\end{cases} \qquad (5.6)$$

A variant of the Saint–Venant principle from [JM95] implies that the problem (5.6) is uniquely solvable and there exist constants $\gamma_0 \in]0, 1[$ and C_π^j and a constant vector $\vec{C}^{j,bl} = (C_1^{j,bl}, A_{j2})$, such that $e^{\gamma_0|y_2|}\nabla_y \vec{w}^{j,bl} \in L^2(Z^+\cup Z^-)^4$, $e^{\gamma_0|y_2|}\vec{w}^{j,bl} \in$

FIGURE 3.2. The cell problem for the boundary layer.

$L^2(Z^-)^2$, $e^{\gamma_0|y_2|}\pi^{j,bl} \in L^2(Z^-)$, and

$$\begin{cases} \mid \vec{w}^{j,bl}(y_1, y_2) - \vec{C}^{j,bl} \mid \le Ce^{-\gamma_0 y_2}, & y_2 > y_*, \\ \mid \pi^{j,bl}(y_1, y_2) - C_\pi^j \mid \le Ce^{-\gamma_0 y_2}, & y_2 > y_*. \end{cases}$$

The auxiliary problem correcting the values of the normal stress of the free fluid at the interface:
We are looking for $(\vec{\beta}^{bl}, \omega^{bl})$ satisfying

$$\begin{cases} -\Delta_y \vec{\beta}^{bl} + \nabla_y \omega^{bl} = 0 & \text{in } Z^+ \cup Z^-, \\ \nabla_y \cdot \vec{\beta}^{bl} = 0 & \text{in } Z^+ \cup Z^-, \\ [\vec{\beta}^{bl}]_X(\cdot, 0) = 0 & \text{on } X, \\ [(\nabla \vec{\beta}^{bl} - \omega^{bl} I)\vec{e}_2]_X(\cdot, 0) = \vec{e}_1 & \text{on } X, \\ \vec{\beta}^{bl} = 0 & \text{on } \cup_{k=1}^\infty \{\partial \mathfrak{S} - (0, k)\}, \\ (\vec{\beta}^{bl}, \omega^{bl}) & \text{is } y_1 - \text{periodic.} \end{cases} \tag{5.7}$$

Once again, results from [JM95] imply that problem (5.7) is solvable and there are constants $\gamma_0 \in]0, 1[$, C_1^{bl} and C_ω^{bl}, such that

$$e^{\gamma_0|y_2|}\nabla_y \vec{\beta}^{bl} \in L^2(Z_{BL})^4,$$

$$e^{\gamma_0|y_2|}\vec{\beta}^{bl} \in L^2(Z^-)^2, \quad e^{\gamma_0|y_2|}\omega^{bl} \in L^2(Z^-),$$

and

$$\begin{cases} \mid \vec{\beta}^{bl}(y_1, y_2) - (C_1^{bl}, 0) \mid \le Ce^{-\gamma_0 y_2}, & y_2 > y_*, \\ \mid \omega^{bl}(y_1, y_2) - C_\omega^{bl} \mid \le Ce^{-\gamma_0 y_2}, & y_2 > y_*. \end{cases} \tag{5.8}$$

The auxiliary problem describing Stokes flow in the free fluid domain:
We search for (u_0, π_0) satisfying

$$\begin{cases} -\Delta u_0 + \nabla \pi_0 = f & \text{in } \Omega_1, \\ \nabla \cdot u_0 = 0 & \text{in } \Omega_1, \\ u_0 = 0 & \text{on } \Sigma, \\ (u_0, \pi_0) & \text{is } L\text{-periodic in } x_1. \end{cases} \tag{5.9}$$

The auxiliary problem describing Darcy flow in the porous medium:
We search for p satisfying

$$\begin{cases} \nabla \cdot \{A(f - \nabla p)\} = 0 & \text{in } \Omega_2, \\ p = 0 & \text{on } \Sigma, \\ p & \text{is } L\text{-periodic in } x_1. \end{cases} \tag{5.10}$$

The counterflow effects caused by the stabilization of $\vec{w}^{j,bl}$ to $\vec{C}^{j,bl}$ in Ω_1:
Let

$$\tilde{f}_j(x) = (f_j - \frac{\partial p}{\partial x_j})(x_1, -0) \exp(-\delta x_2^2), \quad \delta > 0, \ j = 1, 2.$$

Then, we look for (u^{jk}, π^{jk}) satisfying

$$\begin{cases} -\Delta u^{jk} + \nabla \pi^{jk} = 0 & \text{in } \Omega_1, \\ \nabla \cdot u^{jk} = 0 & \text{in } \Omega_1, \\ u^{jk}(x_1, +0) = \tilde{f}_j(x_1, 0)\vec{e}_k & \text{on } \Sigma, \\ (u^{jk}, \pi^{jk}) & \text{is } L\text{-periodic in } x_1. \end{cases}$$

Now we are in position to formulate the results, using the following procedure:

a) We eliminate the external force f. This step corresponds to subtracting Darcy's approximation in the porous solid and, respectively, the solution to Stokes system in Ω_1 with the no-slip condition at the interface Σ and with the right-hand side f from the solution. This correction creates a jump at Σ.

b) We eliminate the jump on Σ by subtracting the corresponding boundary layer velocities. This step requires a precise study of the Stokes system in the unbounded, half-perforated strips and establishing the corresponding Saint–Venant's principle describing stabilization of corresponding velocities and pressures when $y_2 \to +\infty$. Those results are in [JM95]. Stabilization to constants differing from zero creates counterflow effects in the free fluid.

c) We eliminate the compressibility effects arising from the previous corrections. This step creates a number of auxiliary problems not contributing to the effective velocity. For details, one can consult [JM95].

d) We estimate the pressure correction through the velocity correction.

e) We find an estimate for the velocity correction the L^2-norm of which is of order $C\varepsilon$ in the porous solid.

f) The L^2-norm of the velocity correction at the interface is now of order $C\sqrt{\varepsilon}$, and we use a very weak solution of the Stokes system to obtain an estimate for the L^2-norm of the velocity correction.

This long and relatively complicated procedure is developed in detail in [JM95], and the corresponding theorems are stated and proven. Because we are interested, mainly, in effective models, we present only some weak convergence results. Investigating the behavior of solutions to problems (5.2)–(5.5), as $\varepsilon \to 0$, requires extending $\vec{v}^\varepsilon$ and p^ε to the whole Ω. Let $\bar{g}$ denote the extension by zero of a function g to $\Omega \setminus \mathfrak{P}^\varepsilon$, and let $\tilde{p}^\varepsilon$ be the extension of the pressure described in section 3.1.3. Finally, let H denote Heaviside's function.

Then, we can state the following theorem:

Theorem 5.1 *Let $\gamma = 2$. Then,*

$$\frac{\bar{u}^\varepsilon}{\varepsilon^2} \rightharpoonup H(x_2)(u_0 + \sum_{j,k} C_k^{j,bl} u^{jk}) + H(-x_2)A(f - \nabla p) \text{ in } L^2_{loc}(\Omega)^2 \text{ weakly,}$$

$$\tilde{p}^\varepsilon \longrightarrow H(-x_2)p \qquad \text{in } L^2_{loc}(\Omega) \text{ strongly,}$$

$$\nabla(\frac{\bar{u}^\varepsilon}{\varepsilon^2} - \sum_j \vec{w}_j^\varepsilon(f_j - \frac{\partial p}{\partial x_j}))$$

$$\rightharpoonup H(x_2)\nabla(u_0 + \sum_{j,k} C_k^{j,bl} u^{jk}) + \sum_j \tilde{f}_j(C_1^{j,bl} - A_{1j})\vec{e}_1 \otimes \vec{e}_2 \delta_\Sigma$$

weak* in $M_p(K)^4$ for every bounded domain $K \subset \Omega$, (5.11)

where $M_p(K)^4$ denotes the dual of the space $C(K)$.

Remark We define the rescaled effective velocity of the free fluid due to the counterflow effects by

$$\vec{v}_F = u_0 + \sum_{j,k} C_k^{j,bl} u^{jk} \qquad \text{in } \Omega_1$$

and

$$\vec{v}_D = A(f - \nabla p) \qquad \text{in } \Omega_2.$$

Then, using Theorem 5.1, we easily find the conditions at the interface Σ between the two different flows:

$$\vec{v}_F(x_1, +0)\vec{e}_2 = \vec{v}_D(x_1, -0)\vec{e}_2 \qquad \text{on } \Sigma \qquad (5.12)$$

(i.e., the normal effective velocity is continuous on Σ),

$$\vec{v}_F(x_1, 0)\vec{e}_1 - \vec{v}_D(x_1, -0)\vec{e}_1 = (\sum_j (C_1^{j,bl} - A_{1j})\tilde{f}_j(x_1, +0))\vec{e}_1 \qquad \text{on } \Sigma \quad (5.13)$$

(i.e., there is a jump in the tangential effective velocity on Σ, and it is given by (5.13)),

$$p(x_1, -0) = 0 \qquad \text{on } \Sigma. \qquad (5.14)$$

As already mentioned, the conditions on Σ coupling our two effective flows were derived in a number of papers in the physics literature.

In particular, it is possible to compare our results with those in the paper [LSP75]. One of the cases discussed in that paper is when the pressure gradient on the side of a porous body is normal to the interface. This corresponds to our choice $\gamma = 2$. Relationships (5.12) and (5.14) were derived, but the formula (5.13) causing a jump in tangential velocity was not found, because it requires a constant $C_1^{j,bl}$ corresponding to the boundary layer introduced by (5.6).

As for the Beavers and Joseph law from [BJ67] and [Saf71], the result (5.11) indicates its validity. However, it does not rigorously establish that law. It seems that it can be established only after obtaining a corrector for the gradients of velocities, which is presently an open problem.

Now we switch to the case $\gamma = 0$. Let us introduce new auxiliary problems very similar to those discussed above.

We start with a new problem for the Darcy pressure:

$$\nabla \cdot \left(A(f - \nabla p^0) \right) = 0 \qquad \text{in } \Omega_2,$$

$$p^0(x_1, -0) = (\sigma_0 e_2 e_2)(x_1, +0) - C_\omega^{bl}(\sigma_0 e_2 e_1)(x_1, +0) \text{ on } \Sigma, \qquad (5.15)$$

$$p^0 \text{ is } L\text{-periodic in } x_1$$

(cf. Eq. (5.10)), where $\sigma_0 = \pi_0 I - \nabla u_0$, (u_0, π_0) are given by (5.9) and $\vec{C}_\omega^{bl}$ is given by (5.8).

Let us state the effective behavior of the velocity and the pressure:

Theorem 5.2

$$\bar{u}^\varepsilon \longrightarrow H(x_2)u_0 \text{ in } H_{loc}^1(\Omega)^2 \text{ strongly,}$$

$$\bar{p}^\varepsilon \longrightarrow H(-x_2)p^0 + H(x_2)\pi_0 \qquad \text{in } L_{loc}^2(\Omega) \text{ strongly,}$$

and

$$\frac{\bar{u}^\varepsilon}{\varepsilon^2} \rightharpoonup A(f - \nabla p^0) \text{ in } L_{loc}^2(\Omega_2)^2 \text{ weakly.}$$

Remark As one expects, the velocity of the free flowing fluid is dominant for $\gamma = 0$. The free fluid flow behaves as if it is in contact with a rigid wall, i.e., in the leading order of approximation, we have the no-slip condition on Σ. This corresponds to the results from the paper [ESP75] covering the case b) when the pressure gradient on the side of porous solid is not normal to the interface.

However, the relation (5.15) between the pressures is far removed from any intuition. Because $\Sigma \subset \{x_2 = 0\}$,

$$p^0(x_1, -0) = \pi_0(x_1, +0) + C_\omega^{bl} \frac{\partial (u_0)_1}{\partial x_2}(x_1, -0) \qquad \text{on } \Sigma, \qquad (5.16)$$

involving the contribution from the geometry in C_ω^{bl}. For general periodic geometry, pressure continuity cannot be expected. In [ESP75], the second term on the right-hand side of (5.16) does not appear. It should be noted that, in [ESP75], the boundary layer determining C_ω^{bl} was not explicitly constructed.

4

Non-Newtonian Flow

Andro Mikelić

4.1   Introduction

In the preceding chapter, the flow of Newtonian fluids through porous media was discussed in detail. However, non-Newtonian fluids are extensively involved in a number of applied problems involving the production of oil and gas from underground reservoirs. There are at least two typical situations, the flow of heavy oils and enhanced oil recovery (EOR). In EOR applications, non-Newtonian fluids, such as low concentration polymer solutions are injected to increase the viscosity of agents that displace the oil. Similarly, many heavy oils behave as Bingham viscoplastic fluids.

Therefore, it is important to establish filtration laws governing non-Newtonian flows through porous media. As in the Newtonian case in Chapter 3, the homogenization method should enable us to simplify solving the non-Newtonian flow equations in extremely complicated geometric structures, such as porous media. This can be accomplished by deriving filtration laws keeping only some overall properties of the porous medium within the averaged coefficients.

For single-phase Newtonian fluid flow, the method of homogenization has confirmed the well-known Darcy law, giving us, in addition, formulas for permeability and porosity. With even the simplest non-Newtonian flow the situation changes, and the filtration laws are not clear. In engineering literature, porous media are frequently modeled as a bundle of capillary tubes. Then, for an individual capillary tube, a sort of "dual" functional relationship between shear stress and shear rate is usually found (see, e.g., Bird et al. [BSL60], Christopher, Middleman [CM65], Shah, Yortsos [SY93] and references therein for more details). Such a "local" law is, then, used to obtain nonlinear Darcy-type laws at a macro level.

To start we should choose a class of fluids to be considered. In general, due to the complicated geometry of porous media, the flow is not really a shearing flow, and we should deal with the equations for viscoelastic fluids (see Bird et al. [BAH87]). However, in this survey chapter, we will consider only steady-state flows. Supposing that the characteristic strain rate of the system porous medium/fluid is much smaller than the characteristic strain rate of the fluid filling the pores, we conclude that the *Deborah number* is small, and, consequently, we consider only quasi-Newtonian fluids. This approach is widely used, but it should be noted that, in porous media with very nonuniform cross-sectional areas, many pseudoplastic fluids exhibit extensional viscosity effects, especially, when the pressure drop is high enough. Dealing with equations of viscoelastic flow in porous media is an open mathematical problem.

For quasi-Newtonian fluids, the viscosity depends on shear rate. The three typical examples are

(a) "Power law" or *Ostwald–de Waele model*,

$$\eta_r(\xi) = \lambda|\xi|^{r-2}, \quad \xi \in \mathbb{R}^4, \tag{1.1}$$

with $\lambda > 0$ and $\infty > r > 1$, where λ and r are parameters characteristic for each fluid. The matrix norm $|\cdot|$ is defined by $|\xi|^2 = Tr(\xi\xi^*)$.

(b) *Carreau–Yasuda law*,

$$\eta_r(\xi) = (\eta_0 - \eta_\infty)(1 + \lambda|\xi|^2)^{\frac{r}{2}-1} + \eta_\infty, \quad \xi \in \mathbb{R}^4 \tag{1.2}$$

with $\eta_0 > \eta_\infty > 0, \lambda > 0, r > 1$, where η_0 is the zero-shear-rate viscosity, λ is a time constant, and $(r-1)$ is a dimensionless constant describing the slope in the "power law region" of $\log \eta_r$ versus $\log (D(v))$.

(c) *Bingham law*,

$$\eta_r(\xi) = \begin{cases} \infty, & |\xi| \leq \tau_0 \\ \mu_0 + \tau_0|\xi|^{-1}, & |\xi| \geq \tau_0 \end{cases}, \quad \xi \in \mathbb{R}^4, \tag{1.3}$$

where τ_0 is the "yield stress" and $\mu_0 > 0$.

4.2   Equations Governing Creeping Flow of a Quasi-Newtonian Fluid

For simplicity, we will restrict ourselves to the case where the porous medium Ω is a square in $\mathbb{R}^2$, i.e., $\Omega = (0, a)^2$. The medium consists of a subdomain $\mathfrak{P}^\varepsilon$ (the part filled with the fluid) and a closed set $\mathfrak{S}^\varepsilon$, given by $\mathfrak{S}^\varepsilon = \overline{\Omega} \setminus \mathfrak{P}^\varepsilon$ (the solid part of Ω). Here, ε denotes the typical pore size.

We consider the creeping flow of a quasi-Newtonian fluid with injection at the boundary. The mathematical model is given by (see Bird et al. [BAH87] for details)

$$\begin{cases} -\nabla \cdot \{\eta_r(D(\vec{v}^\varepsilon))D(\vec{v}^\varepsilon)\} + \nabla p^\varepsilon = 0 & \text{in } \mathfrak{P}^\varepsilon, \\ \nabla \cdot \vec{v}^\varepsilon = 0 & \text{in } \mathfrak{P}^\varepsilon, \\ \vec{v}^\varepsilon = 0 & \text{on } \partial\mathfrak{P}^\varepsilon \setminus \partial\Omega, \\ \vec{v}^\varepsilon = \mathbf{Re}^\varepsilon u & \text{on } \partial\Omega, \end{cases} \tag{2.1}$$

where ε is a characteristic length parameter for the porous medium, $\mathbf{Re}^\varepsilon$ is the dimensionless Reynolds number, $\vec{v}^\varepsilon$ is the velocity, $D(\vec{v}^\varepsilon) = (\nabla\vec{v}^\varepsilon +^t \nabla\vec{v}^\varepsilon)/2$ is the *rate-of-strain tensor* (where the symbol t denotes the transpose), p^ε is the pressure, and $\vec{u}$ is a given injection velocity. The nonlinear function η_r is given by (1.1) or (1.2). The Bingham law (1.3) can be treated similarly, but because of space limitations we will omit it completely and refer to Lions, Sanchez-Palencia [LSP81] and Bourgeat, Mikelić [BM95c].

Assume that $\mathfrak{P}^\varepsilon \in C^\infty$ and that $\vec{u}$ is the trace of a smooth, solenoidal, i.e., divergence-free, function defined on Ω. Let

$$V_r(\mathfrak{P}^\varepsilon) = \{\vec{v}|\ \vec{v} \in W_0^{1,r}(\mathfrak{P}^\varepsilon)^2, \quad \nabla \cdot \vec{v} = 0 \text{ in } \mathfrak{P}^\varepsilon\}.$$

Then, using the Browder–Minty theorem, we directly deduce the existence of a unique weak solution $(\vec{v}^\varepsilon, p^\varepsilon) \in (\mathbf{Re}^\varepsilon u + V_r(\mathfrak{P}^\varepsilon)) \times L_0^{r'}(\mathfrak{P}^\varepsilon)$ for (2.1) in the case of the power law (1.1). For the Carreau–Yasuda law (1.2), there is a unique solution $(\vec{v}^\varepsilon, p^\varepsilon) \in (\mathbf{Re}^\varepsilon u + V_2(\mathfrak{P}^\varepsilon)) \times L_0^2(\mathfrak{P}^\varepsilon)$ for (2.1), if $1 < r \leq 2$, and $(\vec{v}^\varepsilon, p^\varepsilon) \in (\mathbf{Re}^\varepsilon u + V_r(\mathfrak{P}^\varepsilon)) \times L_0^{r'}(\mathfrak{P}^\varepsilon)$, if $r > 2$.

Remark In general, we should consider that Eq. (2.1) also contains the inertial term $(\vec{v}^\varepsilon \nabla)\vec{v}^\varepsilon$. Then, the monotonicity is lost, and, in the case of the Carreau–Yasuda law (1.1) we will use the classical theory of the Navier–Stokes equations from Lions [Lio69], Cattabriga [Cat61], and Temam [Tem79] to establish the existence of at least one weak solution in the same function spaces as in the creeping flow case. The same observation applies to the power law (1.1) with $r \geq 2$. For $r < 2$, we refer to Blavier, Mikelić [BM95b] where, for a Reynolds number not too large, the existence of at least one weak or very weak solution is established for $r \geq 3/2$ and $3/2 > r > 1$, respectively. Anyhow, we will give only the results for Reynolds numbers not too large. Otherwise there would be some inertial effects, as in the Newtonian case (see section 3.2).

Remark Nonstationary variants of the system (2.1), including the inertial term but with somewhat regularized η_r, are considered in Cioranescu [Cio92] (also see the references therein). It should be noted that the presence of time could produce memory effects, as in the Newtonian case (see section 3.2.1).

Remark In nonisothermal non-Newtonian flow, the effects of viscous heating are important. Mathematical results in that direction are in Baranger and Mikelić [BM94a] and [BM95a]. We are not going to consider such flows in porous media here.

4.3   Description of a Periodic ε-Geometry, Construction of the Restriction Operator, and Review of the Results of Two-Scale Convergence in L^q-Spaces

Due to the complexity of the problem, we will restrict ourselves to the case of a periodic porous medium. It should be noted that the results of Beliaev and Ko-zlov [BK95] on homogenization of flow in a random porous medium could be extended to our case. However, we will pass over those questions in this short review, because, in most random cases, the conclusion follows that from the periodic case.

We will consider a periodic, porous medium, defined in section 1.4, as an intersection of the domain Ω with a periodic lattice in $\mathbb{R}^2$. The lattice consists of cubes $\varepsilon Y_k \equiv \varepsilon(Y + k)$, where $k \in \mathbb{Z}^2$ and $Y = (0, 1)^2$. Inside Y an open set $\mathfrak{P}$ is given (it represents the fluid part). For simplicity, we assume that $\mathfrak{P}$ is a complement of an open set $\mathfrak{S}$, strictly contained in Y and locally placed on one side of its boundary Γ, which is assumed to be a smooth curve. The set $\mathfrak{S}$ represents the solid particle. Then, the domain Ω has a smooth boundary and, for sufficiently small $\varepsilon > 0$, we will consider the sets

$$Z^\varepsilon = \{k \in \mathbb{Z}^2 : \varepsilon Y_k \subset \Omega\},$$

$$\mathfrak{S}^\varepsilon = \bigcup_{k \in Z^\varepsilon} \varepsilon \mathfrak{S}_k,$$

$$\Gamma^\varepsilon = \partial \mathfrak{S}^\varepsilon,$$

and

$$\mathfrak{P}^\varepsilon = \Omega \setminus \overline{\mathfrak{S}}^\varepsilon.$$

The set $\mathfrak{S}^\varepsilon$ represents the solid matrix, Γ^ε is the solid-liquid interface, and $\mathfrak{P}^\varepsilon$ represents the liquid part of the medium. Obviously,

$$\overline{\mathfrak{P}}^\varepsilon = \bigcup_{k \in Z^\varepsilon} (\varepsilon \overline{\mathfrak{P}}_k),$$

$$\partial \mathfrak{P}^\varepsilon = \partial \Omega \cup \Gamma^\varepsilon.$$

It is important to note that, for simplicity, we have assumed that the fluid part of the porous medium is connected and the solid part is not. This is realistic for two-dimensional flow problems in porous media but not for a three-dimensional situation.

To investigate the behavior of solutions of (2.1), as $\varepsilon \to 0$, we need to extend $\vec{v}^\varepsilon$ and p^ε to the whole of Ω. We extend $\vec{v}^\varepsilon$ by zero in $\Omega \setminus \mathfrak{P}^\varepsilon$. It is well-known that extension by zero preserves L^q and $W_0^{1,q}$ norms for $1 < q < \infty$.

Extending the pressure is a much more difficult task. The extension is closely related to the construction of the restriction operator $R_q \in \mathcal{L}\big(W^{1,q}(Y)^2, W_\Gamma^{1,q}(\mathfrak{P})^2\big)$, where $W_\Gamma^{1,q}(\mathfrak{P}) = \{z \in W^{1,q}(\mathfrak{P}) : z = 0 \text{ on } \Gamma\}$.

Generalizing the results of Tartar [Tar80], Allaire [All89], and Mikelić [Mik91],

Lemma 3.1 *There exists an operator* $R_q \in \mathcal{L}\big(W^{1,q}(Y)^2, W_\Gamma^{1,q}(\mathfrak{P})^2\big)$, $1 < q < +\infty$, *with the properties*

$$R_q \vec{\psi} = \vec{\psi} \text{ for } \vec{\psi} = 0 \text{ on } \Gamma$$

and

$$\nabla \cdot (R_q \vec{\psi}) = 0 \text{ for } \nabla \cdot \vec{\psi} = 0.$$

Now, for $\vec{\psi} \in W^{1,q}(\Omega)^2$, $1 < q < +\infty$, we define

$$\vec{\psi}_k^\varepsilon(y) = \vec{\psi}(\varepsilon y), \quad y \in Y_k, \quad k \in Z^\varepsilon.$$

Lemma 3.2 *The operator R_q^ε, defined by the formula*

$$(R_q^\varepsilon \vec{\psi})(x) = (R_q \vec{\psi}_k^\varepsilon)(\frac{x}{\varepsilon}), \quad x \in \varepsilon \mathfrak{P}_k, \quad k \in Z^\varepsilon,$$

has the properties $R_q^\varepsilon \in \mathcal{L}(W^{1,q}(\Omega)^2, W^{1,q}_{\Gamma^\varepsilon}(\mathfrak{P}^\varepsilon)^2)$,

$$R_q^\varepsilon \vec{\psi} = \vec{\psi} \text{ for } \vec{\psi} = 0 \text{ on } \Gamma^\varepsilon,$$

$$\nabla \cdot (R_q^\varepsilon \vec{\psi}) = 0 \text{ for } \nabla \cdot \vec{\psi} = 0,$$

$$\|R_q^\varepsilon \vec{\psi}\|_{L^q(\mathfrak{P}^\varepsilon)^2} \le C\big(\|\vec{\psi}\|_{L^q(\Omega)^2} + \varepsilon\|\nabla\vec{\psi}\|_{L^q(\Omega)^4}\big),$$

and

$$\|\nabla(R_q^\varepsilon \vec{\psi})\|_{L^q(\mathfrak{P}^\varepsilon)^4} \le C\big(\frac{1}{\varepsilon}\|\vec{\psi}\|_{L^q(\Omega)^2} + \|\nabla\vec{\psi}\|_{L^q(\Omega)^4}\big).$$

Because of these properties of the extension operators, it is natural to consider $R_q\vec{\psi}$ ($R_q^\varepsilon\vec{\psi}$, respectively) extended by zero to Y (Ω, respectively).

The following variant of Korn's inequality from Mjasnikov, Mosolov [MM71] will be very useful:

Lemma 3.3 *For each $\vec{\psi} \in W_0^{1,q}(\mathfrak{P}^\varepsilon)^2$, we have the inequality*

$$\|\vec{\psi}\|_{L^q(\mathfrak{P}^\varepsilon)^2} \le C\varepsilon\|D(\vec{\psi})\|_{L^q(\mathfrak{P}^\varepsilon)^4}, \quad 1 < q < +\infty.$$

In the next subsection, a priori estimates for pressure will be derived. After getting the a priori estimates, we will extend the pressure to the solid part of the porous medium. Following the idea of Lipton, Avellaneda [LA90], we define the extension of pressure p^ε by

$$\tilde{p}^\varepsilon = \begin{cases} p^\varepsilon & \text{in } \mathfrak{P}^\varepsilon \\ \frac{1}{|Y_{F_i}^\varepsilon|} \int_{Y_{F_i}^\varepsilon} p^\varepsilon & \text{in the } Y_{\mathfrak{S}_i}^\varepsilon \text{ for each } i \end{cases} \tag{3.1}$$

where $Y_{F_i}^\varepsilon$ is the fluid part of the cell Y_i^ε. Note that the solid part of the porous medium is the union of all $Y_{\mathfrak{S}_i}^\varepsilon$.

Tartar's construction together with appropriate a priori estimates from the next subsection and techniques from Mikelić [Mik91] guarantee the compactness of the family $\{\tilde{p}^\varepsilon\}$ in $L_0^{r'}(\Omega)$.

We recall properties of two-scale convergence in L^q-spaces, which will be crucial for our convergence proof (see section A.3).

Definition 3.4 *The sequence $\{\varphi^\varepsilon\} \subset L^q(\Omega)$, $1 < q < +\infty$ is said to two-scale converge in L^q to a limit $w \in L^q(\Omega \times Y)$ if, for any $\sigma \in C_0^\infty(\Omega; C_\#^\infty(Y))$ ("#" denotes Y-periodicity),*

$$\lim_{\varepsilon \to 0} \int_\Omega \varphi^\varepsilon(x)\sigma\left(x, \frac{x}{\varepsilon}\right) dx = \int_\Omega \int_Y \varphi(x, y)\sigma(x, y) \, dy \, dx.$$

Next, we will give various useful properties of two-scale convergence in an L^q-setting. For the proof, we refer to section A.3 and Allaire [All92a].

Proposition 3.5 *(i). From each bounded sequence in $L^q(\Omega)$, one can extract a subsequence which two-scale converges in L^q to a limit $w \in L^q(\Omega \times Y)$.*

(ii). Let φ^ε and $\varepsilon\nabla\varphi^\varepsilon$ be bounded sequences in $L^q(\Omega)$. Then, there exists a function $\varphi \in L^q\big(\Omega; W_\#^{1,q}(Y)\big)$ and a subsequence such that both φ^ε and $\varepsilon\nabla\varphi^\varepsilon$ two-scale converge in L^q to φ and $\nabla_y\varphi$, respectively.

(iii). Let φ^ε two-scale converge in L^q to $\varphi \in L^q(\Omega \times Y)$. Then, φ^ε converges weakly in L^q to $\int_Y \varphi(x, y)\, dy$.

After recalling these basic properties, we give a sequential, lower semicontinuity result for two-scale convergence in L^q. Let Φ be a continuous functional on $\mathbb{R}^N$ satisfying

$$0 \le \Phi(\xi) \le C[\,1 + |\xi|^q\,], \qquad \forall \xi \in \mathbb{R}^N. \tag{3.2}$$

Proposition 3.6 *Let Φ be a convex function on $\mathbb{R}^N$ satisfying (3.2). Let $\vec{\psi}^\varepsilon$ be a bounded sequence from $L^q(\Omega)^N$ which two-scale converges in L^q to $\vec{\psi} \in L^q(\Omega \times Y)^N$. Then,*

$$\lim_{\varepsilon \to 0} \inf \int_\Omega \Phi(\vec{\psi}^\varepsilon)\, dx \ge \int_\Omega \int_Y \Phi(\vec{\psi})\, dy\, dx.$$

Proof The proof of the proposition is given in Bourgeat and Mikelić [BM93]. In somewhat different form it is also given in Allaire [All92a].   Q.E.D.

Proposition 3.7 *(i). Let $\sigma \in L_\#^\infty(Y)$, $\sigma^\varepsilon = \sigma(x/\varepsilon)$, and let a sequence $\{\varphi^\varepsilon\} \subset L^q(\Omega)$ two-scale converge in L^q to a limit $\varphi \in L^q(\Omega \times Y)$. Then, $\sigma^\varepsilon\varphi^\varepsilon$ two-scale converges in L^q to the limit $\sigma\varphi$.*

(ii). Let Φ and Φ_1 be continuous functionals defined on $\mathbb{R}^N$ and $\mathbb{R}^{N^2}$, respectively, and let $\vec{\psi} \in C_0^\infty\big(\Omega; C_\#^\infty(Y)\big)^N$. Then,

$$\int_\Omega \Phi(\vec{\psi}^\varepsilon)\, dx \to \int_\Omega \int_Y \Phi(\vec{\psi})\, dy\, dx,$$

and

$$\int_\Omega \Phi_1(\varepsilon\nabla\vec{\psi}^\varepsilon)\, dx \to \int_\Omega \int_Y \Phi_1(\nabla_y\vec{\psi})\, dy\, dx,$$

as $\varepsilon \to 0$, where $\vec{\psi}^\varepsilon(x) = \vec{\psi}(x, \frac{x}{\varepsilon})$.

4.4   Statement of the Principal Results

In this section, we present the principal results of this paper. We will formulate the homogenized problems, state corresponding uniqueness results, and, then, give the convergence theorems.

We start with the case of the Carreau law (1.2) leading to a number of homogenized problems. The first and simplest possibility is $\gamma > 1$ (i.e., the Reynolds number $\mathbf{Re}^\varepsilon$ is of order smaller than ε). Let us formulate the auxiliary problem.

Let $(\vec{w}_i, \pi_i) \in H^1(\mathfrak{P})^2 \times L_0^2(\mathfrak{P})$, $(i = 1, 2)$ be the Y-periodic solution of the cell problem,

$$\begin{cases} -\Delta \vec{w}_i + \nabla \pi_i = \vec{e}_i & \text{in } \mathfrak{P}, \\ \nabla \cdot \vec{w}_i = 0 & \text{in } \mathfrak{P}, \\ \vec{w}_i = 0 & \text{on } \Gamma, \end{cases}$$

and let

$$K = (K_{ij})_{i,j=1,2},$$

$$K_{ij} = \int_{\mathfrak{P}} w_{ij}(y)\, dy.$$

The matrix K (permeability tensor) is symmetric and positive definite.

Let $(\vec{v}, p)$ be the solution of the (homogenized) problem,

$$\begin{cases} \nabla \cdot \vec{v} = 0 & \text{in } \Omega, \\ \vec{v} = -\dfrac{K}{\eta_0} \nabla p & \text{in } \Omega, \\ \vec{v} \cdot \vec{v} = \vec{u} \cdot \vec{v} & \text{on } \partial\Omega. \end{cases} \qquad (4.1)$$

Then,

Theorem 4.1 *Let* $\mathbf{Re}^\varepsilon = \varepsilon^\gamma$ *with* $\gamma > 1$, *let* $(\vec{v}^\varepsilon, p^\varepsilon)$ *be the weak solutions of problem (1.2), (2.1), and let* $(\vec{v}, p)$ *be the solution of (4.1). Let* $\vec{v}^\varepsilon$ *be extended by zero to* Ω, *and let* $\tilde{p}^\varepsilon$ *be given by (3.1). Then, in the limit* $\varepsilon \to 0$,

$$\frac{1}{\mathbf{Re}^\varepsilon} \vec{v}^\varepsilon \to \vec{v} \text{ weakly in } L^2(\Omega)^2,$$

and

$$\frac{\varepsilon^2}{\mathbf{Re}^\varepsilon} \tilde{p}^\varepsilon \to p \text{ strongly in } L_0^{\min\{2,r'\}}(\Omega).$$

Proof It is along the same lines as the proof of Theorem 1 in Bourgeat and Mikelić [BM95c]. Because we are very close to the Newtonian case already considered in Chapter 3, we will omit it.   Q.E.D.

Remark It is important to note that Theorem 4.1 and (4.1) imply that, for $\gamma > 1$, i.e., for Reynolds number $\mathbf{Re}^\varepsilon < \varepsilon$, the Carreau law leads to Darcy's law.

Now, let us consider the Carreau law (1.2) with $\gamma = 1$. First, we formulate the homogenized problem.

Find

$$(\vec{v}_0, p, \pi) \in L^{\max\{2,r\}}(\Omega; W_\Gamma^{1,\max\{2,r\}}(\mathfrak{P}))^2$$
$$\times L_0^{\min\{2,r'\}}(\Omega) \times L^{\max\{2,r\}}\left(\Omega; L_0^{\min\{2,r'\}}(\mathfrak{P})\right),$$

such that

$$\begin{cases} -\nabla_y \cdot \{\eta_r(D_y(\vec{v}_0))D_y(\vec{v}_0)\} + \nabla_y \pi = -\nabla_x p(x) & \text{in } \mathfrak{P}, \\ \nabla_y \cdot \vec{v}_0(x, y) = 0 & \text{in } \mathfrak{P}, \\ (\vec{v}_0, \pi) \text{ is } Y - \text{periodic} & \text{a.e. on } \Omega, \\ \vec{v}_0 = 0 & \text{on } \Gamma, \\ \nabla_x \cdot \int_{\mathfrak{P}} \vec{v}_0(x, y)\, dy = 0 & \text{in } \Omega, \\ \vec{v}(x) \cdot \int_{\mathfrak{P}} \vec{v}_0(x, y)\, dy = v(x) \cdot \vec{u} & \text{on } \partial\Omega. \end{cases} \qquad (4.2)$$

We have the following result:

Theorem 4.2 *Problem (4.2) has a unique solution. Moreover,*

$$p \in W^{1,max\{2,r\}}(\Omega).$$

Theorem 4.3 *("Theorem A") Let* $\mathbf{Re}^\varepsilon = \varepsilon$, *let* $(\vec{v}^\varepsilon, p^\varepsilon)$ *be the weak solutions of problem (1.2), (2.1), and let* $(\vec{v}_0, p, \pi)$ *be the weak solution of the differential equation in (4.2). Furthermore, let* $\vec{v}^\varepsilon$ *be extended by zero to* Ω, *and let* $\tilde{p}^\varepsilon$ *be given by (3.1). Then, in the limit* $\varepsilon \to 0$

$$\varepsilon^{-1}\vec{v}^\varepsilon \to \vec{v}_0 \text{ in the two-scale sense in } L^{max\{2,r\}},$$

$$\varepsilon^{-1}\vec{v}^\varepsilon \rightharpoonup \vec{v} = \int_{\mathfrak{P}} \vec{v}_0(x, y) \, dy \quad \text{weakly in } L^{max\{2,r\}}(\Omega)^2,$$

$$\nabla \vec{v}^\varepsilon \to \nabla_y \vec{v}_0 \quad \text{in the two-scale sense in } L^{max\{2,r\}},$$

and

$$\varepsilon \tilde{p}^\varepsilon \to p \quad \text{strongly in } L_0^{min\{2,r'\}}(\Omega).$$

The proof of Theorem 4.3 is given in section 4.8.

Remark It turns out that there is a critical Reynolds number of order ε (i.e., $\gamma = 1$) for which the linear regime of flow (4.1) changes to a nonlinear and nonlocal one, described by system (4.2). Linearity holds only if $r = 2$.

The last case of the Carreau law (1.2) that we will consider is with $\gamma < 1$. It exhibits an interesting behavior: the flow regime is linear as in the Newtonian case for $1 < r \leq 2$ (i.e., for pseudoplastic fluids) and nonlinear for $r > 2$. Moreover, we find different linear laws for $1 < r < 2$ and for $r = 2$. Note that the result for $r = 2$, i.e., the Newtonian case, is already known from Mikelić, Aganović [MA87].

Let $\{\bar{v}, \bar{p}\} \in L^2(\Omega)^2 \times H^1(\Omega) \cap L_0^2(\Omega)$ be the solution of the (homogenized) problem

$$\begin{cases} \nabla \cdot \bar{v} = 0 & \text{in } \Omega, \\ \bar{v} = -\dfrac{K}{\eta_\infty}\nabla\bar{p} & \text{in } \Omega, \\ \bar{v} \cdot \vec{v} = \vec{u} \cdot \vec{v} & \text{on } \partial\Omega. \end{cases} \tag{4.3}$$

Then,

Theorem 4.4 *Let* $\gamma < 1, \mathbf{Re}^\varepsilon = \varepsilon^\gamma$ *and* $1 < r < 2$. *Let* $(\vec{v}^\varepsilon, p^\varepsilon)$ *be the weak solutions of the Carreau law system (1.2) and (2.1), and let* $(\bar{v}, \bar{p})$ *be the weak solution of (4.3). Furthermore, let* $\vec{v}^\varepsilon$ *be extended by zero to* Ω, *and let* $\tilde{p}^\varepsilon$ *be given by (3.1). Then, in the limit* $\varepsilon \to 0$,

$$\frac{1}{\mathbf{Re}^\varepsilon}\vec{v}^\varepsilon \rightharpoonup \bar{v} \text{ weakly in } L^2(\Omega)^2,$$

and

$$\frac{\varepsilon^2}{\mathbf{Re}^\varepsilon}\tilde{p}^\varepsilon \to \bar{p} \text{ strongly in } L_0^2(\Omega).$$

The next theorem describes the situation for $r = 2$.

Theorem 4.5 *Let $\gamma < 1$, $\mathbf{Re}^\varepsilon = \varepsilon^\gamma$, and $r = 2$. Let $(\vec{v}^\varepsilon, p^\varepsilon)$ be the weak solutions of (1.2) and (2.1), and let $(\vec{v}, p)$ be the weak solution of (4.1). Then, for the extension $\tilde{p}^\varepsilon \in L_0^2(\Omega)$ of the function p^ε, defined by (3.1),*

$$\frac{1}{\mathbf{Re}^\varepsilon} \vec{v}^\varepsilon \rightharpoonup \vec{v} \ \text{weakly in } L^2(\Omega)^2,$$

and

$$\frac{\varepsilon^2}{\mathbf{Re}^\varepsilon} \tilde{p}^\varepsilon \to p \ \text{strongly in } L_0^2(\Omega),$$

as $\varepsilon \to 0$.

Proof See Mikelić and Aganović [MA87].   Q.E.D.

Remark It is important to note that Theorem 4.4 (i.e., $1 < r < 2$) and Theorem 4.5 (i.e., $r = 2$) in the same range of Reynolds numbers ($\varepsilon < \mathbf{Re}^\varepsilon$) lead to different Darcy's laws (4.1) and (4.3).

Finally, let us describe the nonlinear law occurring in the case $\gamma < 1, r > 2$. We formulate the homogenized problem.

Find $(\vec{v}_0, p, \pi) \in L^r\left(\Omega; W_\Gamma^{1,r}(\mathfrak{P})\right)^2 \times L_0^{r'}(\Omega) \times L^r\left(\Omega; L_0^{r'}(\mathfrak{P})\right)$, such that

$$
\begin{cases}
-\nabla_y \cdot \left((\eta_0 - \eta_\infty)\lambda^{\frac{r}{2}-1}|D_y(\vec{v}_0)|^{r-2}D_y(\vec{v}_0)\right) & \\
\quad +\nabla_y \pi = -\nabla_x \, p(x) & \text{in } \mathfrak{P}, \\
\nabla_y \cdot \vec{v}_0(x, y) = 0 & \text{in } \mathfrak{P}, \\
(\vec{v}_0, \pi) \text{ is } Y - \text{periodic} & \text{a.e. on } \Omega, \quad (4.4) \\
\vec{v}_0 = 0 & \text{on } \Gamma, \\
\nabla_x \cdot \int_{\mathfrak{P}} \vec{v}_0(x, y)\, dy = 0 & \text{in } \Omega, \\
\vec{v}(x) \cdot \int_{\mathfrak{P}} \vec{v}_0(x, y)\, dy = \vec{v}(x) \cdot \vec{u} & \text{on } \partial\Omega.
\end{cases}
$$

We have the following results:

Theorem 4.6 *Problem (4.4) has a unique solution. Furthermore, $p \in W^{1,r}(\Omega)$.*

Theorem 4.7 *Let $\gamma < 1$, $\mathbf{Re}^\varepsilon = \varepsilon^\gamma$, and $r > 2$. Let $(\vec{v}^\varepsilon, p^\varepsilon)$ be the weak solutions of problem (1.2) and (2.1), and let $(\vec{v}_0, p, \pi)$ be the weak solution of problem (4.4). Furthermore, let $\vec{v}^\varepsilon$ be extended by zero, and let $\tilde{p}^\varepsilon$ be defined by (3.1).*
Then, in the limit $\varepsilon \to 0$,

$$\frac{1}{\mathbf{Re}^\varepsilon} \vec{v}^\varepsilon \to \vec{v}_0 \quad \text{in the two-scale sense in } L^r,$$

$$\frac{\varepsilon}{\mathbf{Re}^\varepsilon} \nabla \vec{v}^\varepsilon \to \nabla_y \vec{v}_0 \quad \text{in the two-scale sense in } L^r,$$

$$\frac{1}{\mathbf{Re}^\varepsilon} \vec{v}^\varepsilon \rightharpoonup \vec{v} = \int_{\mathfrak{P}} \vec{v}_0(x, y)\, dy \quad \text{weakly in } L^r(\Omega)^2,$$

and

$$\varepsilon\left(\frac{\varepsilon}{\mathbf{Re}^\varepsilon}\right)^{r-1} \tilde{p}^\varepsilon \to p \quad \text{strongly in } L_0^{r'}(\Omega).$$

Completing the statement of our results for the case of the Carreau law (1.2), we turn our attention to the power law (1.1). Because of the simple algebraic formula for the law, we can formulate the results in one theorem.

The homogenized problem is stated as follows:

Find $(\vec{v}_0, p, \pi) \in L^r\left(\Omega; W_\Gamma^{1, r}(\mathfrak{P})\right)^2 \times L_0^{r'}(\Omega) \times L^r(\Omega; L_0^{r'}(\mathfrak{P}))$ such that

$$
\begin{cases}
-\lambda \, \nabla_y \cdot \left(|D_y(\vec{v}_0)|^{r-2} D_y(\vec{v}_0)\right) + \nabla_y \pi = -\nabla_x \, p(x) & \text{in } \mathfrak{P}, \\
\nabla_y \cdot \vec{v}_0(x, y) = 0 & \text{in } \mathfrak{P}, \\
(\vec{v}_0, \pi) \text{ is } Y - \text{periodic} & \text{for (a.e.) } x \in \Omega, \\
\vec{v}_0 = 0 & \text{on } \Gamma, \\
\nabla_x \cdot \int_{\mathfrak{P}} \vec{v}_0(x, y) \, dy = 0 & \text{in } \Omega, \\
\vec{v}(x) \cdot \int_{\mathfrak{P}} \vec{v}_0(x, y) \, dy = \vec{v}(x) \cdot \vec{u} & \text{on } \partial\Omega.
\end{cases}
$$

$$(4.5)$$

Then, we have the following results:

Theorem 4.8 *Problem (4.5) has a unique solution. Furthermore, $p \in W^{1,\, r}(\Omega)$.*

Remark Note that, after proper scaling, problems (4.4) and (4.5) are identical. Furthermore, proofs of Theorem 4.7 and Theorem 4.9 are very similar, and we give only the proof of Theorem 4.9 in section 4.9.

Theorem 4.9 *("Theorem B") Let $(\vec{v}^\varepsilon, p^\varepsilon)$ be the weak solutions of the "power law" system (1.1) and (2.1), and let $(\vec{v}_0, p, \pi)$ be the weak solution of problem (4.5). Furthermore, let $\vec{v}^\varepsilon$ be extended by zero to Ω, and let $\tilde{p}^\varepsilon$ be given by (3.1).*
Then, in the limit $\varepsilon \to 0$,

$$
\frac{1}{\mathrm{Re}^\varepsilon} \vec{v}^\varepsilon \to \vec{v}_0 \quad \text{in the the two-scale sense in } L^r,
$$

$$
\frac{\varepsilon}{\mathrm{Re}^\varepsilon} \nabla \vec{v}^\varepsilon \to \nabla_y \vec{v}_0 \quad \text{in the two-scale sense in } L^r,
$$

$$
\frac{1}{\mathrm{Re}^\varepsilon} \vec{v}^\varepsilon \rightharpoonup \vec{v} = \int_{\mathfrak{P}} \vec{v}_0(x, y) \, dy \quad \text{weakly in } L^r(\Omega)^2,
$$

and

$$
\varepsilon \left(\frac{\varepsilon}{\mathrm{Re}^\varepsilon}\right)^{r-1} \tilde{p}^\varepsilon \to p \text{ strongly in } L_0^{r'}(\Omega).
$$

Remark A direct comparison between Eqs. (4.4) in Theorem 4.7 and Eqs. (4.5) in Theorem 4.9 shows that the homogenized problems in the case of dilatant fluid ($r > 2$) for both the power law and Carreau's law have the same form for some range of Reynolds numbers and $r > 2$. In fact, for such a range of parameters, the flow of a dilatant fluid through porous media is governed by a nonlinear "power-like" law, no matter whether an empiricism employed to describe the fluid behavior in the pores was like the Carreau law or the power law.

Remark The result in Theorem 4.9 is not of a Darcy law type and, generally, does not lead to the usual filtration law used in standard engineering treatment (e.g., as in Wu et al. [WPW91]):

$$\vec{v} = (\frac{K}{\mu_{eff}}[-\frac{\partial p}{\partial x}])^{\frac{1}{r-1}}. \tag{4.6}$$

This filtration law is obtained by modeling a porous medium as a collection of long capillary tubes through which fully developed laminar flow occurs. If we assume flow only in the x_1 direction, then, the variables x and y in the nonlinear Stokes system (4.5) can be separated. Then, solving (4.5) leads to a nonlinear, one-dimensional power-like law, identical to the law in the engineering literature. In our notation, it reads

$$\int_{\mathfrak{P}} \vec{v}(x, y)\, dy = |\frac{dp}{dx_1}|^{r'-2} \cdot (-\frac{dp}{dx_1}) \int_{\mathfrak{P}} \vec{w}(y)\, dy,$$

where $\vec{w}$ is the solution of the nonlinear Stokes system (4.5) when the right-hand side of (4.5) is $\vec{e}_1$. For classical, nonrigorous derivation of the law (4.6), we refer to Bird et al. [BSL60] and Christopher and Middleman [CM65].

However, it should be noted that this argument holds only in the one-dimensional case. Our laws for $N > 1$ are nonlocal, and they cannot be reduced to the N-dimensional variants of (4.6), connecting the Darcy velocity $\vec{v}$ and some power of ∇p. Moreover, we should note that, because most experiments are performed for one-dimensional flows, from an engineering point of view, it is difficult to observe any effect that depends on the dimension.

4.5   Inertia Effects for Non-Newtonian Flows through Porous Media

The inertia term $(\vec{v}^\varepsilon \nabla)\vec{v}^\varepsilon$ should be contained in Eq. (2.1). It is negligible for small Reynolds number $\mathbf{Re}^\varepsilon$, but, otherwise, it plays a significant role. The case of flow governed by a forcing term $\vec{f}$ with $\vec{u} = 0$ is considered in Bourgeat and Mikelić [BM95c]. It was found that the results stated in section 4.4 are valid for $\gamma < \gamma_{crit}$. For $r = 2$, the value of γ_{crit} is equal to -1.

For a flow governed by injection, the situation is even more complicated. First, as shown in Blavier and Mikelić [BM95b] for $1 < r \le 2$, we have existence only for $\mathbf{Re}^\varepsilon$ not too large. The situation is analogous for $r > 2$. However, using the two-scale asymptotic expansion, we are able to determine the critical values of γ.

Let us assume the power law (1.1). Then, the expansions for the viscous energy term and for the inertial term are of the same order, if $(\frac{\mathbf{Re}^\varepsilon}{\varepsilon})^{r-1} = (\mathbf{Re}^\varepsilon)^2$, which gives $\gamma = (r-1)/(r-3)$.

In the case $r = 2$, there is a rigorous mathematical theory for the corresponding nonlinear and nonlocal, two-scale system of partial differential equations established in Mikelić [Mik95]. In the non-Newtonian case, there is no theory for $\gamma = \gamma_{crit}$.

4.6   Proof of the Uniqueness Theorems

In this section, we will prove Theorems 4.2, 4.6, and 4.8.

Proof of Theorem 4.2.

Let us assume that problem (1.3) has two solutions $(\vec{v}_0^1, p^1, \pi^1)$ and $(\vec{v}_0^2, p^2, \pi^2)$ in the class

$$L^{\max\{2,r\}}\left(\Omega; W_\Gamma^{1,\max\{2,r\}}(\mathfrak{P})\right)^2 \times L_0^{\min\{2,r'\}}(\Omega) \times L^{\max\{2,r\}}\left(\Omega; L_0^{\min\{2,r'\}}(\mathfrak{P})\right).$$

Let $\vec{v}_0 = \vec{v}_0^1 - \vec{v}_0^2$, $p = p^1 - p^2$, and $\pi = \pi^1 - \pi^2$. We introduce the function space

$$X = \{\vec{z} \in W_\Gamma^{1,\max\{2,r\}}(\mathfrak{P}) : \; \nabla_y \cdot \vec{z} = 0 \text{ in } \mathfrak{P} \text{ and } \vec{z} \text{ is } Y - \text{periodic}\}.$$

Let the operator A be given by

$$A(\vec{z}) = -\nabla \cdot \{\eta_r(D_y(\vec{z})) \, D_y(\vec{z})\}, \quad \vec{z} \in X.$$

Then, A is a strictly monotone and continuous operator, $A : X \to X'$.

After simple manipulations with (4.2),

$$\langle A(\vec{v}_0^1) - A(\vec{v}_0^2), \vec{v}_0 \rangle_{X',X} = -\nabla_x p \cdot \int_{\mathfrak{P}} \vec{v}_0.$$

By integrating over Ω and using the strict monotonicity,

$$\alpha \int_\Omega \|\vec{v}_0(x)\|_X^2 \, dx \le \int_\Omega p(x) \, \nabla_x \cdot \int_{\mathfrak{P}} \vec{v}_0 = 0.$$

Therefore, $\vec{v}_0 = 0$ (a.e.) on $\Omega \times \mathfrak{P}$. It implies $\nabla_y \pi + \nabla_x p = 0$, and, consequently,

$$-\nabla_x p \cdot \int_{\mathfrak{P}} \vec{\sigma}(y) \, dy = 0, \quad \forall \vec{\sigma} \in H_\#^1(\mathfrak{P})^2, \; \nabla_y \cdot \vec{\sigma} = 0 \text{ in } \mathfrak{P}, \; \vec{\sigma} = 0 \text{ on } \Gamma.$$

Now, using Lemma 2.10 from Allaire [All92a], p. 1499–1500, $\nabla_x p = 0$ which implies $p = 0$ in $L_0^{\min\{2,r'\}}(\Omega)$. Finally, $\nabla_y \pi = 0$ and $\pi = 0$ in $L^{\max\{2,r\}}\left(\Omega; L_0^{\min\{2,r'\}}(\mathfrak{P})\right)$.

Analogously,

$$-\nabla_x p \cdot \int_{\mathfrak{P}} \vec{\sigma}(y) \, dy \in L^{\max\{2,r\}}(\Omega)$$

for all $\vec{\sigma} \in H_\#^1(\mathfrak{P})^2$ with $\nabla_y \cdot \vec{\sigma} = 0$ in $\mathfrak{P}$ and $\vec{\sigma} = 0$ on Γ. Hence $p \in W^{1,\max\{2,r\}}(\Omega)$ and $\pi \in L^{\max\{2,r\}}\left(\Omega; L_0^{\min\{2,r'\}}(\mathfrak{P})\right)$.   Q.E.D.

Proofs of Theorems 4.6 and 4.6.

They are analogous to the preceding proof.   Q.E.D.

Remark This intrinsic monotone structure is applicable to the numerical solutions of those nonlocal problems.

4.7   Uniform A Priori Estimates

In this section, we will derive uniform a priori estimates for solutions of (1.2) and (2.1) and (1.1) and (2.1), respectively. The a priori estimates for the velocities are similar to those corresponding for the case of Newtonian flow. Following Mikelić and Aganović [MA87], we take $\mathbf{Re}^\varepsilon R_q^\varepsilon u$ as a lift for the trace and obtain a priori estimates for the velocities:

Lemma 7.1 *Let $\vec{v}^\varepsilon$ be a weak solution of (1.2) and (2.1), and $1 < r < +\infty$. Then,*

$$\|\vec{v}^\varepsilon\|_{L^{\max\{2,r\}}(\mathfrak{P}^\varepsilon)^2} \le C\mathbf{Re}^\varepsilon,$$

and

$$\|\nabla \vec{v}^\varepsilon\|_{L^{\max\{2,r\}}(\mathfrak{P}^\varepsilon)^4} \le C\mathbf{Re}^\varepsilon/\varepsilon.$$

Lemma 7.2 *Let $\vec{v}^\varepsilon$ be a weak solution for (1.1) and (2.1), $1 < r < +\infty$. Then,*

$$\|\vec{v}^\varepsilon\|_{L^r(\mathfrak{P}^\varepsilon)^2} \le C\mathbf{Re}^\varepsilon,$$

and

$$\|D(\vec{v}^\varepsilon)\|_{L^r(\mathfrak{P}^\varepsilon)^4} \le C\mathbf{Re}^\varepsilon/\varepsilon.$$

In the previous two lemmas, we have derived a priori estimates for the velocity. To derive a priori estimates for pressure, we use the momentum equation:

$$\langle \nabla p^\varepsilon, \vec{\varphi} \rangle_{\mathfrak{P}^\varepsilon} = -\int_{\mathfrak{P}^\varepsilon} \eta_r(D(\vec{v}^\varepsilon))D(\vec{v}^\varepsilon)D(\vec{\varphi}), \quad \forall \vec{\varphi} \in H_0^1(\Omega)^2 \cap W_0^{1,r}(\Omega)^2. \quad (7.1)$$

Now we use (7.1) to derive estimates for pressure.

We start with the Carreau law.

Proposition 7.3 *Let η_r be given by formula (1.2) (Carreau's law). Then, for $r \le 2$, the pressure p^ε satisfies the inequality*

$$\|\nabla p^\varepsilon\|_{H^{-1}(\mathfrak{P}^\varepsilon)^2} \le C\mathbf{Re}^\varepsilon/\varepsilon. \quad (7.2)$$

For $r > 2$, we again have the inequality

$$\|\nabla p^\varepsilon\|_{W^{-1,r'}(\mathfrak{P}^\varepsilon)^2} \le C\frac{\mathbf{Re}^\varepsilon}{\varepsilon}\left(1 + \left(\mathbf{Re}^\varepsilon/\varepsilon\right)^{r-2}\right).$$

Now we are able to extend pressure.

Lemma 7.4 *Let all the assumptions of Proposition 7.3 hold. Then, there is an extension $\tilde{p}^\varepsilon \in L_0^{\min\{2,r'\}}(\Omega)$ of the function p^ε, such that*

$$\|\tilde{p}^\varepsilon\|_{L_0^{\min\{2,r'\}}(\Omega)}$$
$$+\|\nabla\tilde{p}^\varepsilon\|_{W^{-1,\min\{2,r'\}}(\Omega)^2} \le C\frac{\mathbf{Re}^\varepsilon}{\varepsilon^2}\left(1 + \max\{0, r-2\}\left(\mathbf{Re}^\varepsilon/\varepsilon\right)^{r-2}\right).$$

Now we turn to the system (1.1) and (2.1) (the case of the power law).

Lemma 7.5 *Let η_r be given by formula (1.1) (power law). Then the pressure p^ε satisfies the inequality*

$$\|\nabla p^\varepsilon\|_{W^{-1,r'}(\mathfrak{P}^\varepsilon)^2} \leq C\left(\mathbf{Re}^\varepsilon/\varepsilon\right)^{r-1}.$$

In the next step, we extend the pressure as in Mikelić [Mik91] and get the following estimate.

Lemma 7.6 *Let all assumptions of Lemma 7.5 hold. Then, there exists an extension $\tilde{p}^\varepsilon \in L_0^{r'}(\Omega)$ of the function p^ε such that*

$$\|\tilde{p}^\varepsilon\|_{L_0^{r'}(\Omega)} + \|\nabla \tilde{p}^\varepsilon\|_{W^{-1,r'}(\Omega)^2} \leq C\left(\mathbf{Re}^\varepsilon/\varepsilon\right)^{r-1}/\varepsilon.$$

The special properties of the restriction operator R_q^ε and the derived properties of the extension of the pressure permit the following result:

Proposition 7.7 *Let $\tilde{p}^\varepsilon$ be the extension of the pressure from Lemma 7.4 or Lemma 7.6, and let $\{\vec{\psi}^\varepsilon\} \subset L_0^{\max\{2,r\}}(\Omega)^2$ be an arbitrary sequence which converges weakly to 0. Then,*

$$\int_\Omega \tilde{p}^\varepsilon \vec{\psi}^\varepsilon \to 0 \quad \text{as } \varepsilon \to 0.$$

Proof It is essentially contained in the proof of Lemma 4.4, pp. 175–176, of Mikelić [Mik91]. Q.E.D.

Lemma 7.8 *The extensions $\tilde{p}^\varepsilon$ of the pressures p^ε from Lemmas 7.4 and 7.6 are given by (3.1), i.e., $\tilde{p}^\varepsilon = p^\varepsilon$ in the fluid part $\mathfrak{P}^\varepsilon$, and $\tilde{p}^\varepsilon = \frac{1}{|Y_{F_i}^\varepsilon|}\int_{Y_{F_i}^\varepsilon} p^\varepsilon$ in $Y_{\mathfrak{S}_i}^\varepsilon$ for each i, where $Y_{\mathfrak{S}_i}^\varepsilon$ and Y_{F_i} are the solid and the fluid parts of Y_i^ε, respectively.*

Proof See the articles Lipton and Avellaneda [LA90] and Allaire [All89]. Q.E.D.

4.8   Proof of Theorem A

In this section we will consider the problem of Theorem 4.3, i.e., Carreau's law for the viscosity with $\gamma = 1$. It is important to note that we cannot conclude the convergence $\eta_r^\varepsilon \to \eta_0$ for $\gamma = 1$. This gives rise to a completely different law relating the homogenized velocity and the gradient of the homogenized pressure.

We start with the following result which is a straightforward consequence of the results from section 4.3, Proposition 3.5 (i), Lemma 7.4, and Proposition 7.7:

Proposition 8.1 *Let $\gamma = 1$, and let $(\vec{v}^\varepsilon, p^\varepsilon)$ be the solutions of problem (1.2) and (2.1). Then there exist subsequences of $\{\vec{v}^\varepsilon\}$ and $\{\tilde{p}^\varepsilon\}$ (again denoted by the same symbols) and functions $\vec{v}_0^* \in L^{\max\{2,r\}}(\Omega \times Y)^2$, $p^* \in L_0^{\min\{2,r'\}}(\Omega)$, such that, in the limit $\varepsilon \to 0$,*

$$\varepsilon^{-1}\vec{v}^\varepsilon \to \vec{v}_0^* \text{ in the two-scale sense in } L^{\max\{2,r\}}, \tag{8.1}$$

$$\nabla \vec{v}^\varepsilon \rightarrow \nabla_y \vec{v}_0^* \in L^{\max\{2,r\}}(\Omega \times Y)^4 \text{ in the two-scale sense in } L^{\max\{2,r\}},$$

$$\varepsilon^{-1}\vec{v}^\varepsilon \rightharpoonup \vec{v}^* = \int_Y \vec{v}_0^*(x, y) \, dy \text{ weakly in } L^{\max\{2,r\}}(\Omega)^2,$$

and

$$\varepsilon \tilde{p}^\varepsilon \rightarrow p^* \text{ in } L_0^{\min\{2,r'\}}(\Omega). \tag{8.2}$$

The function $\vec{v}^*$ satisfies the equations

$$\nabla \cdot \vec{v}^* = 0 \quad in \ \Omega$$

and

$$\vec{v}^* \cdot \vec{v} = \vec{u} \cdot \vec{v} \quad on \ \Gamma.$$

The properties of the two-scale limit $\vec{v}_0^*$ are more precisely described by the following lemma:

Lemma 8.2 $\vec{v}_0^* \in L^{\max\{2,r\}}\big(\Omega; W_\Gamma^{1,\max\{2,r\}}(\mathfrak{P})\big)^2$ *and* $\nabla_y \cdot \vec{v}_0^* = 0$.

Proof See Bourgeat and Mikelić [BM95c].   Q.E.D.

Our next goal is to derive the law giving the relationship between $\vec{v}_0^*$ and ∇p. Let the functional Φ_r be defined by

$$\Phi_r(x) = \frac{\eta_\infty}{2}|x|^2 + \frac{\eta_0 - \eta_\infty}{r\lambda}\Big\{\big[1 + \lambda|x|^2\big]^{r/2} - 1\Big\},$$

for symmetric 2×2 matrices x. Note that $\nabla \Phi_r(x) = \eta_r(x)x$ and Φ_r is a strictly convex, proper functional defined on symmetric matrices, $1 < r < +\infty$.

Proposition 8.3 *The functions $\vec{v}_0^*$ and p^* defined by (8.1) and (8.2), respectively, satisfy the equations*

$$- \nabla_y \cdot \{\eta_r(D_y(\vec{v}_0^*))D_y(\vec{v}_0^*)\} + \nabla_y \pi(x, y) = -\nabla_x p^*(x) \text{ in } \mathfrak{P}, \tag{8.3}$$

$$\nabla_y \cdot \vec{v}_0^* = 0 \text{ in } \mathfrak{P}, \quad (\vec{v}_0^*, \pi) \text{ is } Y - periodic, \quad \vec{v}_0^* = 0 \text{ on } \Gamma. \tag{8.4}$$

Proof We write Eq. (2.1) in the following equivalent form:

$$\int_\Omega \Phi_r(\varepsilon D(\vec{z})) - \int_\Omega \Phi_r(\varepsilon D(\vec{w}^\varepsilon))$$

$$\geq \varepsilon \int_\Omega \nabla p^\varepsilon \cdot (\vec{w}^\varepsilon - \vec{z}), \ \forall \vec{z} \in H_0^1(\mathfrak{P}^\varepsilon)^2 \cap W_0^{1,r}(\mathfrak{P}^\varepsilon)^2 + R_q^\varepsilon \vec{u} \tag{8.5}$$

where $\vec{w}^\varepsilon = \varepsilon^{-1}\vec{v}^\varepsilon$.

Now, let $\vec{\psi}(x, y) \in C_0^\infty(\Omega; C_\#^\infty(\mathfrak{P})^2)$, $\vec{\psi}(x, y)\big|_\Gamma = 0$, $\nabla_y \cdot \vec{\psi}(x, y) = 0$. We define $\vec{\psi}^\varepsilon(x) = \vec{\psi}(x, \frac{x}{\varepsilon})$. Inserting $\vec{z} = \vec{\psi}^\varepsilon + R_q^\varepsilon$ in (8.5) and using propositions 3.6 and 3.7,

$$\int_\Omega \int_Y \Phi_r(D_y(\vec{\psi} + R_q)) \, dx \, dy - \int_\Omega \int_Y \Phi_r(D_y(\vec{v}_0^*)) \, dx \, dy \geq \lim_{\varepsilon \to 0} \varepsilon \int_\Omega \tilde{p}^\varepsilon \nabla_x \cdot \vec{\psi}^\varepsilon,$$

as $\varepsilon \to 0$. The elementary properties of the two-scale convergence imply that

$$\varepsilon \int_\Omega \tilde{p}^\varepsilon \nabla_x \cdot \vec{\psi}^\varepsilon \to \int_\Omega \int_Y p^* \nabla_x \cdot \vec{\psi}(x, y)\, dx\, dy \text{ as } \varepsilon \to 0$$

and we get

$$\langle -\nabla_y \cdot \{\eta_r(D_y(\vec{v}_0^*))D_y(\vec{v}_0^*)\} + \nabla_x p^*(x),\ \vec{\psi}(y)\rangle_{X',X} = 0,$$

$$\forall \vec{\psi} \in X = \{\vec{z} \in W_\Gamma^{1,\max\{2,r\}}(\mathfrak{P})^2 : \vec{z} \text{ is } Y\text{-periodic and } \nabla_y \cdot \vec{z} = 0 \text{ in } \mathfrak{P}\}. \quad (8.6)$$

It is easy to see that (8.6) implies (8.3) by the variant of de Rham's formula in a periodic setting (see Temam [Tem79]). Formula (8.4) has been proven in Lemma 8.2. $\hspace{4cm}$ Q.E.D.

Now we are able to prove Theorem 4.3:

Proof of Theorem 4.3: The convergence for the subsequence is a direct consequence of the preceding propositions. Theorem 4.2 implies that problem (4.2) has a unique solution, and, therefore, the limits do not depend on subsequences, i.e., $v = \vec{v}_0^*$ and $p = p^*$. $\hspace{4cm}$ Q.E.D.

4.9   Proof of Theorem B

In this section, we will prove Theorem 4.9.

Proposition 9.1 *Let* $(\vec{v}^\varepsilon, p^\varepsilon)$ *be corresponding solutions of the power law system (1.1) and (2.1). Then there exist subsequences of* $\{\vec{v}^\varepsilon\}$ *and* $\{\tilde{p}^\varepsilon\}$ *(again denoted by the same symbols) and functions* $\vec{v}_0^* \in L^r(\Omega \times Y)^2$, $p^* \in L_0^{r'}(\Omega)$, *such that*

$$\left(\frac{\varepsilon}{\mathrm{Re}^\varepsilon}\right)^{r-1} \vec{v}^\varepsilon \to \vec{v}_0^* \text{ in the two-scale sense in } L^r, \tag{9.1}$$

$$\left(\frac{\varepsilon}{\mathrm{Re}^\varepsilon}\right)^{r-1} \nabla \vec{v}^\varepsilon \to \nabla_y \vec{v}_0^* \in L^r(\Omega \times Y)^4 \text{ in the two-scale sense in } L^r, \tag{9.2}$$

$$\left(\frac{\varepsilon}{\mathrm{Re}^\varepsilon}\right)^{r-1} \vec{v}^\varepsilon \rightharpoonup \vec{v}^* = \int_Y \vec{v}_0^*(x, y)\, dy \text{ weakly in } L^r(\Omega)^2, \tag{9.3}$$

and

$$\varepsilon\left(\frac{\varepsilon}{\mathrm{Re}^\varepsilon}\right)^{r-1} \tilde{p}^\varepsilon \to p^* \text{ in } L_0^{r'}(\Omega), \tag{9.4}$$

as $\varepsilon \to 0$. *The average* $\vec{v}^*$ *satisfies the equations*

$$\nabla_x \cdot \vec{v}^* = 0 \text{ in } \Omega,$$

and

$$\vec{v}^* \cdot \vec{v} = \vec{u} \cdot \vec{v} \text{ on } \Gamma.$$

Analogously to Lemma 8.2,

Lemma 9.2 $\vec{v}_0^* \in L^r\left(\Omega; W_\Gamma^{1,\,r}(\mathfrak{P})^2\right)$, and $\nabla_y \cdot \vec{v}_0^* = 0$ in $\mathfrak{P}$.

Now we are going to establish the law giving a relationship between $\vec{v}_0^*$ and ∇p^*

Proposition 9.3 *The functions $\vec{v}_0^*$ and p^*, defined by (9.1) and (9.4), respectively, satisfy the equations*

$$-\lambda\, \nabla \cdot \left\{ |D_y(\vec{v}_0^*)|^{r-2} D_y(\vec{v}_0^*) \right\} + \nabla_y \pi(x, y) = -\nabla_x p^*(x) \text{ in } \mathfrak{P},$$

$$\nabla_y \cdot \vec{v}_0^* = 0 \text{ in } \mathfrak{P},$$

$$(\vec{v}_0^*, \pi) \text{ is } Y\text{-periodic,}$$

and

$$\vec{v}_0^* = 0 \text{ on } \Gamma.$$

Proof We use Eq. (2.1) and formula (1.1) and write, in an equivalent form,

$$\int_\Omega \frac{\lambda}{r} |\varepsilon D(\vec{\varphi})|^r \, dx - \int_\Omega \frac{\lambda}{r} |\varepsilon D(\vec{w}^\varepsilon)|^r \, dx \geq -\langle \nabla \tilde{\omega}^\varepsilon,\ \vec{\varphi} - \vec{w}^\varepsilon \rangle_\Omega, \tag{9.5}$$

for all $\varphi \in W_0^{1,\,r}(\mathfrak{P}^\varepsilon)^2 + R_r^\varepsilon u$, where $\vec{w}^\varepsilon = \vec{v}^\varepsilon \left(\frac{\varepsilon}{\mathbf{Re}^\varepsilon}\right)^{r-1}$ and $\omega^\varepsilon = p^\varepsilon \varepsilon \left(\frac{\varepsilon}{\mathbf{Re}^\varepsilon}\right)^{r-1}$.

Now we choose $\vec{\psi} \in C_0^\infty\left(\Omega; C_\#^\infty(\mathfrak{P})\right)^2$, such that $\vec{\psi}(x, y) = 0$ on Γ for (a.e.) $x \in \Omega$, $\nabla_y \cdot \vec{\psi} = 0$ in $\mathfrak{P}$ and define $\vec{\psi}^\varepsilon(x) = \vec{\psi}(x, \frac{x}{\varepsilon})$. We insert $\varphi = \vec{\psi}^\varepsilon + R_r^\varepsilon u$ in (9.5). Then,

$$-\langle \nabla \tilde{\omega}^\varepsilon, \vec{\psi}^\varepsilon \rangle_\Omega = \int_\Omega \tilde{\omega}^\varepsilon \nabla_x \cdot \vec{\psi}^\varepsilon \to \int_\Omega \int_Y p^* \nabla_x \cdot \vec{\psi}(x, y) \, dx \, dy, \text{ as } \varepsilon \to 0.$$

Now the results from section 4.3 imply that

$$\int_\Omega \int_Y \frac{\lambda}{r} |D_y(\vec{\psi} + R_r u)|^r \, dx \, dy$$

$$- \int_\Omega \int_Y \frac{\lambda}{r} |D_y(\vec{v}_0^*)|^r \, dx \, dy \geq -\langle \nabla p^*(x), \int_Y (\vec{\psi} - \vec{v}_0^*) \, dy \rangle_\Omega.$$

Consequently,

$$-\lambda\, \nabla_y \cdot \left\{ |D_y(\vec{v}_0^*)|^{r-2} D_y(\vec{v}_0^*) \right\} + \nabla_y \pi(x, y) = -\nabla p^*(x) \text{ in } \mathfrak{P},$$

$$\nabla_y \cdot \vec{v}_0^* = 0 \text{ in } \mathfrak{P},$$

$$\vec{v}_0^* = 0 \text{ on } \Gamma,$$

and $(\vec{v}_0^*, \pi)$ is Y-periodic for (a.e.) $x \in \Omega$ by de Rham's formula.   Q.E.D.

Proof of Theorem 4.9: It is a direct consequence of the preceding propositions. Q.E.D.

4.10   Conclusion

In this chapter, we applied the method of homogenization in modeling non-Newtonian flow through porous media. Starting from the Stokes (and incompressible Navier–Stokes) equations with viscosity dependent on shear rate we have obtained both nonlinear and linear relationships between filtration velocity and effective pressure.

When modeling viscosity by the "power law" or Ostwald–de Waele model (1.1), we found that filtration velocity is the average of a velocity $\vec{v}_0$. This velocity depends on a fast variable y and a slow variable x, and we determine $\vec{v}_0(x, y)$ and the effective pressure $p(x)$ by solving the system (4.5). This system of partial differential equations no longer involves the extremely complicated geometry of our porous medium and should be used instead of the original flow equations. Let us, however, point out that it has an intrinsic, nonlinear, two-scale coupling which poses some challenges in the numerical simulation. This structural complexity comes from nonlinear viscosity and should not be seen as a drawback. We cannot always expect to have laws as simple as Darcy's, describing complex natural phenomena.

As already pointed out in the last remark of section 4.4, our system (4.5) reduces to the widely used, nonlinear, Darcy-like law (4.6) in the case of a directional flow. That example also confirms a "dual" functional relationship between shear stress and shear rate usually found when modeling porous media as a bundle of capillary tubes. However, our conclusion is that direct generalizations of (4.6) are incorrect. Contrary to the case of the Darcy law, when homogenization only justified something already well-established, here, homogenization was used to derive an a priori unknown model.

The case of the Carreau–Yasuda law (1.2) is somewhat different. Depending on the relation between the Reynolds number $\mathbf{Re}^\varepsilon = \varepsilon^\gamma$ and the pore size ε, we obtain several effective models. More precisely, if $\gamma > 1$ (slow injection), our effective model is the well-known Darcy law. For $\gamma = 1$, we obtain the nonlinear two-scale (4.2). For higher injection rates ($\gamma < 1$), we get the Darcy law for pseudoplastic fluids and system (4.4), equivalent to (4.5), for dilatant fluids. If the injection rate is too high, we can expect some inertial effects and, the hypothesis of creeping flow is no longer valid.

Other modeling problems, such as inertial, thermal, and memory effects for non-Newtonian flow were not considered in this short chapter. Using the homogenization method described in this book and material from other chapters, all of those problems can be addressed similarly.

On the other hand, numerical solution of systems of the type (4.5) is still an open problem. This class of problems appeared after finding the homogenized systems of type (4.5). Furthermore, a theory of systems of type (4.5), corresponding to the random case, is still to be developed.

Finally, let us point out that, apart from its role in justifying various effective models in physics, homogenization is a very powerful tool in modeling.

5

Two-Phase Flow

Alain Bourgeat

5.1 Derivation of the Generalized Nonlinear Darcy Law

5.1.1 Introduction

In practical applications, the problem of modeling two-phase flow behavior is of greatest importance. In petroleum engineering the so-called enhanced oil recovery (EOR) process is based on displacing a fluid (oil) by another one (gas or water, for example). In soil science, water and contaminant or water and air are involved in many underground flows in the unsaturated zone. Domestic gas, before delivery, is kept in natural reservoirs and moved by water and a cushion gas.

For single-phase flow through a porous medium, by using homogenization theory, we are now able to rigorously derive filtration laws according to the porosity range (see for instance Chapter 3) or according to the Reynolds number as for instance in [BMP95] or in [Mik95].

Unlike one-phase flow, deriving the filtration law from the behavior of the two-phase mixture at the pore level is still not clearly and totally rigorously understood, although there have been several attempts (see for instance [ASP86], [EP87], [Mar82], [Whi86]) leading to partial results.

However, engineers from all fields (petroleum engineering, soil science, environment...) use the model, introduced on empirical grounds and on experiments by Wyckoff–Botset [WB36] and Leverett [Lev38], to describe the two-phase flow behavior at the macroscopic scale (the scale of rock properties). The key to this macroscopic model is the concept of relative permeability, i.e., the assumption that each phase has its own Darcy velocity, depending on saturation, i.e., the presence of the other phase:

$$\vec{q}_\xi = -K \frac{k_{r\xi}(S_\xi)}{\mu_\xi}(\nabla p_\xi - \rho_\xi g \nabla \mathcal{Z}), \qquad \xi = \alpha, \beta, \tag{1.1}$$

where the subscripts α and β stand for the wetting and the nonwetting phase fluid. Each phase also has its own density ρ_ξ and its own viscosity μ_ξ, and we denote, by (p_α, S_α) and (p_β, S_β), the pressures and saturations for the wetting and nonwetting phases, respectively. In simultaneous flow of two fluids, it is assumed that each fluid interferes with the flow of the other. If the one-phase rock permeability were a diagonal tensor $K = kI$, the effective permeabilities $Kk_{r\xi}$ would be less than or equal to the single-phase permeability k (i.e., $kk_{r\xi} \leq k$). The functions of the saturation $k_{r\xi}(S_\xi)$, which take values between 0 and 1, are, then, called relative permeability curves.

Two mathematical approaches have been explored to develop multiphase flow equations derived empirically.

The first approach is based on the method used in the thermodynamic theory of mixtures and irreversible processes (as, for instance, in [Bow76] and [GM62]). It has been developed essentially in [Mar82] and in [Whi86]. As in statistical mechanics, it is assumed that averaging over all sets of microscopically identical systems will define macroscopic quantities and their behavior. Existence of macroscopic behavior, in the context of randomly distributed pores, has been rigorously justified in the case of one-phase flow in [Mat67] and in [BK95]. On the other hand, for computing effective parameters, it has been proven that averaging is equivalent to the usual stochastic homogenization (see for instance [Sab92] and [BB95]), provided that the medium (or the mixture) is statistically homogeneous (i.e., shift-invariant in probability) and ergodic. But due to the complexity of the phenomena involved in describing a multiphase multicomponent flow at the pore level, until now there has been no mathematical proof of the validity for volume averaging, and some heuristics have to be used to obtain a macroscopic model from microscopic behavior. The procedure is as follows. First, choose the macroscopic quantities which would be involved in describing the macroscopic state of the system and of the phenomena leading to its evolution. Second, write the balance equations (for mass, momentum, and energy) according to the kinematical, dynamical, and thermodynamic principles for each phase at the pore level. Third, from spatial volume averaging, develop a macroscopic version of the previous balance equations. At this step, it should be noticed that there are some microscopic "additive" quantities like the mass or mass flux of a component in a fluid phase, which, by volume averaging, will naturally define macroscopic quantities

$$\mathbf{U}(t, x) = (u * m)(t, x) = \int_{\mathbb{R}^3} u(t, x - y)m(y)dy;$$

for the definition of the weight function m see below.

Some other "nonadditive" quantities, for instance, pressure, temperature, or velocity cannot be obtained directly from volume averaging. They are derived, at the macroscopic level, by a "closure equation", based on periodicity assumption as in [Whi86] and in [BQW88], or by a total macroscopic entropy equation as in [Mar82]. This entropy equation is built so that the macroscopic quantities, involved in describing the macroscopic state, obey macroscopic thermodynamic principles. In the following, we will adopt the second point of view which is close to the physical meaning of the equations.

The second way of obtaining macroscopic behavior from microscopic behavior at the pore level is by two-scale asymptotic expansion as in the standard homogenization theory (see, for instance, [SP80] or [BP89]). One attempt has been made in this direction for the stationary case, i.e., with no motion of the interface (see [ASP86] and [ST91]).

In section 5.1.2 we will start with the first method, based on volume averaging; the second method, based on asymptotic expansion will be presented in section 5.1.3.

5.1.2   Obtaining Macroscopic Laws by Volume Averaging

This method has been used for complicated phenomena (see, for instance, [Mar82] and the bibliography therein), like multiphase fluid flow with heat and diffusion fluxes, chemical reaction, mass adsorption, and desorption. But, for simplicity, following [EP87], we will restrict the description of this method to a system of two incompressible Newtonian fluid phases (α and β) with several components $i = 1, \ldots, n$, and we will neglect only the phenomena taking place at the interfaces. We will also assume that the solid is perfectly rigid and at rest, neglecting the gravitational acceleration and assuming that the external forces f^i are uniform in all phases.

We start by recalling the microscopic balance equations written at the pore level.

Equations at the Pore Level

In each ξ-fluid phase, $\xi = \{\alpha, \beta\}$, the following "microscopic" quantities are chosen to describe the phenomena at the pore scale. We denote the *mass per unit volume* by ρ_ξ, the *mass concentration* of the ith component for $i = 1, \ldots, n$ by $c_{\xi i}$ (with $\sum_{i=1}^{n} c_{\xi i} = 1$), the *fluid phase velocity* and *pressure* by (v^ξ, p_ξ), the *diffusion mass flux* of the ith component in the ξ-fluid phase by $j^{\xi i}$ (with $\sum_{i=1}^{n} j^{\xi i} = 0$), the *internal energy* by e_ξ, and the *external forces per unit mass* of component i by $\vec{f}^{\xi i}$.

Now, for each phase $\xi = \{\alpha, \beta, s\}$ including the solid phase (denoted by the subscript s), we write the mass, momentum and energy balance equations:

$$\frac{\partial}{\partial t}(\rho_\xi c_{\xi i}) + \nabla \cdot \omega^{\xi i} = 0 , \qquad i = 1, \ldots, n, \tag{1.2}$$

$$\frac{\partial}{\partial t}\omega^\xi + \nabla \cdot (\omega^\xi \otimes v^\xi - \sigma^\xi) - \rho_\xi c_{\xi i} f^{\xi i} = 0, \tag{1.3}$$

and

$$\frac{\partial}{\partial t}(\rho_\xi e_\xi) + \nabla \cdot (\rho_\xi e_\xi v^\xi + Q^\xi) - \sigma^\xi : \nabla v^\xi + j^{\xi i} \cdot f^{\xi i} = 0. \tag{1.4}$$

Unless stated otherwise, throughout, we are using the repeated indices Einstein notation ($a_i b_i \equiv \sum_{i=1}^{n} a_i b_i$).

Summing (1.2) over $i = 1, \ldots, n$ yields the microscopic *continuity equation*,

$$\frac{\partial \rho_\xi}{\partial t} + \nabla \cdot \omega^\xi = 0. \tag{1.5}$$

In the equations above, we have denoted, by $\omega^{\xi i}$ the ith component of the *mass flux* in the ξ fluid phase, Q^ξ the *heat flux*, and ω^ξ the *total mass flux* with

$$\sum_{i=1}^{n} \omega^{\xi i} = \omega^\xi = \rho_\xi v_\xi \qquad \text{and} \qquad \omega^{\xi i} = \rho_\xi c_{\xi i} \omega^\xi + j^{\xi i}. \qquad (1.6)$$

The momentum tensor, in the momentum balance equation for any phase ξ, is composed of the convective momentum flux $\omega^\xi \otimes v^\xi$ and the symmetric stress tensor $\sigma^\xi = -p_\xi I + \varepsilon^\xi$ (ε^ξ being the viscosity tensor).

Due to the assumptions on the solid part, (assumed to be perfectly rigid and at rest) equations (1.2)–(1.4) for the solid phase s reduce to the energy balance equation,

$$\frac{\partial}{\partial t}(\rho_s e_s) + \nabla \cdot Q^s = 0. \qquad (1.7)$$

Remark All quantities appearing in (1.2)–(1.7) with subscripts or superscripts $\xi \in \{\alpha, \beta, s\}$ are assumed to be continuous with continuous derivatives in each ξ-phase but, except for the velocities, they may have jump discontinuities across the phase interfaces. Consequently, derivatives appearing in the equations above must be taken as distributions.

Averaging Process

Let m be a positive function of class C^∞ with compact support in $\mathbb{R}^3$ such that

$$\int_{\mathbb{R}^3} m(x)dx = 1.$$

Let $\mathcal{X}_\xi(x, t)$, for $\xi \in \{\alpha, \beta\}$, and, let $\mathcal{X}_s(x)$ be the functions equal to one inside the ξ-phase, for $\xi \in \{\alpha, \beta, s\}$, and zero elsewhere.

We may then define the macroscopic quantities for the porous medium, the porosity ϕ and the ξ-fluid phase saturation S_ξ, by

$$\phi = (1 - \mathcal{X}_s) * m,$$

and

$$\phi S_\xi = \mathcal{X}_\xi * m, \qquad \xi \in \{\alpha, \beta\}.$$

Now by convolution of Eqs. (1.2)–(1.4) and the function m, we obtain the macroscopic balance equations in any ξ-fluid phase associated with the corresponding macroscopic quantities.

For instance, the macroscopic mass balance obtained in this way is

$$\frac{\partial}{\partial t}(\phi S_\xi \rho_\xi C_{\xi i}) + \nabla \cdot (\phi S_\xi \rho_\xi C_{\xi i} v^\xi + j^{\xi i}) + b_{\xi si} + \sum_{\substack{\xi_1, \xi_2 \in \{\alpha, \beta\} \\ \xi_1 \neq \xi_2}} b_{\xi_1 \xi_2 i} = 0 \qquad (1.8)$$

for $i = 1, \ldots, n$, where, due to their additive property, the following quantities have been defined for each ξ-fluid phase:

ρ_ξ, the macroscopic mass per unit volume, $\phi S_\xi \rho_\xi = \rho_\xi * m$;

$C_{\xi i}$, the macroscopic mass concentration of the ith component in the ξ-phase, $\phi S_\xi \rho_\xi C_{\xi i} = \rho_\xi c_{\xi i} * m$;

$\mathbf{v}^\xi$, the macroscopic ξ-fluid phase velocity, $\phi S_\xi \rho_\xi \mathbf{v}^\xi = \rho_\xi v^\xi * m$;

$\mathbf{j}^{\xi i}$, the macroscopic ξ-fluid phase mass flux diffusion of the ith component, $\mathbf{j}^{\xi i} = j^{\xi i} * m + \rho_\xi (c_{\xi i} - C_{\xi i})(v^\xi - \mathbf{v}^\xi) * m$;

$\omega^{\xi i}$, the macroscopic ith component mass flux in the ξ-fluid phase, $\omega^{\xi i} = \phi S_\xi \rho_\xi C_{\xi i} \mathbf{v}^\xi$;

$b_{\xi_1 \xi_2 i}$, the ith component mass leaving the ξ_1-phase through the interface $\Sigma_{\xi_1 \xi_2}$, per unit time and unit volume of porous medium,

$$b_{\xi_1 \xi_2 i} = ([[\rho_{\xi_1} C_{\xi_1 i}(v^{\xi_1} - V) + j^{\xi i}]] \cdot \vec{v} \, \delta_{\xi_1 \xi_2}) * m, \quad \xi_1 \in \{\alpha, \beta\}, \quad \xi_2 \in \{\alpha, \beta, s\}$$

(V is the velocity of the interface $\Sigma_{\xi_1 \xi_2}$ between the ξ_1 and the ξ_2-phase, $\vec{v}$ is the outward ξ_1 unit normal to the interface, $\delta_{\xi_1 \xi_2}$ is the $\mathbb{R}^3$ Dirac measure with support $\Sigma_{\xi_1 \xi_2}$, and $[[a]]$ denotes the jump of the quantity a through $\Sigma_{\xi_1 \xi_2}$).

To relate the macroscopic quantities by a macroscopic continuity equation like (1.5), we add equations (1.8) for $i = 1, \ldots, n$ and using

$$\sum_{i=1}^{n} C_{\xi i} = 1,$$

$$\sum_{i=1}^{n} \mathbf{j}^{\xi i} = 0,$$

$$\mathbf{b}_{\xi_1 \xi_2} = \sum_{i=1}^{n} b_{\xi_1 \xi_2 i},$$

and

$$\sum_{i=1}^{n} \omega^{\xi i} = \omega^\xi = \phi S_\xi \rho_\xi \mathbf{v}^\xi,$$

we, finally, obtain

$$\frac{\partial}{\partial t}(\phi S_\xi \rho_\xi) + \nabla \cdot \omega^\xi + \mathbf{b}_{\xi s} + \sum_{\substack{\xi_1, \xi_2 \in \{\alpha, \beta\} \\ \xi_1 \neq \xi_2}} \mathbf{b}_{\xi_1 \xi_2} = 0, \quad \xi = \alpha, \beta. \tag{1.9}$$

Similarly convolution of the other microscopic balance equations (1.3)–(1.4) by m leads to the definition of the following macroscopic quantities:

the macroscopic stress tensor $\sigma^\xi = \sigma^\xi * m + (\rho_\xi(v^\xi - \mathbf{v}^\xi) \otimes (v^\xi - \mathbf{v}^\xi)) * m$,

the macroscopic energy $\mathbf{e}_\xi$, $\phi S_\xi \rho_\xi \mathbf{e}_\xi = (\rho_\xi e_\xi) * m$,

the macroscopic heat flux

$$\mathbf{Q}^\xi = Q^\xi * m + \rho_\xi(e_\xi - \mathbf{e}_\xi)(v^\xi - \mathbf{v}^\xi) * m - (v^\xi \cdot (\sigma^\xi - \sigma^\xi)) * m.$$

For the solid phase we obtain the macroscopic quantities in the same way:

$$(1 - \phi)\rho_s = \rho_s * m,$$

$$(1 - \phi)\rho_s \mathbf{e}_s = (\rho_s e_s) * m,$$

and

$$\mathbf{Q}^s = Q^s * m.$$

Obtaining Constitutive Equations

The problem, now, is to define the macroscopic "nonadditive" quantities like temperature θ_ξ or pressure $\mathbf{p}_\xi$. As in [EP87], the first step is to impose the first *constitutive equation* for the macroscopic tensor:

$$\sigma^\xi = -\phi S_\xi \mathbf{p}_\xi I + \varepsilon^\xi, \quad \xi = \alpha, \beta. \tag{1.10}$$

The second step is to make the averaged quantities in previous macroscopic balance equations satisfy the fundamental thermodynamic principles. It is assumed that every thermodynamic relationship for a medium at rest under equilibrium thermodynamic conditions remains valid at each point of the thermodynamic macroscopic level for any fluid or solid phase and for interfaces.

From this, a total macroscopic entropy balance equation is derived by adding the macroscopic entropy balance equations for any fluid or solid phases and for interfaces. According to the principle of thermodynamics for an irreversible process, [GM62], this equation should have the form of an entropy balance equation

$$\frac{\partial}{\partial t} \mathbf{E} + \nabla \cdot S = \bar{\omega} \tag{1.11}$$

where $\mathbf{E}$ and S are an entropy density and an entropy flux. Moreover, the entropy source density $\bar{\omega}$ should be, as usual, a product of generalized forces and generalized fluxes. These generalized forces and fluxes have been obtained by adding the microscopic entropy sources of phases and interfaces and averaging them by convolution with m.

We start, at the macroscopic level, from the Gibbs–Duhem equation

$$\mathbf{e}_\xi + \frac{\mathbf{p}_\xi}{\rho_\xi} - \theta_\xi E_\xi - \mu_{\xi i} C_{\xi i} = 0, \quad \xi = \alpha, \beta, \tag{1.12}$$

where θ_ξ is the temperature, E_ξ is the ξ-fluid entropy and $\mu_{\xi i}$ is the macroscopic thermodynamic potential of the ith component in the phase ξ, and, from the relations between differentials

$$d(\rho_\xi \mathbf{e}_\xi) = \theta_\xi d(\rho_\xi E_\xi) + \mu_{\xi i} d(\rho_\xi C_{\xi i}), \quad \xi = \alpha, \beta, \tag{1.13}$$

and

$$d(\rho_s \mathbf{e}_s) = \theta_s d(\rho_s E_s). \tag{1.14}$$

Using the previous macroscopic mass equation (1.8) with the macroscopic continuity equation (1.9), we have to compute the right-hand side of the expected macroscopic entropy balance equation (1.11), i.e., to compute

$$\frac{\partial}{\partial t}(\phi S_\xi \rho_\xi E_\xi) + \nabla \cdot (\phi S_\xi \rho_\xi E_\xi \mathbf{v}^\xi + \frac{Q^\xi}{\theta_\xi} - \frac{\mu_{\xi i}}{\theta_\xi}\mathbf{j}^{\xi i}), \quad \xi = \alpha, \beta, \quad (1.15)$$

and

$$\frac{\partial}{\partial t}((1 - \phi)\rho_s E_s) + \nabla \cdot (\frac{Q^s}{\theta_s}). \quad (1.16)$$

Those computations lead to a macroscopic entropy source density in (1.11) composed of a sum of generalized fluxes and generalized forces. Imposing linear relations between these generalized fluxes and forces finally leads to the missing *constitutive equations*.

Assuming interfacial areas between fluids and solid depending only on ϕS_ξ, in the case of two-phase two components flow, but taking into account the dynamic of the interfaces (as in [Pav90] or in [Kal86]), yields a generalized, nonlinear Darcy law

$$\mathbf{v}^\xi = K_{\xi\xi}[-\nabla \mathbf{p}_\xi] + K_{\xi\zeta}[-\nabla \mathbf{p}_\zeta] + K_{\xi Q}\nabla(1/\theta_\xi), \quad \xi = \alpha, \beta, \quad \zeta \neq \xi, \quad (1.17)$$

where $K_{\xi\xi}$, $K_{\xi\zeta}$ and $K_{\xi Q}$ are three two-rank tensors, and leads to a dynamic capillary pressure relation,

$$\mathbf{p}_\alpha - \mathbf{p}_\beta = (\cos \varphi)f(S_\alpha) + g(S_\alpha) + \pi \frac{\partial(\phi S_\alpha)}{\partial t}, \quad (1.18)$$

where φ is the wettability angle and $\pi \geq 0$ is a constant.

Remark Relationships (1.17) and (1.18) are different from the usual generalized Darcy's law as in Eq. (1.1),

$$\mathbf{v}^\xi = Kk_{r\xi}(S_\xi)\nabla \mathbf{p}_\xi, \quad (1.19)$$

and, from the usual capillary pressure relation,

$$\mathbf{p}_c = \mathbf{p}_\alpha - \mathbf{p}_\beta = (\cos \varphi)f(S_\alpha). \quad (1.20)$$

In (1.17) there is coupling between the two phases and the temperature gradient. Even if we write $K_{\xi\xi} = Kk_{r\xi}/\mu_\xi$ in (1.17), where K is the two-rank tensor for the rock permeability, there is no way to ensure $k_{r\xi}$ to depend only on S_ξ, as in (1.19).

In (1.18) functions, f and g are given by interfacial areas, and the last term takes into account the motion of the fluids to reach the equilibrium corresponding to a new curvature due to a change in one of the pressures.

5.1.3   Obtaining Macroscopic Laws by Homogenization

Introduction and Assumptions

In this paragraph, we are following the work of Auriault and Sanchez-Palencia [ASP86], assuming no interface motion but periodicity of the porous medium.

This work is based on the two-scale asymptotic expansion technique for obtaining macroscopic or homogenized equations. A similar result has been obtained later in [ST91] by the energy method. The periodicity assumption for the porous medium is made to use the two-scale homogenization technique for explicitly solving the so-called auxiliary cell problem.

According to the definition in Chapter 3, the porous medium Ω is made of repeated shrunken unit cells $\Omega = \bigcup_{k \in \mathbb{Z}^3} \varepsilon(Y + k)$ where Y is the unit cell $]0, 1[^3$. The pore volume $\mathfrak{P}$ in each cell Y is filled with two fluids ($\mathfrak{P}_\alpha$ and $\mathfrak{P}_\beta$), separated by an interface $\Sigma_{\alpha\beta}$, $\mathfrak{P} = \mathfrak{P}_\alpha \cup \mathfrak{P}_\beta \cup \Sigma_{\alpha\beta}$, and Γ_s is the boundary between the fluid and the solid. In addition to the usual assumption made on $\mathfrak{P}$ (see Chapter 3), $\mathfrak{P}_\alpha$ and $\mathfrak{P}_\beta$ are also assumed to be connected. The ith fluid part in Ω is denoted by

$$\mathfrak{P}_\xi^\varepsilon = \bigcup_{k \in \mathbb{Z}^3} \varepsilon(\mathfrak{P}_\xi + k), \quad \xi = \alpha, \beta,$$

and the solid part of Ω is $\mathfrak{S}^\varepsilon = \bigcup_{k \in \mathbb{Z}^3} \varepsilon\left[(Y \setminus \overline{\mathfrak{P}}) + k\right].$

We make the following assumptions in each cell Y:

H_1: The flow is assumed to be stationary. Then, the tangential component of the velocity v is continuous across the interface $\Sigma_{\alpha\beta}$, and the normal component is zero.

H_2: $\Sigma_{\alpha\beta}$ intersects Γ_s with a wetting angle φ.

H_3: According to the Laplace relation, the jump of the normal component of the stress tensor across $\Sigma_{\alpha\beta}$ is given by $[[(pI + \sigma)\vec{v}]] = T(\frac{1}{R_1} + \frac{1}{R_2})\vec{v}$ where σ is the viscous stress tensor, R_1 and R_2 are the main curvature radii, and $\frac{1}{R_1} + \frac{1}{R_2} = C$ is called the mean curvature. The tangential component is continuous.

H_4: The surface tension T is of order ε, $T = \varepsilon\kappa$ (to make the surface tension effects of order 1 at the level of the pore, because, then, the curvature in a capillary tube is of order ε).

H_5: There is a relationship between the stress tensor jump through the interface and the saturation S_ξ in the cell. These saturations are defined by the ratio between the fluid on one side of the interface and the total volume of the fluid, $|\mathfrak{P}_\xi|/|\mathfrak{P}_\alpha + \mathfrak{P}_\beta|$.

H_6: The flow in each phase is slow enough (creeping flow) to neglect inertial effects.

Remark These assumptions are not very realistic according to experiments; φ depends on the history of the fluid and leads to hysteresis phenomena. It is also easy to imagine different shapes of the interface $\Sigma_{\alpha\beta}$ and, then, different saturations leading, however, to the same stress tensor jump, according to the Laplace formula. However, as in the previous method of averaging (see section 5.1.2 and [Pav90]), this type of assumption is necessary to develop a final, general Darcy law.

Microscopic Equations Governing Creeping, Two-Phase Flow in a Porous Medium

According to the previous assumptions, the following equations describe the stationary incompressible two-phase flow in the porous medium $\mathfrak{P}_\xi^\varepsilon$. The momentum balance equation is

$$\left.\begin{array}{c} -\nabla p_\varepsilon + \nabla \cdot \sigma^\varepsilon = f \quad \text{in } \mathfrak{P}_\xi^\varepsilon, \quad \xi = \alpha, \beta, \\ \sigma^\varepsilon = 2\mu_\xi D(v^\varepsilon), \end{array}\right\} \tag{1.21}$$

with $D(v^\varepsilon) = 1/2(\nabla v^\varepsilon + {}^t\nabla v^\varepsilon)$, and the mass balance equation is given by

$$\left.\begin{array}{c} \nabla \cdot v^\varepsilon = 0 \quad \text{in } \mathfrak{P}_\xi^\varepsilon, \quad \xi = \alpha, \beta, \\ v^\varepsilon = 0 \quad \text{on } \Gamma^\varepsilon, \quad \text{and} \\ -[[(p_\varepsilon I + \sigma^\varepsilon]]\vec{v} = \kappa\gamma^\varepsilon\vec{v}, \\ v^\varepsilon \cdot \vec{v} = 0, \quad [[v^\varepsilon \cdot t]] = 0, \\ \text{on the fluid interface } \Sigma_{\alpha\beta}^\varepsilon = \bigcup_{k \in \mathbb{Z}^3} \varepsilon(\Sigma_{\alpha\beta} + k), \end{array}\right\} \tag{1.22}$$

where the superficial tension $\mathcal{T} = \varepsilon\kappa$, and the mean curvature is $c^\varepsilon = \varepsilon^{-1}\gamma^\varepsilon$.

Asymptotic Expansion

As for one-phase flow (see [SP80]) the asymptotic expansion for the pressure and the velocity is sought in the following form:

$$\begin{array}{rl} p_\varepsilon(x) & = p_0(x, y) + \varepsilon p_1(x, y) + \dots, \quad \text{and} \\ \varepsilon^{-2}v^\varepsilon(x) & = v^0(x, y) + \varepsilon v^1(x, y) + \dots. \end{array} \tag{1.23}$$

Moreover, it is assumed that there is an expansion of the curvature

$$\gamma^\varepsilon = \gamma^0 + \varepsilon\gamma^1 + \dots. \tag{1.24}$$

Identifying terms of the same order gives the first problem in Y at order ε^{-1}:

$$\left.\begin{array}{cl} (i) & \nabla_y p_0 = 0 \quad \text{in } \mathfrak{P}_\xi, \quad \xi = \alpha, \beta, \\ (ii) & [[p_0 I]] \cdot \vec{v} = \kappa\gamma^0\vec{v} \quad \text{on } \Sigma_{\alpha\beta}, \\ & p_0, \gamma^0 \, Y - \text{periodic}, \end{array}\right\}, \tag{1.25}$$

and the second problem in Y at order ε^0:

$$\left.\begin{array}{cl} (i) & -\nabla_y p_1 + \nabla_y \cdot \sigma_y^0 = \nabla_x p_0 + f \quad \text{in } \mathfrak{P}_\xi, \quad \xi = \alpha, \beta, \\ & \sigma_y^0 = 2\mu_\xi D_y(v^0), \quad D_y(v^0) = 1/2(\nabla_y v^0 + {}^t\nabla_y v^0), \\ (ii) & \nabla_y \cdot v^0 = 0 \quad \text{in } \mathfrak{P}_\xi, \quad \xi = \alpha, \beta, \\ (iii) & v^0 = 0 \quad \text{on } \Gamma_s, \\ (iv) & -[[p_1 I + \sigma_y^0]]\vec{v} = \kappa\gamma^1\vec{v}, \\ (v) & v^0 \cdot \vec{v} = 0, \quad [[v^0 \cdot t]] = 0 \quad \text{on } \Sigma_{\alpha\beta} \text{ in } \mathfrak{P}, \\ & p_1, v^0 \, Y\text{-periodic}. \end{array}\right\}. \tag{1.26}$$

Because p^0 is constant in $\mathfrak{P}_\alpha$ and also in $\mathfrak{P}_\beta$ from $(1.25)_{(i)}$, $(1.26)_{(i)}$ could be rewritten as:

$$- \nabla_y \cdot \sigma_y^0 + \nabla_y p_1 = -\lambda(y) \quad \text{in } \mathfrak{P}, \tag{1.27}$$

where

$$\lambda(y) = \nabla_x p_0 - f(x) = \left\{ \begin{array}{ll} \lambda^\alpha(x), & \text{if } y \in \mathfrak{P}_\alpha, \\ \lambda^\beta(x), & \text{if } y \in \mathfrak{P}_\beta \end{array} \right. .$$

Then, problem (1.27) with $(1.26)_{(ii)}$–$(1.26)_{(v)}$, is well-posed and determines v^0 and p_1 (up to a constant) in each $\mathfrak{P}_\xi$ as a function of λ^ξ, $\xi = \alpha, \beta$. More precisely, it has been shown in [ASP86], Proposition 5.1, that the variational formulation of this local problem has a unique Y-periodic solution (p_1, v^0) and that the macroscopic velocity of each fluid,

$$\mathbf{v}^\xi(x) = \frac{1}{|Y|} \int_{\mathfrak{P}_\xi} v^0(x, y) dy, \tag{1.28}$$

verifies a generalized Darcy filtration law,

$$\mathbf{v}^\xi = -K_{\alpha\alpha} \lambda^\alpha + K_{\alpha\beta} \lambda^\beta, \quad \xi = \alpha, \beta, \tag{1.29}$$

where the tensors $K_{\xi_1 \xi_2}$ depend on $\mathfrak{P}_\alpha$, $\mathfrak{P}_\beta$, μ_α, and μ_β.

Remark In the case of a stationary, isothermal flow, the filtration law obtained in (1.29) is the same as the law obtained in (1.17).

5.2   Upscaling Two-Phase Flow Characteristics in a Heterogeneous Reservoir with Capillary Forces (Finite Peclet Number)

5.2.1   Introduction

According to the results of the previous section, we now assume that the behavior of immiscible, two-phase flow through a porous domain at a low flow rate, with no mass transfer between the fluid phases, is described at any time $t \in]0, \tau[$ by a continuity equation in each phase $\xi = \alpha, \beta$:

$$\phi(x) \frac{\partial(S_\xi \rho_\xi)}{\partial t} + \nabla \cdot (\rho_\xi q_\xi) = f_\xi \quad \text{in} \quad \Omega \times]0, \tau[. \tag{2.1}$$

f_ξ is the ξ mass flow rate per volume injected into Ω or produced from Ω, where $\Omega \subset \mathbb{R}^3$ is the reservoir domain, $\phi(x)$ is the porosity of the rock, denoting the fraction of the volume available for flow, and ρ_ξ is the density per unit volume of the ξ-phase. The saturation of the ξ-fluid phase S_ξ is the fraction of the space available to flow occupied by that phase.

The subscripts α or β denote the wetting or the nonwetting phase, and q_ξ is the superficial Darcy velocity, depending on the other phase,

$$q_\xi = -K \frac{k_{r\xi}(S_\xi)}{\mu_\xi}(\nabla p_\xi - \rho_\xi g \nabla \mathcal{Z}), \qquad \xi = \alpha, \beta, \tag{2.2}$$

where $K(x)$ is the rock absolute permeability tensor which quantifies the ability of the rock to transmit any single-phase fluid, $k_{r\xi}(S_\xi)$ is the ξ-phase relative permeability, which depends on the saturation to account for the interference of flow between the phases. μ_ξ is the ξ-phase viscosity, $\Lambda_\xi(S_\alpha) = \dfrac{k_{r\xi}(S_\xi)}{\mu_\xi}$ is the ξ-phase mobility, g is the magnitude of acceleration due to gravity, and $\mathcal{Z}$ is the vector pointing in the direction of gravity. Relative permeability curves are usually obtained from laboratory experiments or from empirical polynomial formulas (see Fig. 5.1).

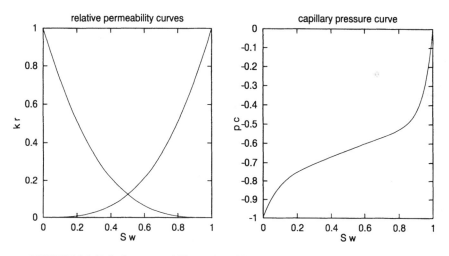

FIGURE 5.1. Relative permeability and capillary pressure ($p_c = p_\alpha - p_\beta$) curves.

Because both phases are flowing,

$$S_\alpha + S_\beta = 1. \tag{2.3}$$

Each phase also has separate pressures, p_α and p_β, and the difference between these pressures is the capillary pressure p_c. We shall assume that, as in usual empirical formulas (see, for instance, [RB49] or [Lev40]), the capillary pressure is a unique, nonlinear, monotone decreasing function of saturation (see Fig. 5.1):

$$p_c(x, S_\alpha) = p_\alpha - p_\beta. \tag{2.4}$$

For a more general discussion of the physical principles used to establish Eqs. (2.1)–(2.3), we refer to [Mar81], [Sch57] and [Ewi83].

As we already stated in this section, in the following, we will not use special notations either for vectors, for instance, the velocity q_ξ, or for tensors, for instance, the permeability K. Assuming incompressibility, substituting equation (2.2) into Eq. (2.1), using (2.3) and the capillary pressure law (2.4) leads to

$$\phi \frac{\partial S}{\partial t} - \nabla \cdot \left(a(S)K(x)\nabla S - b_g(S)\nabla P_g - b(S)q_T \right) = f_\alpha, \qquad (2.5)$$

and

$$\nabla \cdot q_T = 0 \qquad \text{in} \qquad \Omega \times]0, \tau[, \qquad (2.6)$$

where $S = S_\alpha$ is the saturation of the wetting fluid, $q_T = q_\alpha + q_\beta$ is the total fluid velocity, $d(S) = \Lambda_\alpha(S) + \Lambda_\beta(S)$ is the total mobility, and $a(S) = \dfrac{\Lambda_\alpha(S)\Lambda_\beta(S)}{d(S)} \dfrac{\partial p_c(S)}{\partial S}$, $b(S) = \dfrac{\Lambda_\alpha(S)}{d(S)}$ is the α-fractional flow,

$$b_g(S) = \frac{2\Lambda_\alpha(S)\Lambda_\beta(S)}{d(S)} \frac{\rho_\alpha - \rho_\beta}{\rho_\alpha + \rho_\beta},$$

and $P_g = \dfrac{1}{2}(\rho_\alpha + \rho_\beta)g\,Z$ is the gravity potential.

Finally, introducing the concept of *"reduced pressure"* P, defined, for instance, in [AD85], [AKM90], or [CJ86], Eq. (2.6) can be expanded as follows:

$$q_T = -K(x)d(S)(\nabla P + \gamma_2(S)\nabla P_g) \qquad (2.7)$$

with

$$\gamma_2(S) = \frac{2(\rho_\alpha \Lambda_\alpha(S) + \rho_\beta \Lambda_\beta(S))}{d(S)(\rho_\alpha + \rho_\beta)}$$

and

$$P = \frac{1}{2}(p_\alpha + p_\beta) + \int_{S_m}^{S} \left[b(\xi) - \frac{1}{2} \right] \frac{\partial p_c}{\partial \xi}\, d\xi,$$

where S_m is the wetting fluid's irreducible saturation $(k_{r\alpha}(S_\alpha) = 0, \forall S_\alpha \leq S_m)$.

The main motivation for homogenizing two-phase flows comes from the need for upscaling the geological characteristics of the reservoir under two conditions.

The first occurs when it is necessary to build data for numerical simulators from geological data. The entire reservoir size is usually of the order of several kilometers, and the mesh size in simulators is of the order of 10 meters. Thus, there are clearly two scales, the geological scale and the mesh scale, and for effective parameters on each mesh, we need a way to upscale the geological data.

The second might happen during numerical simulations when, for numerical purposes, it is necessary to coarsen meshes. In this situation there are also two scales, and it is necessary to build the data for the coarse mesh from the fine mesh.

5.2.2 Definition of the Homogenization Problem

We now assume that ε is the nondimensional scale of the heterogeneities or the dimensionless period diameter if the medium is assumed to be periodic. In each period, we assume, for simplicity, that there are only two types of rock, and then, each coefficient and curve depend on both the period ε and the type of rock $i = 1, 2$. We may then associate a typical cell $Y =]0, 1[^3$ with this porous medium, obtained by magnifying any period of the domain Ω by the change of variable $y = \dfrac{x}{\varepsilon}$. Then, $Y_i, i = 1, 2$, is the subdomain of Y corresponding to ith rock type.

We denote the subdomain of Ω corresponding to the ith rock type by $\Omega_i^\varepsilon, i = 1, 2$, with the interface $\Gamma_{1,2}^\varepsilon = \partial\Omega_1^\varepsilon \cap \partial\Omega_2^\varepsilon$ assumed to be sufficiently smooth. $\vec{\nu}_{1,2}^\varepsilon$ is the unit normal to $\Gamma_{1,2}^\varepsilon$ exterior to Ω_1^ε. In any subdomain Ω_i^ε, there are different functions depending on S, such as Λ_ξ, b, d, and P_c. To emphasize the difference, the subscript or superscript i will be added to each of them.

The rock permeability tensor K^ε and the rock porosity ϕ^ε change rapidly as x varies, with period ε, and they are of the form,

$$K^\varepsilon(x) = K\left(\frac{x}{\varepsilon}\right)$$

and

$$\phi^\varepsilon(x) = \phi\left(\frac{x}{\varepsilon}\right) ,$$

with K and ϕ Y-periodic.

Because of the continuity of the capillary pressure, the pressure and flow rate, p_ε^i and $q_\xi^{\varepsilon,i} \cdot \vec{\nu}$, in each phase at the interface between different rock types, are both continuous (see [Mar81]). Consequently, the saturation and the "reduced pressure", S_ε^i and $P^{\varepsilon,i}$, are both discontinuous at the interface between different rock types.

If we denote the capillary pressure by u^ε, the conditions at $\Gamma_{1,2}^\varepsilon$ are

$$\begin{aligned}
q_\alpha^{\varepsilon,1} \cdot \vec{\nu}_{1,2}^\varepsilon &= q_\alpha^{\varepsilon,2} \cdot \vec{\nu}_{1,2}^\varepsilon , \\
q_T^{\varepsilon,1} \cdot \vec{\nu}_{1,2}^\varepsilon &= q_T^{\varepsilon,2} \cdot \vec{\nu}_{1,2}^\varepsilon, \quad \text{and} \\
P^{\varepsilon,1} &= P^{\varepsilon,2} + \mathcal{G}(u^\varepsilon).
\end{aligned} \tag{2.8}$$

$\mathcal{G}(u)$ is the jump of the "reduced pressure", coming from both the jump of saturation and the change of the mobility curve at the interface $\Gamma_{1,2}^\varepsilon$:

$$\mathcal{G}(u) = \int_{p_c^1(S_m^1)}^u \left[B_1(s) - \frac{1}{2}\right] ds - \int_{p_c^2(S_m^2)}^u \left[B_2(s) - \frac{1}{2}\right] ds \tag{2.9}$$

with $B_i = b_i \circ (p_c^i)^{-1}, i = 1, 2$.

Due to the discontinuity of saturation at the interface $\Gamma_{1,2}^\varepsilon$, instead of S, we now use, the capillary pressure u^ε, which is continuous at this interface. This change of unknown is possible, as in [ABEA90] or in [BH95b], assuming $p_c^1(S_k^1) = p_c^2(S_k^2)$, $k = m, M$ with $(1 - S_M)$ being the nonwetting irreducible saturation.

In most geological situations, the imprecision of the data and the asymptotic character of the capillary pressure curves near S_m^i and S_M^i make it possible to extrapolate these capillary pressure curves and, consequently, to have this assumption verified.

With this new unknown u^ε, under the previous definition and assumptions, the Eqs. (2.5) and (2.6), in $\Omega_i^\varepsilon \times]0, \tau[; i = 1, 2$, become

$$\frac{\partial \lambda^\varepsilon}{\partial t}(x, u^\varepsilon) + \nabla \cdot q_\alpha^{\varepsilon,i} = 0 \tag{2.10}$$

with the wetting fluid velocity $q_\alpha^{\varepsilon,i}$,

$$q_\alpha^{\varepsilon,i} = -A_i(u^\varepsilon)K^\varepsilon(x)\nabla u^\varepsilon + B_i(u^\varepsilon)q_T^{\varepsilon,i}, \tag{2.11}$$

and the total velocity $q_T^{\varepsilon,i}$,

$$q_T^{\varepsilon,i} = -D_i(u^\varepsilon)K^\varepsilon(x)\nabla P^{\varepsilon,i}, \tag{2.12}$$

and

$$\nabla \cdot q_T^{\varepsilon,i} = 0. \tag{2.13}$$

In the equations above, the functions $\lambda(x, u) = \phi(x)S = \phi(x)(p_c^i)^{-1}(u)$, $B_i(u) = b_i \circ (p_c^i)^{-1}(u)$, and $D_i(u) = d_i \circ (p_c^i)^{-1}(u)$ are, respectively, the wetting fluid saturation times the porosity, the wetting fluid fractional flow, and the total mobility considered as functions of the capillary pressure u^ε. The function $A_i(u) = \dfrac{\Lambda_\alpha^i(S)\Lambda_\beta^i(S)}{d^i(S)} \circ (p_c^i)^{-1}(u)$ is zero for $S = S_m$ or S_M. $P^\varepsilon, q_\alpha^\varepsilon$, and q_T^ε are, respectively, the *"reduced pressure"*, the wetting fluid velocity, and the total velocity, extended to the whole domain $\Omega = \Omega_1^\varepsilon \bigcup \Gamma_{1,2}^\varepsilon \bigcup \Omega_2^\varepsilon$.

The interface conditions at $\Gamma_{1,2}^\varepsilon$ are given by (2.8), and the boundary conditions are given on the boundary of Ω, $\partial\Omega = \Gamma_{in} \bigcup \Gamma_{pr} \bigcup \Gamma_{nf}$. Where $\vec{v}$ is the unitary outward normal and Γ_{in}, Γ_{pr}, and Γ_{nf} denote, respectively, the injection, the production, and the no-flux boundaries:

$$u^\varepsilon = u_{max} \text{ and } q_T^\varepsilon \cdot \vec{v} = -q_e \text{ on } \Gamma_{in} \times]0, \tau[, \tag{2.14}$$

$$u^\varepsilon = u_{min} \text{ and } P^\varepsilon = 0 \text{ on } \Gamma_{pr} \times]0, \tau[, \tag{2.15}$$

and

$$q_\alpha^\varepsilon \cdot \vec{v} = 0 \text{ and } q_T^\varepsilon \cdot \vec{v} = 0 \text{ on } \Gamma_{nf} \times]0, \tau[\tag{2.16}$$

with $u_{min} = p_c^{-1}(S_m)$ and $u_{max} = p_c^{-1}(S_M)$.

The initial condition is given by

$$\lambda^\varepsilon(x, u(x, 0)) = \lambda_0^\varepsilon(x) \quad \text{in } \Omega. \tag{2.17}$$

Under suitable assumptions, (see assumptions A1–A6 in [BH95b] and assumptions A1–A8 in [BH95a]), existence of the solution $(u^\varepsilon, q_T^\varepsilon, P^\varepsilon)$ of (2.10)–(2.17) is proven in [BH95b], Theorem 5.1, using a mixed variational formulation (see Eqs. (2.3)–(2.8) in [BH95b] and Eqs. (2.6)–(2.12) in [BH95a]).

Uniqueness can be obtained with arguments from [AKM90], [CJ86] [Gag80], [KS77], and [Mik89], under additional assumptions to decouple the capillary pressure equations from the *"reduced pressure"* equations and with additional regularity assumptions.

5.2.3   General Homogenization Result

In the very general case, assuming a small enough jump $\mathcal{G}$ of the *"reduced pressure"* and using the two-scale convergence (see section A.3 for definition), we have the following convergence results (see Proposition (4.1) in [BH95a] for the exact mathematics):

$$u^\varepsilon \longrightarrow u \quad \text{strongly in } L^p(\Omega \times]0, \tau[) , \quad \forall p \in [1, +\infty[, \quad (2.18)$$

$$\frac{\partial \beta_0(u^\varepsilon)}{\partial x_j} \text{ two-scale converges to } \frac{\partial \beta_0(u)}{\partial x_j} + \frac{\partial u_1}{\partial y_j}, \ j = 1, 2, 3, \quad (2.19)$$

$$q_{T,j}^\varepsilon \text{ two-scale converges to } q_{T,j}^0 , \quad j = 1, 2, 3, \quad (2.20)$$

$$\begin{cases} \tilde{P}^\varepsilon \rightharpoonup \tilde{P} \text{ weakly in } L^2(0, T; H^1(\Omega)) \\ \tilde{P}^\varepsilon \text{ two-scale converges to } \tilde{P}, \end{cases} \quad (2.21)$$

$$\frac{\partial \tilde{P}^\varepsilon}{\partial x_j} \text{ two-scale converges to } \frac{\partial \tilde{P}}{\partial x_j} + \frac{\partial \tilde{P}_1}{\partial y_j}, \quad j = 1, 2, 3, \quad (2.22)$$

where $(u, q_T^0, \tilde{P}, u_1, \tilde{P}_1)$ is solution of the homogenized problem

$$\frac{\partial \bar{\lambda}}{\partial t}(u) + \nabla_x \cdot \{q_\alpha^0\}_Y = 0 , \quad \text{in } \Omega \times]0, \tau[, \quad (2.23)$$

$$\nabla_y \cdot q_\alpha^0 = 0 , \quad \text{in } \Omega \times]0, \tau[\times Y, \quad (2.24)$$

where the wetting fluid velocity q_α^0 and the total velocity q_T^0 are defined by

$$q_\alpha^0 = -F(y, u)K(y)[\nabla_x \beta_0(u) + \nabla_y u_1] + B(y, u)q_T^0 , \quad (2.25)$$

$$\begin{aligned} q_T^0 = {} & -D(y, u)K(y)[\nabla_x \tilde{P} + \nabla_y \tilde{P}_1] \\ & + \mathcal{X}_{Y_2}(y)D(y, u)K(y)(\mathcal{G} \circ \beta_0^{-1})' \\ & \circ \beta_0(u)[\nabla_x \beta_0(u) + \nabla_y u_1], \end{aligned} \quad (2.26)$$

$$\nabla_y \cdot q_T^0 = 0 \text{ in } \Omega \times]0, \tau[\times Y, \quad (2.27)$$

and

$$\nabla_x \cdot \{q_T^0\}_Y = 0 \text{ in } \Omega \times]0, \tau[, \quad (2.28)$$

with $u_1(x, t, \cdot)$ and $\tilde{P}_1(x, t, \cdot)$ Y-periodic and with the boundary and initial conditions

$$\beta_0(u) = \beta_0(u_{\max}) \quad \text{and} \quad \{q_T^0\}_Y \cdot \vec{v} = -q_e \quad \text{on } \Gamma_{\text{in}} \times]0, \tau[, \tag{2.29}$$

$$\beta_0(u) = \beta_0(u_{\min}) \quad \text{and} \quad \tilde{P} = 0 \quad \text{on } \Gamma_{\text{pr}} \times]0, \tau[, \tag{2.30}$$

$$\{q_\alpha^0\}_Y \cdot \vec{v} = 0 \quad \text{and} \quad \{q_T^0\}_Y \cdot \vec{v} = 0 \quad \text{on } \Gamma_{\text{nf}} \times]0, \tau[, \tag{2.31}$$

and

$$\bar{\lambda}(u(x, 0)) = \bar{\lambda}(u_0(x)) \text{ in } \Omega . \tag{2.32}$$

In Eqs. (2.23)–(2.32) above, we have denoted

$$\bar{\lambda}(z) = \frac{1}{|Y|} \int_Y \lambda(y, z) dy \quad \text{and}$$

$$\{f\}_Y = \frac{1}{|Y|} \int_Y f(y) dy, \tag{2.33}$$

and, in the convergence results (2.21),

$$\tilde{P}^\varepsilon = P^\varepsilon + \mathcal{G}(u^\varepsilon) \mathcal{X}_{\Omega_2^\varepsilon}, \tag{2.34}$$

is a *"continuous reduced pressure"*, where $\mathcal{X}_{\Omega_i^\varepsilon}, i = 1, 2$ and $\mathcal{X}_Y$, are the characteristic functions of Ω_i^ε and Y.

Moreover, if we denote $A_0(\eta) = \text{Min}\{A_1(\eta), A_2(\eta)\}, \eta \in]u_{\min}, u_{\max}[$, then, in (2.19), (2.25) and (2.29), (2.30),

$$\beta_0(z) = \int_{u_{\min}}^z A_0(s) ds, \tag{2.35}$$

and

$$F_i(z) = A_i(z)/A_0(z). \tag{2.36}$$

Remark The appearance of $\beta_0(u)$, instead of u, in (2.19)–(2.30), as in the standard two-scale convergence, is due to the possible degeneracy of the nonlinear terms $A_i(u)$.

Due to the strong nonlinarity of the homogenized problem in (2.23)–(2.32), there is no way, except in some special cases, to decouple and separately solve the auxiliary local problem in Y from the global problem in Ω. In the next part of this section, we present some special cases where it is possible to decouple the local problem from the global one.

In [Hid93], numerical simulations are presented for these special cases, to compare the behavior of a heterogeneous quarter five spot to the behavior of the homogenized corresponding quarter five spot. (A five spot consists of a production well surrounded by four regularly spaced injection wells. Due to symmetry, for modeling purposes, it suffices to consider one quarter of it.)

5.2.4   Some Special Cases

Stratified Porous Media with Orthogonal Flow to the Stratification

When the total flow q_T^ε is orthogonal to the stratification, the stratified case, reduces to the one-dimensional case (see [BH95a] section 6, Theorem 6.1) and, by solving (2.23)–(2.25) first, allows decoupling them from (2.26)–(2.28) which then reduces, finally, to $q_T^0 = q_e$. In the one-dimensional case, the global (or homogenized) behavior is described by the system of Eqs. (2.23) and (2.24) with

$$q_\alpha^0 = \mathcal{F}^*(u)\partial_x\beta_0(u) + \mathcal{B}^*(u)q_e. \tag{2.37}$$

$$\mathcal{F}^*(z) = \left\{\frac{1}{F(\cdot, z)K}\right\}_Y^{-1}, \tag{2.38}$$

and

$$\mathcal{B}^*(z) = \left\{\frac{\mathcal{B}(\cdot, z)}{F(\cdot, z)K(\cdot)}\right\}_Y \mathcal{F}^*(z), \tag{2.39}$$

with F and β_0 defined by (2.35) and (2.36).

Remark In this case, we have kept the possibility that the *"reduced pressure"* may have jumps and, then, oscillations can happen in the convective term in (2.37).

Constant Fractional Flow in the Whole Domain

If we assume that $B_1 = B_2 = B$, then, $\mathcal{G} \equiv 0$ and $\tilde{P} \equiv P$, and, finally, the auxiliary local problem (2.24) and equation (2.28) can be decoupled and solved in the following way:

First, we define two tensors $\mathcal{F}^* = (\mathcal{F}_{i,j}^*)_{1\leq i,j\leq 3}$ and $\mathcal{D}^* = (\mathcal{D}_{i,j}^*)_{1\leq i,j\leq 3}$ by

$$\mathcal{F}_{ij}^*(z) = \int_Y F(y, z)K_{ik}(y)\left[\delta_{kj} + \frac{\partial\theta_j}{\partial y_k}(y, z)\right]dy \tag{2.40}$$

and

$$\mathcal{D}_{ij}^*(z) = \int_Y \mathcal{D}(y, z)K_{ik}(y)\left[\delta_{kj} + \frac{\partial\omega_j}{\partial y_k}(y, z)\right]dy, \tag{2.41}$$

where δ_{ij} is Kronecker's symbol and where θ_k and ω_k, $k = 1, 2, 3$, are the unique solutions in $H_\#^1(Y)$ (up to a constant), for any $z \in [u_{\min}, u_{\max}]$, of

$$-\nabla_y \cdot [F(y, z)K(y)\nabla_y\theta_k(y, z)] = \nabla_y \cdot [F(y, z)K(y)e_k] \tag{2.42}$$

and

$$-\nabla_y \cdot [D(y, z)K(y)\nabla_y\omega_k(y, z)] = \nabla_y \cdot [D(y, z)K(y)e_k] \tag{2.43}$$

with e_k the kth standard basis vector of $\mathbb{R}^3$.

Then, we have (see Proposition 5.1 in [BH95a]) the global (or homogenized) equations in $Q_\tau = \Omega\times]0, \tau[$:

$$\frac{\partial\bar{\lambda}}{\partial t}(u) + \nabla \cdot q_\alpha = 0 \tag{2.44}$$

with

$$q_\alpha = -\mathcal{F}^*(u)\nabla\beta_0(u) + B(u)q_T \qquad (2.45)$$

and the total velocity q_T verifying

$$q_T = -\mathcal{D}^*(u)\nabla P \qquad (2.46)$$

and

$$\nabla \cdot q_T = 0. \qquad (2.47)$$

The boundary and initial conditions are the same as in (2.29)–(2.32) with $q_T = \{q_T^0\}_Y$ and $q_\alpha = \{q_\alpha^0\}_Y$.

Remark Assuming only a jump of order ε at the interface, i.e., assuming in (2.34), that $\mathcal{G}(u^\varepsilon) \simeq O(\varepsilon)$, leads, in (2.26), to a simpler, homogenized total flux

$$q_T^0 = -D(y, u)K(y)\left[\nabla_x \tilde{P} + \nabla_y \tilde{P_1}\right]. \qquad (2.48)$$

Remark The case, where the capillary pressure curves in each media type have been approximated by linear functions and $a_i^\varepsilon(s)$ by α_i^ε, α_i^ε constant in any Ω_i^ε, has been studied first (see [AB88] or [Bou87]), using the energy method and asymptotic expansions, and also leads to global homogenized equations decoupled from the auxiliary local problem.

Nonuniformly, Periodically Oscillating, Rock Absolute Permeability and Porosity

In this case, capillary pressure and relative permeabilities curves are assumed to be the same in the whole domain Ω. There is only one rock type, i.e., $P_c(s)$, $k_{ra}(s)$ and $k_{r\beta}(s)$ do not change in any part of the domain Ω, and all quantities are directly defined on the whole domain Ω. But porosity and permeability of the rock are ε-periodic, i.e., of the form $\phi(x/\varepsilon, x)$ and $K(x/\varepsilon, x)$.

The system of equations (2.10)–(2.13) reduces to the following equations in $\Omega \times]0, \tau[$:

$$\phi\left(\frac{x}{\varepsilon}, x\right)\frac{\partial S^\varepsilon}{\partial t} + \nabla \cdot q_\alpha^\varepsilon = f_\alpha, \qquad (2.49)$$

$$q_\alpha^\varepsilon = -a(S^\varepsilon)K\left(\frac{x}{\varepsilon}, x\right)\nabla S^\varepsilon + b(S^\varepsilon)q_T^\varepsilon, \qquad (2.50)$$

$$q_T^\varepsilon = -d(S^\varepsilon)K\left(\frac{x}{\varepsilon}, x\right)\nabla P^\varepsilon, \qquad (2.51)$$

and

$$\nabla \cdot q_T^\varepsilon = 0 \qquad (2.52)$$

with the boundary conditions

$$q_T^\varepsilon \cdot \vec{v} = -q_e \quad \text{and} \quad S^\varepsilon = S_M \quad \text{on } \Gamma_{in} \times]0, \tau[, \qquad (2.53)$$

$$P^\varepsilon = 0 \quad \text{and} \quad S^\varepsilon = S_m \quad \text{on } \Gamma_{pr} \times]0, \tau[, \qquad (2.54)$$

$$q_T^\varepsilon \cdot \vec{v} = 0 \quad \text{and} \quad q_\alpha^\varepsilon \cdot \vec{v} \quad \text{on } \Gamma_{nf} \times]0, \tau[, \qquad (2.55)$$

and with the initial condition $S^\varepsilon = S_0$ in Ω.

Such a system has first been theoretically investigated in [Bou84a], [Bou84b], [Bou85], [Bou86], and [Mik89], by the energy method (see definition in Appendix A, section A.2) or by asymptotic expansion (see the definition, for instance, in [BLP78], [BP89], and [SP82]). In these systems, it has been proven that

- *There is a global (homogenized) system of equations exactly like the original (2.1)–(2.4) (see, for instance, Theorem 2 in [Bou86] or Theorem 3.2 and 3.3 in [Mik89]).*

- *There is an effective permeability tensor $K^{\text{hom}}(x)$ and an effective porosity $\phi^{\text{hom}}(x)$ given by the same auxiliary local problem as in the one-phase flow linear equation (see, e.g., Appendix A, Theorem 2.1), i.e.,*

$$\phi^{\text{hom}}(x) = \{\phi(x, y)\}_Y,$$

$$K^{\text{hom}}(x) = \{K(y, x) + \alpha(y, x)\}_Y \qquad (2.56)$$

$$\alpha_{i,j}(y) = K_{ik}(y, x)\frac{\partial W^j}{\partial y_k}(y, x), \qquad (2.57)$$

and

$$\begin{cases} -\nabla_y \cdot (K(y, x)\nabla W^j(y, x)) = \nabla_y \cdot (K(y, x)e_j) \text{ in } \mathbb{R}^3 \\ W^j \qquad Y\text{-periodic function.} \end{cases} \qquad (2.58)$$

Finally in [AB88] and [ABK91], numerical simulations are presented, comparing the behavior (saturation and pressure contours, breakthrough time, production curves) of a heterogeneous quarter five spot to the corresponding homogenized one.

Theorem 2.1 *The homogenized equations, corresponding to the limit of equations (2.49)–(2.55) as $\varepsilon \to 0$, are*

$$\phi^{\text{hom}}(x)\frac{\partial S_\xi}{\partial t} + \nabla \cdot q_\xi = f_\xi, \quad in \quad \Omega \times]0, \tau[, \quad and$$

$$q_\xi = -K^{\text{hom}}(x)\frac{k_{r\xi}(S_\xi)}{\mu_\xi}\nabla p_\xi, \quad \xi = \alpha, \beta,$$

with ϕ^{hom} and K^{hom} given by (2.56).

Remark It should be noted that the above results cover not only the case $K(x/\varepsilon)$ but also the case where $K(x/\varepsilon, x)$ is nonuniformly oscillating (i.e., depends also on the location x in the reservoir), for instance, when there are several different geological rock facies.

5.2.5   Randomly Heterogeneous Porous Media

In sections 5.2.2–5.2.4, we have assumed a periodic distribution for the heterogeneities. However, using the recent extension of the two-scale method to the stochastic context (see [BMW94]) because the periodic case defines an ergodic dynamical system (on Y, considered with the Lebesgue measure dy as a probability space) by $T(x)y = y + x$ (mod 1), the previous results could be extended to some type of randomly distributed heterogeneities.

However, for simplicity, in this paragraph we are presenting the results for a randomly heterogeneous porous medium similar to the one of section 5.2.4. We are assuming that the rock absolute permeability and the porosity define stochastically homogeneous random fields (for a more complete definition of a homogeneous random field, see, for instance, [JKO94]). But we assume only one rock type, i.e., the capillary pressure and relative permeabilities curves do not vary in the reservoir domain Ω.

In the last decade, we have seen rapid developments in research literature treating multiphase flow in reservoirs in a probabilistic framework. But, due to the complexity of this theory, the natural tendency of most of the research in this area has been to emphasize computational refinements using pragmatic averaging techniques without developing an accurate mathematical theory.

Using the tools of stochastic homogenization, it is proven in [BKM95] that a nonlinear effective system of deterministic equations exists which governs flow behavior in a homogeneous medium, equivalent, in the sense of homogenization theory, to the original. When ε is small enough (i.e., when the number of heterogeneities is sufficiently large), the randomly heterogeneous porous medium behaves like a deterministic medium with an effective permeability tensor denoted K^0 (Theorem 4.5 in [BKM95]). This effective permeability tensor K^0 is obtained from the realizations of an auxiliary stochastic problem (Proposition 3.1 in [BKM95]). Using the rescaling parameter ε, corresponding to the characteristic scale of heterogeneities, the convergence of the homogenization process for $\varepsilon \to 0$ is proven. Furthermore, by using regularity results for the nonlinear effective equations, the correctors are constructed, and strong convergence is established (see Theorems 5.3 and 5.7 in [BKM95]).

The naturally heterogeneous porous medium is considered as a realization of a statistically homogeneous three-dimensional random field. More precisely, the rock permeability tensor $K(x)$ and the rock porosity $\phi(x)$ are random functions of position, changing rapidly as x varies. The parameter $\varepsilon > 0$ is defined as the ratio between a typical length scale associated with the variation in permeability and porosity and the length scale of the reservoir Ω.

The permeability tensor K and the porosity ϕ are statistically homogeneous random functions of the form $K(x/\varepsilon, \omega)$ and $\phi(x/\varepsilon, \omega)$, $\omega \in \mathcal{M}$ (the probability space). The saturation S^ε and the pressure P^ε in the system of equations (2.49)–(2.52) are now realizations of random variables. They are, then, denoted by S_ε^ω, p_ε^ω, and they satisfy the stochastic partial differential equations

$$
\left.
\begin{aligned}
q_\varepsilon^\omega &= -d(S_\varepsilon^\omega)K\left(\frac{x}{\varepsilon}, \omega\right)\nabla p_\varepsilon^\omega, \\[4pt]
\nabla \cdot \, q_\varepsilon^\omega &= 0, \\[4pt]
\phi\left(\frac{x}{\varepsilon}, \omega\right)\frac{\partial S_\varepsilon^\omega}{\partial t} - \nabla \cdot \left(K\left(\frac{x}{\varepsilon}, \omega\right)a(S_\varepsilon^\omega)\nabla S_\varepsilon^\omega - b(S_\varepsilon^\omega)q_\varepsilon^\omega\right) &= 0, \\[4pt]
\text{for} \quad (t, x, \omega) \in {]0, \tau[} \times \Omega \times \mathcal{M},
\end{aligned}
\right\}
\tag{2.59}
$$

where q_ε^ω stands for the total velocity.

Principal Theoretical Results

In this paragraph we recall the main results previously obtained in [BKM95], concerning the homogenization of Eqs. (2.59), i.e., the behavior of $(S_\varepsilon^\omega, P_\varepsilon^\omega)$ as $\varepsilon \to 0$.

First of all, we recall the probabilistic assumptions (for more details see, for instance, [BKM95], [BMW94], or [Koz80]):

- $(\mathcal{M}, \mathcal{F}, \pi)$ is a probability space.

- T is a three-dimensional dynamical system given on $\mathcal{M}$, i.e., a family $\{T(x) \, ; \, x \in \mathbb{R}^3\}$, $T(x) \, : \, \mathcal{M} \to \mathcal{M}$ of invertible maps, such that for all $x \in \mathbb{R}^3$, $T(x)$ and $T(x)^{-1}$ are measurable. Moreover, T is an additive group and it preserves the measure π.

- T is ergodic with respect to the measure π.

- The permeability tensor K and the porosity ϕ are statistically homogeneous random fields, i.e., are functions $v \, : \, \mathbb{R}^3 \times \mathcal{M} \to \mathbb{R}^d$ (with $d = 1$ for ϕ and $d = 3 \times 3$ for K) having the property $v(x + y, \omega) = v(x, T(y)\omega)$, $\forall x, y \in \mathbb{R}^3$. This is similar to assuming shift invariance on the k-point probability functions of any order.

Under these assumptions it has been proven in [BKM95] that $(S_\varepsilon^\omega, P_\varepsilon^\omega)$ tends to a limit (S, P), independent of ω, solving a homogenized system of equations which is deterministic, namely, with constant effective rock porosity ϕ^0 and permeability K^0. This homogenized system is of the same type as the homogenized system obtained in Theorem 2.1 by homogenization of (2.49)–(2.55), assuming dependence only on x/ε for ϕ and K (see Theorem 4.5 in [BKM95]).

At first glance, this final result, i.e., the existence of deterministic behavior on a sufficiently large scale, seems to be contradictory to the well-known inherent instability of such displacement (see, for instance, [CVV59], [Hom87], [TH86], [WMS76]).

In fact, the only way to understand the apparent contradiction between the unstable displacement and our result is to notice that the scale of the fingering is, in some way, macroscopic (as shown in [Hom87]) and bigger than the characteristic size ε ($\varepsilon \to 0$) of the heterogeneities. Consequently, there is no interaction between the fingering scale and the heterogeneities scale.

With adequate assumptions of regularity, the usual a priori estimates (see Proposition 3.5 and 3.7 in [BKM95]), for a.e. ω, lead to the following convergence results.

Theorem 2.2 *For a.e.* $\omega \in \mathcal{M}$, *let* $\{S_\varepsilon^\omega, p_\varepsilon^\omega, q_\varepsilon^\omega\}$ *be the corresponding solutions to problem (2.59). Then, there exists a subsequence, denoted by the same index* ε *and* $\{S, p, q\}$, *such that, as* $\varepsilon \to 0$,

$$S_\varepsilon^\omega \rightharpoonup S \quad \text{weakly in} \quad L^2(0, \tau; V), \tag{2.60}$$

$$S_\varepsilon^\omega \to S \quad \text{strongly in} \quad L^2(Q_\tau), \tag{2.61}$$

$$\phi(T(\varepsilon^{-1}x)\omega)\frac{\partial S_\varepsilon^\omega}{\partial t} \rightharpoonup \mathcal{E}(\phi)\frac{\partial S}{\partial t} \quad \text{weakly in} \quad L^2(0, \tau; V'), \tag{2.62}$$

$$p_\varepsilon^\omega \rightharpoonup p \quad \text{weakly in} \quad L^2(0, \tau; W), \tag{2.63}$$

$$b(S_\varepsilon^\omega)d(S_\varepsilon^\omega)\nabla p_\varepsilon^\omega \rightharpoonup b(S)d(S)\nabla p \quad \text{weakly in} \quad L^2(Q_\tau)^3, \tag{2.64}$$

$$a(S_\varepsilon^\omega)\nabla S_\varepsilon^\omega \rightharpoonup a(S)\nabla S \quad \text{weakly in} \quad L^2(Q_\tau)^3, \tag{2.65}$$

$$d(S_\varepsilon^\omega)\nabla p_\varepsilon^\omega \rightharpoonup d(S)\nabla p \quad \text{weakly in} \quad L^2(Q_\tau)^3, \quad \text{and}$$

$$q_\varepsilon^\omega \rightharpoonup q_T \quad \text{weakly in} \quad L^2(Q_\tau)^3, \tag{2.66}$$

where

$$q_T = -d(S)K^0\nabla p \qquad \text{a.e. in } Q_\tau =]0, \tau[\times\Omega \tag{2.67}$$

(with K^0 *defined in the theorem 2.4). Here,* V' *is the dual of* $V = \{v \in H^1(\Omega); v = 0$ *on* $\Gamma_{in} \bigcup \Gamma_{pr}\}$ *and* $W = \{p \in H^1(\Omega); p = 0 \text{ on } \Gamma_{pr}\}$ *and* $\mathcal{E}(f) = \int_{\mathcal{M}} f(z)d\pi$.

From this theorem, using equations satisfied by the limits, we deduce the following theorem.

Theorem 2.3 *The homogenized equation, corresponding to the limits of Theorem 2.2, is expressed as*

$$\begin{cases} \phi^0\dfrac{\partial S_\xi}{\partial t} + \nabla \cdot q_\xi = f_\xi & \text{in} \quad \Omega\times]0, \tau[, \\ q_\xi = -K^0\dfrac{k_{r\xi}S_\xi}{\mu_\xi}\nabla p_\xi, & \xi = \alpha, \beta, \end{cases}$$

with $\phi^0 = \mathcal{E}(\phi)$ *and* K^0 *given by (2.68)–(2.69) below.*

Denoting $V_{pot}^2 = \{f \in L_{pot}^2(\mathcal{M})^3, \mathcal{E}(f) = 0\}$, where $L_{pot}^2(\mathcal{M})^3$ is the set of all $f \in L^2(\mathcal{M})^n$, such that a.e. realization $f(T(x)\omega)$ is a potential in $\mathbb{R}^3$, we have the following theorem (see section 7.2 in [JKO94] and Proposition 3.1 in [BKM95]):

Theorem 2.4 *The auxiliary problem: find* $v_\xi \in V_{pot}^2$, *such that*

$$-\nabla \cdot \left[K(T(x)\omega)(\xi + v_\xi)\right] = 0 \qquad \text{in } \mathbb{R}^3 \tag{2.68}$$

has a unique solution, and the constant positive definite matrix K^0 in (2.67) is defined from the above solutions v_ξ by

$$K^0\xi = \mathcal{E}(K(\xi + v_\xi)), \quad \forall \xi \in \mathbb{R}^3. \tag{2.69}$$

Moreover, under additional assumptions of regularity on the relative permeability curves $k_i(S)$, the correctors are computed, and convergence theorems including the associated corrections are established (see Theorem 5.3, Corollary 5.4, and Theorem 5.7 in [BKM95]).

The similarity of these results to the results for the periodic case in section 5.2.4 should be noted. More precisely, the global (homogenized) system of equations is deterministic and of the same type as the original one (see Theorem 4.5, Proposition 3.8 and problem (P) in [BKM95]). The effective permeability tensor K^0 is given by an auxiliary problem (2.68)–(2.69) which is very close to (2.56)–(2.58) used for periodic heterogeneities and similar to the auxiliary problem (see [Koz80]) used for the linear one-phase flow equation in the same stochastic context.

But, unlike in the periodic case, the auxiliary problem (2.68) has to be solved in $\mathbb{R}^3$, with boundary conditions at infinity appearing nonexplicitly in the definition of the space V_{pot}^2. The averaging on Y is replaced by computing the expectation of realizations in (2.69), which requires cumbersome computing of realizations of solutions $v_\xi(T(x)\omega)$ in (2.68).

In [BB95] (see Theorem 3 and Algorithm 1), an algorithm has been designed, taking advantage of the ergodicity assumption and avoiding the time consuming computations of realizations of solutions for (2.68).

This algorithm is based on the "Representative Elementary Volume" concept and gives both an approximation of K^0 and the minimal volume Ω on which this approximation is valid. Moreover, the theoretical bases (Theorem 3 in [BB95]) of this algorithm give a rigorously mathematical method for "volume averaging".

As for the periodic case, numerical simulations are presented in [BB95], comparing the behavior of a randomly heterogeneous quarter five spot to the corresponding homogenized one.

5.3   Upscaling Two-Phase Flow Characteristics in a Heterogeneous Core, Neglecting Capillary Effects (Infinite Peclet Number)

5.3.1   Introduction

When capillary pressure effects are neglected, equations become hyperbolic, and we lose uniqueness of the solutions. To keep only the physically meaningful solutions, it is necessary to consider only the entropy solutions, defined, for instance, as the limit of the solution with capillary effects when the capillary forces tend to zero. Having the capillary forces tend to zero leads to a singular perturbation problem which could be mathematically handled through the theory of vanishing viscosity, as in [Kru70], developed only for a scalar equation. In the case of a two-phase flow through a porous core, which is a one-dimensional displacement in a porous medium, the equations become the scalar Buckley–Leverett equations. Assuming that either the porosity $\phi(x)$ and the rock permeability $k(x)$ or the initial saturation $S(t = 0)$ are oscillating at the lower scale, the homogenization, coupled with the singular perturbation technique, leads to global behavior, as in [BM94b].

Most laboratory measurements of porous media characteristics are usually done on a cylindrical core. If we consider the flow parallel to Ox in a cylindrical porous medium with axis Ox, pressure and saturations being uniform in each layer perpendicular to Ox, the displacement would, then, be considered one-dimensional. The two fluids (for instance, oil and water) are injected at the end $x = 0$ of the core and are assumed to be incompressible with constant viscosities.

Neglecting the capillary pressure effects, the equations governing the displacement describe the conservation of the fractional water volume (or water saturation) S and the total (oil plus water) fluid velocity flow q. The equations are

$$\phi(x)\frac{\partial S}{\partial t} + q(t)\frac{\partial}{\partial x}b(S) = 0, \tag{3.1}$$

$$q(t) = \frac{k(x)}{\mu(S)}\frac{\partial p}{\partial x}(x, t), \quad \text{and}$$

$$\frac{\partial q}{\partial x} = 0 \quad (x, t) \in]0, 1[\times]0, +\infty[. \tag{3.2}$$

In these equations, the independent variables x and t are the space and time variables. The quantity $p(x, t)$ is the pressure in both fluids (the capillary pressure effects are neglected). The quantity $1/\mu(S)$ is the total mobility, which has been previously denoted in (2.7) by $d(S)$, and $b(S)$ is the water fractional flow. Both $\mu(S)$ and $b(S)$ are known functions of the saturation S. A distinguishing feature of the immiscible displacement equations is that b is nonconvex. Typically, b is S-shaped and has one inflection point.

Eq. (3.1) (Buckley–Leverett equation) for a given q is hyperbolic. Eq. (3.2) is the incompressibility condition, and (3.2) is Darcy's law.

It should be noted that accounting for gravitational effects leads to the same type of equations as above, with the same S-shaped flux, as long as the gravitational flow term is small compared to the transport term $qb(S)$ (see [CS82], [Ewi83], and [Mar81]).

To the Eqs. (3.1)–(3.2), we add the initial and boundary conditions

$$S(x, 0) = S_0(x) \quad \text{on} \quad]0, 1[, \tag{3.3}$$

$$S(0, t) = S_1(t) \quad \text{on} \quad]0, +\infty[, \tag{3.4}$$

$$p(0, t) = p_0(t), \tag{3.5}$$

and

$$p(1, t) = p_1(t). \tag{3.6}$$

Note that (3.5)–(3.6) corresponds to flow with an imposed pressure drop $P(t) = p_0(t) - p_1(t)$ on $]0, +\infty[$.

Another possibility is to give a flow rate at $x = 0$, but, then, the velocity q is a given constant, and the problem is reduced to the homogenization of a standard scalar conservation law.

A detailed discussion of the Buckley–Leverett equation (3.1) can be found in [CS82], [CJ86], [CCS83], [CP79]. Homogenization of equations (3.1)–(3.2) has been studied for linear fractional flow, $b(S) = S$, in [AHZ90].

Linear fractional flow corresponds to two-component miscible flow without dispersion and reduces to a special case of the nonlinear case studied in [BM94b].

The multidimensional version of (3.1)–(3.6), usually called the Muskat problem, has been open for a long time (see, e.g., [CJ86]), and there are some indications that this problem is not well-posed, because the capillarity effects are neglected.

In the first part of [BM94b], after giving the definition of an entropy solution of (3.1)–(3.2), the existence and uniqueness of such a solution are given in Theorems 2.4–2.8 in [BM94b] under the following assumptions:

$$b' \geq 0 \quad \text{on }]0, 1[\qquad b \in C^2_{loc}(\mathbb{R}), \tag{3.7}$$

$$\mu \in W^{1,\infty}(\mathbb{R}), \qquad \mu \geq \mu_0 > 0, \tag{3.8}$$

$$0 < \phi_- \leq \phi(x) \leq \phi_+ \quad \text{a.e. on }]0, 1[, \tag{3.9}$$

$$0 < k_- \leq k(x) \leq k_+ \quad \text{a.e. on }]0, 1[, \tag{3.10}$$

$$S_0 \in W^{1,\infty}(0, 1), \quad S_1 \in W^{2,\infty}(0, +\infty), \quad 0 \leq S_0, S_1 \leq 1, \tag{3.11}$$

$$p_0, p_1 \in L^\infty(0, +\infty), \quad P(t) = p_0(t) - p_1(t) \geq P_0 > 0 \quad \text{a.e. on }]0, +\infty[. \tag{3.12}$$

Existence and uniqueness are proven after the change of coordinates $(x, t) \to (y, \tau)$

$$y(x) = \left(\int_0^1 \phi(\eta) d\eta \right)^{-1} \int_0^x \phi(\eta) d\eta \tag{3.13}$$

which transforms (3.1)–(3.6) into a nonlocal boundary and initial value problem,

$$\frac{\partial w}{\partial \tau} + \frac{\partial}{\partial y} b(w) = 0 \quad \text{in }]0, 1[\times]0, +\infty[, \tag{3.14}$$

$$w(y, 0) = S_0(X(y)) \quad \text{for } 0 < y \leq 1, \tag{3.15}$$

$$w(0, \tau) = S_1(w(\tau)) \quad \text{for } \tau \geq 0, \tag{3.16}$$

where $w(\tau)$ is defined by

$$\frac{dw(\tau)}{d\tau} = (P(w))^{-1} \left(\int_0^1 \phi(\eta) d\eta \right)^2 \int_0^1 \frac{\mu(w(y, \tau))}{k(X(y))\phi(X(y))} dy, \quad w(0) = 0 \tag{3.17}$$

with

$$\frac{d\tau(t)}{dt} = P(t) \left(\int_0^1 \phi(\eta) d\eta \right)^{-2} \left(\int_0^1 \frac{\mu(w(y, \tau))}{k(X(y))\phi(X(y))} dy \right)^{-1}, \quad \tau(0) = 0, \tag{3.18}$$

and

$$\frac{dX}{dy} = \frac{1}{\phi(x)} \int_0^1 \phi(\xi)d\xi, \quad X(0) = 0. \tag{3.19}$$

Defining $S(x, t) = w(y(x), \tau(t))$, we, then, have $\{S, p, q\}$, the entropy solution of (3.1)–(3.6).

5.3.2 Homogenization of the Buckley–Leverett System

First, we investigate what happens when, in (3.1)–(3.6), $\phi(x)$ and $k(x)$ are highly oscillating. For this, we are assuming only that ϕ^ε and $1/k^\varepsilon$ are oscillating around the constant values ϕ_0 and $1/k_0$, i.e.,

$$\phi^\varepsilon \rightharpoonup \phi_0 \text{ weak } * \text{ in } L^\infty(0, 1), \quad \frac{1}{k^\varepsilon} \rightharpoonup \frac{1}{k_0} \text{ weak } * \text{ in } L^\infty(0, 1), \tag{3.20}$$

as ε (the oscillation scale) tends to zero.

Using the change of coordinates (3.13) and (3.17)–(3.19), it can be proven that $(S^\varepsilon, p^\varepsilon, q^\varepsilon)$ with saturation, pressure, and total velocity being the entropy solution (corresponding to the oscillating ϕ^ε and k^ε functions) are converging (see Theorem 3.6 in [BM94b]) to homogenized quantities which are also the entropy solution of a homogenized system. More precisely,

$$S^\varepsilon \to S^0 \quad \text{in } L^1(]0, 1[\times]0, \tau[), \ \forall \tau < +\infty, \tag{3.21}$$

$$p^\varepsilon \to p^0 \quad \text{in } L^1\left(0, \tau \, ; H^1(0, 1)\right), \quad \text{and}$$

$$q^\varepsilon \to q^0 \quad \text{in } L^1(]0, 1[\times]0, \tau[), \ \forall \tau < +\infty, \tag{3.22}$$

where (S^0, p^0, q^0) is the entropy solution of the homogenized equation

$$\phi_0(x)\frac{\partial S^0}{\partial t} + q^0(t)\frac{\partial}{\partial x}b(S^0) = 0, \tag{3.23}$$

$$\frac{\partial q^0}{\partial x} = 0 ; \quad q^0(t) = -\frac{k_0(x)}{\mu(S^0)}\frac{\partial p^0}{\partial x}(x, t) \quad \text{in }]0, 1[\times]0, +\infty[, \tag{3.24}$$

$$S^0(x, 0) = S_0(x) \qquad \text{on }]0, 1[, \quad \text{and}$$

$$S^0(0, t) = S_1(t) \qquad \text{for } t \geq 0. \tag{3.25}$$

Remark The result above shows that, although there is no capillary effect for smoothing the behavior of the flow, (S^ε is only in $BV(]0, 1[\times]0, \infty[)$, there is still a homogenized equation (3.23)–(3.25) of the same type as the original one, and the upscaling of parameters is given by the usual results (ϕ^ε and k^ε tend to the arithmetic average and the harmonic average, respectively).

5.3.3 Propagation of Nonlinear Oscillations

This section corresponds to the behavior of a two-phase, one-dimensional flow described by (3.1)–(3.4) under a highly oscillating initial condition. The only assumption is that the initial condition $S_0^\varepsilon(x)$ is oscillating around a mean value $\bar{S}_0(x)$, i.e.,

$$S(x, 0) = S_0^\varepsilon(x) = S_0\left(\frac{x}{\varepsilon}\right) \rightharpoonup \bar{S}_0(x) \text{ in } L^\infty(0, 1) \text{ weak } *. \qquad (3.26)$$

It can be proven in the following way (see Theorem 4.2 in [BM94b]) that the entropy solution $(S^\varepsilon, p^\varepsilon, q^\varepsilon)$ corresponding to that oscillating initial condition (3.26) converges to (S^0, p^0, q^0) as $\varepsilon \to 0$.

$$S^\varepsilon \to S^0 \quad \text{in } L^p(]0, 1[\times]0, \tau[), \; \forall p < +\infty, \; \forall \tau < +\infty, \qquad (3.27)$$
$$q^\varepsilon \to q^0 \quad \text{in } L^p(0, \tau), \quad \text{and}$$
$$p^\varepsilon \to p^0 \quad \text{in } L^p\left(0, \tau ; H^1(0, 1)\right), \qquad (3.28)$$

where (S^0, p^0, q^0) is the entropy solution of the homogenized system

$$\phi(x)\frac{\partial S^0}{\partial t} + q^0(t)\frac{\partial}{\partial x}b(S^0) = 0, \qquad (3.29)$$

$$\frac{\partial}{\partial x}q^0(t) = 0 ; \quad q^0(t) = -\frac{k_0(x)}{\mu(S^0)}\frac{\partial p^0}{\partial x} \quad \text{in }]0, 1[\times]0, +\infty[, \qquad (3.30)$$

$$S^0(x, 0) = \bar{S}_0(x) \qquad \text{on } (0, 1), \quad \text{and}$$
$$S^0(0, t) = S_1(t) \qquad \text{for } t \geq 0. \qquad (3.31)$$

Remark The result above shows that the initial oscillations are not propagating in the medium. As the original oscillations become more rapid (as in (3.26)), they tend to disappear inside the medium (see (3.27)–(3.28)).

5.4   The Double-Porosity Model of Immiscible Two-Phase Flow

5.4.1   Introduction

Naturally fractured reservoirs may contain many fractures that permeate different regions of the reservoir and are characterized by the existence of a system of highly conductive fissures together with a large number of matrix blocks containing most of the oil. The fractures serve as highly conductive flow paths for the reservoir's fluid and increase the reservoir's effective permeability significantly beyond the permeability that corresponds only to the rock matrix. The reservoir mechanism of fractured systems is significantly different from that of the so-called single porosity system considered in sections 5.2 and 5.3. To describe the fluid flow in such a fractured reservoir, several authors in the engineering literature, for instance, [BZK60], [DA90], [KMJPZ76], and [WR63], showed that, if there are many well-connected fractures, the fracture network behaves as a porous medium equivalent to the so-called "dual-porosity" model.

In Barenblatt's dual-porosity model (see [BZK60]), the fracture width is considerably greater than the characteristic dimensions of the pores and the permeability

K^* of the fissure system considerably exceeds the permeability k of the individual blocks of porous media.

To obtain the dual-porosity model, the fracture system's local properties are averaged over a volume containing both the fractures and a matrix. The so-called dual-porosity model for a porous medium consists of an equivalent, coarse-grained, porous medium in which the fractures act as "pores" and the blocks act as "grains".

Because flow in the fractures is much more rapid than inside the matrix, the fluid does not flow directly from one matrix block to another. Rather, it first flows into the fracture system and, then, it can pass into a block or remain in the fractures. The porous rock matrix system acts as a global "source term" macroscopically distributed over the entire equivalent porous medium.

Denoting by ε the ratio between the size of one block of porous media and the size of the whole domain of calculation Ω, the characteristic time scale, for any parabolic evolution equation in blocks, will be of order ε^{-2}.

On the other hand, for a permeability ratio of order ε^2 between the system of equations in the blocks of porous media and the system of equations in the fractures, means that the ratio between the characteristic time for the rescaled flow ($y = x/\varepsilon$) in a single block and the characteristic time for the flow through the entire system of fractures is of order ε^{-2}. Roughly speaking, we may say that a ratio of ε^2 between the fractures and the porous blocks is the adequate time scaling allowing the matrix-fracture interaction phenomena described before at the global level (i.e., $\varepsilon \to 0$) and leading to the dual-porosity model. At time $t \ll 1$, a large fraction of the reserves is extracted from the fractures. Then, at time $t \simeq O(1)$, the exchange starts between porous blocks and fissures as described in [Pan90] and [Aur83]. This effect has, initially, been observed in [Vog82] in the case of a diffusion equation coupled with Darcy flow. It should also be noted that this ε^2 scaling is done in the engineering literature, for instance, in [Pea76], [Rei80], [ADH91] and [DA90], but is motivated by introducing a geometrical factor of transmissibility. If one takes the ratio of the two permeabilities of order one, then, by the usual theory of homogenization, the limit model will be, as in [Bou84b] or in [BKM95], a single-porosity model. If the ratio is smaller than order ε^2, the blocks, then, do not contribute to the global system of equations in the limit model, corresponding to homogenization of the system only for the fissures.

In [BLM95], the dual-posity model for incompressible two-phase flow has been rigorously derived, mathematically. A rigorous mathematical proof for this dual-porosity model had been obtained before only for single-phase flow in [ADH90], i.e., when the equations are linear.

To prove the general result, the two-scale convergence method, defined in Appendix A, is used to prove the convergence of total velocity and *"reduced pressure"*, but the difficulty added by the degenerate ellipticity prevents using two-scale convergence for the saturation and leads to use periodic modulation, as introduced in [Vog82].

For simplicity, as usual in engineering literature (see, for instance, [Pan90], [Rei80], and [WR63]), we assume a periodic distribution of the cells Y, each cell containing only one matrix block $\mathcal{B}$ surrounded by a system of fissures $\mathcal{F}$;

$Y = \mathfrak{F} \bigcup \mathfrak{B} \bigcup \partial\mathfrak{B}$. We should note that the two-scale method can also be used with randomly distributed media, as in [BMW94], and, although the periodic modulation is used herein under periodic assumptions, it can certainly be extended to some type of randomly distributed fissures.

5.4.2 The Microscopic Model

A system of small fractures surrounding blocks of a porous medium is considered a porous medium. The fractures quantities are defined on the entire system of fissures $\mathfrak{F}^\varepsilon = \Omega \cap \bigcup_{k \in \mathbb{Z}^3} \varepsilon(\mathfrak{F} + k)$. In $\mathfrak{F}^\varepsilon$, the dimensionless absolute permeability is denoted by K^* and the porosity by ϕ^*. The relative permeability curves are denoted $K_{r\xi}(S)$, $\xi = \alpha, \beta$, and the capillary pressure curve $P_c(S)$.

The matrix blocks have been assumed to possess identical properties, so φ and k are periodic of period $\varepsilon\mathfrak{B}$. On the blocks of porous media, $\mathfrak{B}^\varepsilon = \Omega \cap \bigcup_{k \in \mathbb{Z}^3} \varepsilon(\mathfrak{B}+k)$, in the reservoir. Then, oscillating porosity and permeability $\varphi^\varepsilon(x) = \varphi(x/\varepsilon)$ and $k^\varepsilon(x) = k(x/\varepsilon)$ with unique relative permeability $k_{r\xi}(s)$, $\xi = \alpha, \beta$, and unique capillary pressure curve $p_c(s)$.

Denoting Γ^ε the boundary between $\mathfrak{B}^\varepsilon$ and $\mathfrak{F}^\varepsilon$, with $]0, \tau[$ a time interval, we write the conservation of mass in each phase, combined with the generalized Darcy law for $\xi = \alpha, \beta$:

$$\phi^*(x)\frac{\partial S_\xi^\varepsilon}{\partial t} - \nabla \cdot \left\{ \frac{K^*(x)K_{r\xi}(S_\alpha^\varepsilon)}{\mu_\xi}[\nabla P_\xi^\varepsilon - \rho_\xi g] \right\} = f_\xi(x, t) \quad \text{in } \mathfrak{F}^\varepsilon \times]0, \tau[, \tag{4.1}$$

$$\varphi^\varepsilon(x)\frac{\partial s_\xi^\varepsilon}{\partial t} - \varepsilon^2\nabla \cdot \left\{ \frac{k^\varepsilon(x)k_{r\xi}(s_\alpha^\varepsilon)}{\mu_\xi}[\nabla p_\xi^\varepsilon - \rho_\xi g] \right\} = f_\xi(x, t) \quad \text{in } \mathfrak{B}^\varepsilon \times]0, \tau[, \tag{4.2}$$

$$\frac{K^* K_{r\xi}(S_\alpha^\varepsilon)}{\mu_\xi}[\nabla P_\xi^\varepsilon - \rho_\xi g] \cdot \vec{\nu} = \varepsilon^2 \frac{k^\varepsilon k_{r\xi}(s_\alpha^\varepsilon)}{\mu_\xi}[\nabla p_\xi^\varepsilon - \rho_\xi g] \cdot \vec{\nu} \quad \text{on } \Gamma^\varepsilon \times]0, \tau[, \tag{4.3}$$

and

$$S_\xi^\varepsilon = s_\xi^\varepsilon \quad \text{and} \quad P_\xi^\varepsilon = p_\xi^\varepsilon \quad \text{on} \quad \Gamma^\varepsilon \times]0, \tau[; \tag{4.4}$$

$$S_\xi^\varepsilon(x, 0) = S_\xi^0(x) \text{ in } \mathfrak{F}^\varepsilon \quad ; \quad s_\xi^\varepsilon(x, 0) = s_\xi^0(x) \text{ in } \mathfrak{B}^\varepsilon, \tag{4.5}$$

$$S_\beta^\varepsilon + S_\alpha^\varepsilon = 1 \text{ in } \mathfrak{F}^\varepsilon \times]0, \tau[\quad ; \quad s_\beta^\varepsilon + s_\alpha^\varepsilon = 1 \text{ in } \mathfrak{B}^\varepsilon \times]0, \tau[, \tag{4.6}$$

$$P_\alpha^\varepsilon - P_\beta^\varepsilon = P_c(S_\alpha^\varepsilon) \quad \text{in} \quad \mathfrak{F}^\varepsilon \times]0, \tau[,$$

and

$$p_\alpha^\varepsilon - p_\beta^\varepsilon = p_c(s_\alpha^\varepsilon) \quad \text{in} \quad \mathfrak{B}^\varepsilon \times]0, \tau[\tag{4.7}$$

where $\vec{\nu}$ denotes the unitary outward $\mathfrak{B}^\varepsilon$ normal to Γ^ε.

The boundary conditions on $\partial\Omega = \bar{\Gamma}^1 \bigcup \bar{\Gamma}^2$, are

$$P_\xi^\varepsilon = P_\xi^D \quad \text{on } \Gamma^1 \times]0, \tau[\quad , \quad \xi = \alpha, \beta \tag{4.8}$$

(i.e., Γ^1 represents the part of the external boundary of the reservoir $\partial\Omega$ in contact with a liquid continuum);

$$\frac{K^*K_{r\xi}(S_\alpha^\varepsilon)}{\mu_\xi}[\nabla P_\xi^\varepsilon - \rho_\xi g]\cdot \vec{v} = g_\xi \quad \text{on} \quad \Gamma^2\times]0,\tau[, \quad \xi = \alpha, \beta \qquad (4.9)$$

(i.e., a prescribed flow rate is assumed through the previous external boundary Γ^2).

Now, following [AD85] or [CJ86], as in the previous sections, we transform our system (4.1)–(4.8) to a system formulated as an elliptic-parabolic problem for the *"reduced pressures"* P^ε and p^ε, for the total velocities V^ε and v^ε, and for the saturations S^ε and s^ε of the α-phase:

$$\phi^*\frac{\partial S^\varepsilon}{\partial t} - \nabla\cdot\left\{K^*A(S^\varepsilon)\nabla S^\varepsilon + K^*B(S^\varepsilon)g - D(S^\varepsilon)V^\varepsilon\right\} = 0, \qquad (4.10)$$

$$V^\varepsilon = -K^*\Lambda(S^\varepsilon)[\nabla P^\varepsilon - E(S^\varepsilon)g], \qquad (4.11)$$

$$\nabla\cdot V^\varepsilon = 0 \quad \text{in} \quad \mathfrak{F}^\varepsilon\times]0,\tau[, \qquad (4.12)$$

and

$$\varphi^\varepsilon(x)\frac{\partial s^\varepsilon}{\partial t} - \nabla\cdot\left\{\varepsilon^2 k^\varepsilon a(s^\varepsilon)\nabla s^\varepsilon + \varepsilon^2 k^\varepsilon b(s^\varepsilon)g - d(s^\varepsilon)v^\varepsilon\right\} = 0, \qquad (4.13)$$

$$v^\varepsilon = -\varepsilon^2 k^\varepsilon \lambda(s^\varepsilon)[\nabla p^\varepsilon - e(s^\varepsilon)g], \qquad (4.14)$$

$$\nabla\cdot v^\varepsilon = 0 \quad \text{in} \quad \mathcal{B}^\varepsilon\times]0,\tau[, \qquad (4.15)$$

where the *"reduced pressures"* are given by

$$P^\varepsilon(x,t) = \tfrac{1}{2}(P_\beta^\varepsilon + P_\alpha^\varepsilon) + \int_0^{S_\alpha^\varepsilon}\left(\frac{K_{r\alpha}/\mu_\alpha}{K_{r\beta}/\mu_\beta + K_{r\alpha}/\mu_\alpha} - \frac{1}{2}\right)\frac{\partial P_c}{\partial S}dS,$$

and $$p^\varepsilon(x,t) = \tfrac{1}{2}(p_\beta^\varepsilon + p_\alpha^\varepsilon) + \int_0^{s_\alpha^\varepsilon}\left(\frac{k_{r\alpha}/\mu_\alpha}{k_{r\beta}/\mu_\beta + k_{r\alpha}/\mu_\alpha} - \frac{1}{2}\right)\frac{\partial p_c}{\partial s}ds.$$

The functions of saturation are defined by

$$\Lambda(S) = \frac{K_{r\beta}(S)}{\mu_\beta} + \frac{K_{r\alpha}(S)}{\mu_\alpha}, \quad \lambda(s) = \frac{k_{r\beta}(s)}{\mu_\beta} + \frac{k_{r\alpha}(s)}{\mu_\alpha},$$

$$E(S) = \frac{1}{\Lambda(S)}\left[\frac{K_{r\beta}(S)}{\mu_\beta}\rho_\beta + \frac{K_{r\alpha}(S)}{\mu_\alpha}\rho_\alpha\right],$$

$$e(s) = \frac{1}{\lambda(s)}\left[\frac{k_{r\beta}(s)}{\mu_\beta}\rho_\beta + \frac{k_{r\alpha}(s)}{\mu_\alpha}\rho_\alpha\right],$$

$$A(S) = \frac{K_{r\alpha}(S)K_{r\beta}(S)}{\mu_\beta\mu_\alpha\Lambda(S)}\frac{\partial P_c(S)}{\partial S}, \quad a(s) = \frac{k_{r\alpha}(s)k_{r\beta}(s)}{\mu_\beta\mu_\alpha\lambda(s)}\frac{\partial p_c(s)}{\partial s},$$

$$B(S) = \frac{K_{r\alpha}(S)K_{r\beta}(S)}{\mu_\beta\mu_\alpha\Lambda(S)}(\rho_\beta - \rho_\alpha), \quad b(s) = \frac{k_{r\alpha}(s)k_{r\beta}(s)}{\mu_\beta\mu_\alpha\lambda(s)}(\rho_\beta - \rho_\alpha),$$

and

$$D(S) = \frac{K_{r\alpha}(S)}{\mu_\alpha\Lambda(S)}, \quad d(s) = \frac{k_{rw}(s)}{\mu_\alpha\lambda(s)}.$$

The interface conditions (4.3)–(4.4) become

$$S^\varepsilon = s^\varepsilon \quad \text{and} \quad P^\varepsilon = p^\varepsilon \quad \text{on} \quad \Gamma^\varepsilon \times]0, \tau[, \tag{4.16}$$

$$V^\varepsilon \cdot \vec{\nu} = v^\varepsilon \cdot \vec{\nu} \quad \text{on} \quad \Gamma^\varepsilon \times]0, \tau[, \tag{4.17}$$

and

$$K^*[A(S^\varepsilon)\nabla S^\varepsilon + B(S^\varepsilon)g] \cdot \vec{\nu} = k^\varepsilon[\varepsilon^2 a(s^\varepsilon)\nabla s^\varepsilon + \varepsilon^2 b(s^\varepsilon)g] \cdot \vec{\nu} \tag{4.18}$$

on $\Gamma^\varepsilon \times]0, \tau[$.

For simplicity, the boundary and initial conditions are chosen

$$\begin{cases} V^\varepsilon \cdot \vec{\nu} = g_\beta + g_\alpha = 0 \quad \text{on} \quad \Gamma^2 \times]0, \tau[, \\ K^*[A(S^\varepsilon)\nabla S^\varepsilon + B(S^\varepsilon)g] \cdot \vec{\nu} = g_\alpha \quad \text{on} \quad \Gamma^2 \times]0, \tau[, \\ S^\varepsilon = S_\alpha^\varepsilon(x, P_\alpha^D - P_\beta^D) = 0 \quad \text{on} \quad \Gamma^1 \times]0, \tau[\end{cases} \tag{4.19}$$

$$S^\varepsilon(x, 0) = S_\alpha^0(x) \quad \text{in} \quad \mathfrak{F}^\varepsilon, \tag{4.20}$$

and

$$s^\varepsilon(x, 0) = s_\alpha^0(x) \quad \text{in} \quad \mathfrak{B}^\varepsilon. \tag{4.21}$$

Following [AKM90], defining a weak formulation for the Eqs. (4.10)–(4.21), and applying the existence theory (see [AD85], [AKM90] or [KL84]) lead to the existence of at least one weak solution $\forall \varepsilon > 0$.

5.4.3 Compactness and Convergence Results

Assumptions (A1)–(A12) in ([BLM95], section 3) give a priori estimates, and the concept of two-scale convergence leads to the following compactness results (see Proposition 4.1 in [BLM95]).

Theorem 4.1 *Defining* $Z^\varepsilon = \int_0^{S^\varepsilon} \sqrt{A(\eta)} d\eta$, $z^\varepsilon = \int_0^{s^\varepsilon} \sqrt{a(\eta)} d\eta$ *and* $(\tilde{p}^\varepsilon, \tilde{s}^\varepsilon, \tilde{z}^\varepsilon, \tilde{V}^\varepsilon)$ *either as* $(p^\varepsilon, s^\varepsilon, z^\varepsilon, v^\varepsilon)$ *in* $\mathfrak{B}^\varepsilon$ *or as* $(P^\varepsilon, S^\varepsilon, Z^\varepsilon, V^\varepsilon)$ *in* $\mathfrak{F}^\varepsilon$, *using* Π_ε *the extension operator from* $\mathfrak{F}^\varepsilon$ *to* Ω, *as, for instance, in Chapter 3, section 3.1.2 or in [ACDP92], there exists a subsequence, denoted by the same subscript, such that*

$$\Pi_\varepsilon P^\varepsilon \text{ two-scale converges to } P \in L^2\left(]0, \tau[\,;\, H^1(\Omega)\right) \quad ; \tag{4.22}$$

$$\nabla(\Pi_\varepsilon P^\varepsilon) \text{ two-scale converges to } \nabla P + \nabla_y P_1(t, x, y) \tag{4.23}$$
$$\text{where } P_1 \in L^2(]0, \tau[\times \Omega \,;\, H_\#^1(Y)); $$

$$\tilde{p}^\varepsilon \text{ two-scale converges to } p \in L^2(]0, \tau[\,;\, H^1(\Omega)); \tag{4.24}$$

$$\Pi_\varepsilon Z^\varepsilon \text{ two-scale converges to } Z \in L^2(]0, \tau[\,;\, H^1(\Omega)); \tag{4.25}$$

$$\nabla(\Pi_\varepsilon Z^\varepsilon) \text{ two-scale converges to } \nabla Z + \nabla_y Z_1(t, x, y) \tag{4.26}$$
$$\text{where } Z_1 \in L^2(]0, \tau[\times \Omega \,;\, H_\#^1(Y)); $$

$$\Pi_\varepsilon Z^\varepsilon \longrightarrow Z \text{ strongly in } L^q(]0, \tau[\times \Omega), \;\; \forall q < +\infty; \tag{4.27}$$

$$(Z^\varepsilon)^{-1}(\Pi_\varepsilon Z^\varepsilon) = \overline{S}^\varepsilon \longrightarrow S \text{ strongly in } L^q(]0, \tau[\times \Omega), \tag{4.28}$$

$$\forall q < +\infty;$$

$$\tilde{s}^\varepsilon \text{ two-scale converges to } s \in L^2(]0, \tau[\times\Omega \times Y); \tag{4.29}$$

$$\tilde{z}^\varepsilon \text{ two-scale converges to } z \in L^2(]0, \tau[\times\Omega \; ; \; H^1_\#(Y)); \tag{4.30}$$

$$\varepsilon\nabla\tilde{z}^\varepsilon \text{ two-scale converges to } \nabla_y z \in L^2(]0, \tau[\times\Omega \times Y); \tag{4.31}$$

$$\tilde{V}^\varepsilon \text{ two-scale converges to } V \in L^2(]0, \tau[\times\Omega \times Y); \tag{4.32}$$

$$\nabla_y \cdot V = 0, \quad and \quad \nabla_x \cdot \int_Y V(x, y)dy = 0. \tag{4.33}$$

It would then be natural to try passing to the limit using monotonicity. This approach would work nicely if the nonlinearity was not degenerated. The degeneration makes the standard monotonicity argument (see, e.g., [Lio69], pp. 190–204) fairly complicated if not impossible, so we have chosen a different approach, the periodic modulation. This approach, then, makes possible passing to the limit in the degenerate terms depending nonlinearly on S^ε.

The connection between the periodic modulation and H-measures is discussed in [Tar90], pp. 203–204, and the links between the two-scale convergence and the weak convergence of a periodically modulated sequence are given in Proposition 4.6, Lemma 4.7, and Corollary 4.8 in [BLM95].

After having established the link between the two-scale convergence and the weak convergence of the periodically modulated sequence $\{D^\varepsilon s^\varepsilon\}$, see Proposition 4.6 in [BLM95], the strategy is simple. Find an equation for $D^\varepsilon s^\varepsilon$ and pass to the limit (see Proposition 4.9 in [BLM95]).

Having obtained the equations for $D^\varepsilon s^\varepsilon$, it is, then, possible to establish the a priori estimates for $D^\varepsilon z^\varepsilon$ and $D^\varepsilon s^\varepsilon$, by analogy to those obtained for Z^ε and S^ε. Then, we pass to the limit for fixed $k \in \mathbb{Z}^3$ and, finally, the limit equation for s is obtained by a density argument.

At last, we have the convergence results (see Theorem 4.11 in [BLM95]):

Theorem 4.2 *Let $\{P, S, s, V\}$ be the limits defined in Theorem 4.1. They are solutions of the nonlinear system, in $Q_\tau = \Omega \times]0, \tau[$:*

$$\begin{cases} \dfrac{1}{|Y|} \displaystyle\int_Y V dy = \overline{V} = -K^{*H}\Lambda(S)[\nabla_x P - E(S)g] & in \quad Q_\tau, \\ \nabla_x \cdot \overline{V} = 0 & in \quad Q_\tau, \\ \overline{V} \cdot \vec{\nu} = 0 & on \quad \Gamma^2 \times]0, \tau[, \\ P = \dfrac{P^D_\alpha + P^D_\beta}{2} & on \quad \Gamma^1 \times]0, \tau[, \end{cases} \tag{4.34}$$

and

$$\begin{cases} \dfrac{|\mathfrak{F}|}{|Y|}\phi^*\partial_t S - \nabla_x \cdot \left\{ K^{*H}(A(S)\nabla_x S + B(S)g) - D(S)\overline{V} \right\} \\ \quad = \dfrac{1}{|Y|} \displaystyle\int_{\partial M} k(y)a(s)\nabla_y s \cdot \vec{\nu} d\eta \quad in \quad Q_\tau, \\ K^{*H}(A(S)\nabla_x S + B(S)g) \cdot \vec{\nu} = g_\alpha \quad on \quad \Gamma^2 \times]0, \tau[, \\ S(x, 0) = S^0_\alpha(x) \text{ in } \Omega, \quad S = 0 \text{ on } \Gamma^1 \times]0, \tau[. \end{cases} \tag{4.35}$$

The tensor K^{*H} is defined by

$$K^{*H} = \frac{K^*}{|Y|} \int_{\mathfrak{F}} (I + W(y))dy, \tag{4.36}$$

where $W(y)$ is the tensor whose (i, j) component is $\dfrac{\partial w_j}{\partial y_i}$ with w_j a 1-periodic solution in Y of the auxiliary problem:

$$\begin{cases} \Delta_y w_j = 0 & in \quad \mathfrak{F}, \\ \nabla_y w_j \cdot \vec{v} = -e_j \cdot \vec{v} & on \quad \partial\mathfrak{B}. \end{cases} \tag{4.37}$$

For any $x \in \Omega$, s in the right-hand side of (4.35) is defined from

$$\begin{cases} \varphi^0(y)\partial_t s - \nabla_y \cdot \{k(y)a(s)\nabla_y s\} = 0 & in \quad Q_\tau \times \mathfrak{B}, \\ s(x, y, 0) = s^0_\alpha(x) & in \quad \Omega \times \mathfrak{B}, \\ Z = \int_0^s \sqrt{A(\xi)}d\xi = z = \int_0^s \sqrt{a(\xi)}d\xi & on \quad \partial\mathfrak{B}. \end{cases} \tag{4.38}$$

Remark In the above system of equations

- $|\cdot|$ denotes the measure (or the "volume") of $(\cdot)$.

- There is a macroscopic fracture system driven by equations (4.35) in all $\Omega \times]0, \tau[$, similar to the initial one (4.1) but with an effective absolute rock permeability K^{*H} given by (4.36) and an effective porosity $\frac{|\mathfrak{F}|}{|Y|}\phi^*$, but with an additional right-hand side source-like term

$$\frac{1}{|Y|} \int_{\partial\mathfrak{B}} k(y)a(s)\nabla_y s \cdot \vec{v}d\eta = -\frac{1}{|Y|} \int_M \varphi^0(y)\partial_t sdy.$$

- Associated with each $x \in \Omega$ there is a matrix block $\varepsilon(\mathfrak{B}+k)$. The flow inside this matrix block $(\mathfrak{B}+k)$ is described by (4.38) and produces the source-like term in (4.35).

- If we write

$$K^{*H} = K^* \frac{|\mathfrak{F}|}{|Y|}\left(I + \frac{1}{|\mathfrak{F}|}\int_{\mathfrak{F}} W(y)dy\right)$$

in (4.35), then,

$$I + \frac{1}{|\mathfrak{F}|}\int_{\mathfrak{F}} W(y)dy$$

is a tortuosity factor as in the Kozeny model (see, for instance, [Koz27] or [Car37]).

6

Miscible Displacement

Ulrich Hornung

6.1 Introduction

Fluid flow in porous media was discussed in the preceding chapters. How to derive macro Darcy-type laws from micro laws, such as Stokes equations or its generalizations was demonstrated. It is natural to ask how one can use those results and apply them to miscible displacement problems. In soil physics or soil chemistry, e.g., it is of great importance to combine the homogenization techniques, described so far, with problems of diffusion, dispersion, and convection of chemical species that are transported in the fluid flowing in the pore space of a porous medium. In principle, it turns out that all the mathematical methods developed can easily be applied to such problems. In particular, there is no difficulty in taking chemical reactions into consideration, because, usually, these are described by undifferentiated terms in the differential equations. Therefore, they are basically unmodified during the homogenization process.

The main aspects we are going to discuss in this chapter are twofold: We are going to study the micro to meso upscaling (section 6.2) and meso to macro upscaling problems (section 6.3). The first type of problem starts at the pore scale. Here, we combine Stokes flow with diffusion, convection, and adsorption. For the second type we assume that the micro-meso homogenization has already been carried out and focuses on transport problems in fractured, fissured, or aggregated media.

6.2 Upscaling from the Micro- to the Mesoscale

6.2.1 Diffusion, Convection, and Reaction

The underlying equations for fluid flow are the same as in Chapter 1 (Eq. (4.1)) and Chapter 3 (Eq. (1.1)), which we repeat here for convenience.

$$
\begin{cases}
\varepsilon^2 \mu \Delta \vec{v}^\varepsilon(x) = \nabla p^\varepsilon(x), & x \in \mathfrak{P}^\varepsilon, \\
\nabla \cdot \vec{v}^\varepsilon(x) = 0, & x \in \mathfrak{P}^\varepsilon, \\
\vec{v}^\varepsilon(x) = 0, & x \in \Gamma^\varepsilon.
\end{cases}
\tag{2.1}
$$

We are going to assume that certain chemical species are dissolved in the fluid. For simplicity of notation we write the equation for just one substance, but the reader should keep in mind that this is only a special case which can easily be generalized to any number n of species. Allowing diffusion, convection, and reaction, we get a parabolic equation for the concentration of the substance of the form

$$\partial_t u^\varepsilon(t, x) = \nabla \cdot (a_{\mathfrak{P}}^\varepsilon \nabla u^\varepsilon(t, x) - \vec{v}^\varepsilon(x) u^\varepsilon(t, x)) + R_{\mathfrak{P}}(u^\varepsilon(t, x)), \quad x \in \mathfrak{P}^\varepsilon. \quad (2.2)$$

As long as no interaction takes place with the solid phase, i.e., neither with its interior (of the particles or pellets) nor with its surfaces, we get a homogeneous Neumann boundary condition on Γ^ε, namely,

$$\vec{v} \cdot \nabla u^\varepsilon(t, x) = 0, \quad x \in \Gamma^\varepsilon. \quad (2.3)$$

This condition has that simple form, because, in Eq. (2.1), the fluid velocity $\vec{v}^\varepsilon$ is assumed to be zero on the surfaces Γ^ε. For a linear reaction term $R_{\mathfrak{P}}$ it is now an easy exercise (using asymptotic expansions or two-scale convergence) to show that the homogenized limit of the system (2.1), (2.2) is given by

$$\begin{cases} \vec{v}(x) = -\frac{1}{\mu} K \nabla p(x), & x \in \Omega \\ \nabla \cdot \vec{v}(x) = 0, & x \in \Omega \\ |\mathfrak{P}| \partial_t u(t, x) & \\ \quad = \nabla \cdot (A(x) \nabla u(t, x) - \vec{v}(x) u(t, x)) & \\ \quad + |\mathfrak{P}| R_{\mathfrak{P}}(u(t, x)), & x \in \Omega \end{cases} \quad (2.4)$$

where the tensor K is calculated according to formulas (4.6) Chapter 1 or (1.5) Chapter 3, respectively, and the tensor A is obtained from formula (3.13) Chapter 1. For nonlinear reaction terms, the convergence proof is not straightforward, and special assumptions are needed, such as growth conditions and/or monotonicity (see, e.g., [HJM94]).

6.2.2 Adsorption

Now, we are going to replace the trivial boundary condition (2.3) by an adsorption condition. We assume that the substance in question can be adsorbed on the surfaces of the solid matrix. A mathematical formulation makes it necessary to introduce another concentration, namely, that of the adsorbed material. To be precise, whereas the concentration u^ε is a mass per volume concentration, the one introduced here, denoted by U^ε, is a mass per surface concentration. It is reasonable to allow that the actual adsorption rate b^ε depends on both the concentration u^ε in the fluid and on the U^ε on the surface. Thus we arrive at a boundary condition of the form,

$$a_{\mathfrak{P}}^\varepsilon(x) \vec{v} \cdot \nabla u^\varepsilon(t, x) = \varepsilon b^\varepsilon(x, u^\varepsilon(t, x), U^\varepsilon(t, x)), \quad x \in \Gamma^\varepsilon, \quad (2.5)$$

where $\vec{v}$ is the normal vector on Γ^ε pointing into $\mathfrak{P}^\varepsilon$. In this equation, the dependence of b^ε on the space variable x allows that the adsorption rate varies on each particle surface and depends on the individual site on that particle. The factor ε is used to take the different scalings of surfaces and volumes into account and, thus, to allow for a reasonable and meaningful homogenized model. Obviously, the model is complete only after adding an equation describing the dynamics of the adsorbed concentration $U^\varepsilon(t, x)$ which is

$$\partial_t U^\varepsilon(t, x) = b^\varepsilon(x, u^\varepsilon(t, x), U^\varepsilon(t, x)) + R_\Gamma^\varepsilon(x, U^\varepsilon(t, x)), \quad x \in \Gamma^\varepsilon, \qquad (2.6)$$

if we allow a chemical reaction to take place on the surface whose rate is given by R_Γ^ε. In the paper [HJ91], it has been shown that the system (2.1), (2.5), (2.6) has the following homogenized limit:

$$\begin{cases} \vec{v}(x) & = -\frac{1}{\mu} K \nabla p(x), & x \in \Omega \\ \nabla \cdot \vec{v}(x) & = 0, & x \in \Omega \\ |\mathfrak{P}| \partial_t u(t, x) & \\ \quad + Q(t, x) & = \nabla \cdot (A(x)\nabla u(t, x) - \vec{v}(x)u(t, x)) \\ & \quad + |\mathfrak{P}| R_\mathfrak{B}(u(t, x)), & x \in \Omega \\ Q(t, x) & = \int_\Gamma b(x, y, u(t, x), U(t, x, y)) \, d\Gamma(y), & x \in \Omega \\ \partial_t U(t, x, y) & = b(x, y, u(t, x), U(t, x, y)) \\ & \quad + R_\Gamma(x, y, U(t, x, y)), & x \in \Omega, \ y \in \Gamma. \end{cases}$$
$$(2.7)$$

In the paper cited, the proof was given for the case of a linear adsorption rate b and linear reaction terms $R_\mathfrak{B}$ and R_Γ.

So far, we have assumed that the functions b^ε and R_Γ^ε can be written as

$$b^\varepsilon(x, u, U) = b(x, \frac{x}{\varepsilon}, u, U) \text{ and } R_\Gamma^\varepsilon(x, U) = R_\Gamma(x, \frac{x}{\varepsilon}, U),$$

where b and R_Γ are Y-periodic with respect to the variable $y = \frac{x}{\varepsilon}$. In the special case where there is no dependence on y, the formulas simplify. Then, the solution U does not depend on y, and, therefore, the sink term Q is just a multiple of b, namely,

$$Q(t, x) = |\Gamma| b(x, u(t, y), U(t, x)).$$

Another specialization is neglecting chemical reactions on the surfaces. Then,

$$Q(t, x) = |\Gamma| b(x, u(t, x), U(t, x)) = |\Gamma| \partial_t U(t, x).$$

After dropping the arguments, we obtain the following system for the concentrations u and U:

$$\begin{cases} \partial_t(|\mathfrak{P}|u + |\Gamma|U) = \nabla \cdot (A\nabla u - \vec{v}u) + |\mathfrak{P}| R_\mathfrak{B}, \\ \partial_t U = b. \end{cases} \qquad (2.8)$$

This sytem is called a *first-order kinetic* model. Almost the same macromodel arises in the context of mobile and immobile water (see section 6.3.1). Well-known examples of nonlinear adsorption rates are the *Freundlich isotherm*,

$$b(u, U) = \gamma(u^p - U),$$

with $\gamma > 0$ and $0 < p < 1$ and the *Langmuir isotherm,*

$$b(u, U) = \gamma\left(\frac{\alpha u}{1 + \beta u} - U\right),$$

with $\alpha, \beta, \gamma > 0$. A justification of these isotherms, and their generalization is given in [DK92]. Until now, it has not been quite clear how to generalize the above mentioned convergence proofs of the homogenization process to nonlinear adsorption rates of this kind.

Convolution

Whenever the adsorption rate depends linearly on the adsorbed concentration U, one can give explicit formulas for the adsorbed concentration U in terms of the fluid concentration u. We are going to develop this idea here for

$$b(x, y, u, U) = \alpha(x, u) - \gamma(x, y)U, \tag{2.9}$$

assuming that $R_{\mathfrak{G}}$ vanishes. The solution of $\partial_t U = b$ can now easily be found using the method of "variation of constants." The resulting formula is

$$\begin{aligned} U(t, x, y) \ &= U(0, x, y)\exp(-\gamma(x, y)t) \\ &+ \int_0^t \exp(-\gamma(x, y)(t - \tau))\alpha(x, u(\tau, x))\, d\tau. \end{aligned}$$

If we assume vanishing inital values, i.e., $U(0, x, y) = 0$,

$$b(x, y, u, U) = \partial_t U(t, x, y) = \partial_t \left(\int_0^t \exp(-\gamma(x, y)(t - \tau))\alpha(x, u(\tau, x))\, d\tau \right).$$

After integrating over Γ,

$$\begin{aligned} Q(t, x) \ &= \int_\Gamma b\, d\Gamma = \partial_t \int_\Gamma U\, d\Gamma \\ &= \partial_t \left(\int_0^t \int_\Gamma \exp(-\gamma(x, y)(t - \tau))\, d\Gamma(y)\alpha(x, u(\tau, x))\, d\tau \right). \end{aligned}$$

Using the time *convolution* notation,

$$(\varphi * \psi)(t) = \int_0^t \varphi(t - \tau)\psi(\tau)\, d\tau,$$

$$Q(t, x) = \partial_t \left(\rho(t, x) * \alpha(x, u(t, x)) \right),$$

where the *kernel*

$$\rho(\tau, x) = \int_\Gamma \exp(-\gamma(x, y)\tau)\, d\Gamma(y)$$

is a weighted average of exponential functions. After these manipulations, we can rewrite the global equation for u in (2.7) as

$$\begin{aligned} \partial_t &\left(|\mathfrak{P}|u(t, x) + \rho(t, x) * \alpha(x, u(t, x)) \right) \\ &= \nabla \cdot (A(x)\nabla u(t, x) - \vec{v}(x)u(t, x)) + |\mathfrak{P}|R_{\mathfrak{P}}(u(t, x)), \end{aligned}$$

which makes it clear that we have a memory term in the principal part of a diffusion-convection-reaction equation.

Surface Diffusion

The adsorption model discussed here can be generalized by allowing surface diffusion on the surfaces Γ^ε. This means that we add a diffusion term for U^ε in the differential equation. Let ∇^ε denote the *surface gradient* operator on Γ^ε, and let $\nabla^\varepsilon\cdot$ denote the *surface divergence* operator. Then, the generalization of Eq. (2.6) becomes

$$\begin{aligned}\partial_t U^\varepsilon(t,x) &= \nabla^\varepsilon\cdot(a_\Gamma^\varepsilon(x)\nabla^\varepsilon U^\varepsilon(t,x))\\ &+b^\varepsilon(x,u^\varepsilon(t,x),U^\varepsilon(t,x))+R_\Gamma^\varepsilon(x,U^\varepsilon(t,x)),\ x\in\Gamma^\varepsilon.\end{aligned}\tag{2.10}$$

The paper [HJ91] cited above also covers this case (there a_Γ is taken constant). The homogenized limit is almost the same as before, except that the *local equation* now is no longer an ordinary but a partial differential equation, namely,

$$\begin{aligned}\partial_t U(t,x,y) &= \nabla_y^\Gamma\cdot(a_\Gamma(x,y)\nabla_y^\Gamma U(t,x,y))\\ &+b(x,y,u(t,x),U(t,x,y))+R_\Gamma(x,y,U(t,x,y)),\end{aligned}$$

where ∇^Γ denotes the surface gradient operator, $\nabla^\Gamma\cdot$ denotes the surface divergence operator on Γ, and the subscript y indicates that the differentiation is taken with respect to the *local variable y*.

Making the same assumption of linearity (2.9), as before, and dropping chemical reactions on Γ, we may again obtain a convolution formulation. To this end we have to use the solution $r:\Gamma\to\mathbb{R}$ of the following time-dependent *cell problem*

$$\begin{cases}\partial_t r(t,x,y) &= \nabla_y^\Gamma\cdot(a_\Gamma(x,y)\nabla_y^\Gamma r(t,x,y))\\ &-\gamma(x,y)r(t,x,y), & t>0,\ x\in\Omega,\ y\in\Gamma,\\ r(t,x,y) &= 1, & t=0,\ x\in\Omega,\ y\in\Gamma.\end{cases}\tag{2.11}$$

If the initial values of U do not depend on y, we easily get the representation

$$U(t,x,y)=U(0,x)r(t,x,y)+\int_0^t r(t-\tau,x,y)\alpha(x,u(\tau,x))\,d\tau.$$

Integration over Γ yields

$$\begin{aligned}Q(t,x) &= \int_\Gamma b\,d\Gamma\\ &= \partial_t\int_\Gamma U\,d\Gamma\\ &= \partial_t\left(\int_0^t\int_\Gamma r(t-\tau,x,y)\,d\Gamma(y)\alpha(x,u(\tau,x))\,d\tau\right),\end{aligned}$$

and, thus, again,

$$Q(t,x)=\partial_t\Big(\rho(t,x)*\alpha(x,u(t,x))\Big),$$

where the kernel is now

$$\rho(\tau,x)=\int_\Gamma r(\tau,x,y)\,d\Gamma(y).$$

In the special case where the functions γ and a_Γ do not depend on y, we return to the situation discussed earlier where the cell function r was an exponential function, because, in this case, the solution U and the cell function r are independent of y.

Numerical simulations for this type of model and also for nonlinear versions of it can be found in [HJ87].

Irreversible Adsorption

The adsorption isotherms discussed here and used mostly in the literature are of the form

$$b(u, U) = \gamma(\Phi(u) - U)$$

with a constant $\gamma > 0$ and a strictly increasing function $\Phi : [0, \infty) \rightarrow [0, \infty)$. Processes of this form are reversible, because they allow (positive) adsorption for $U < \Phi(u)$ and desorption (i.e., negative adsorption) for $U > \Phi(u)$. As a consequence, the adsorbed concentration U at a given site will always follow the value $\Phi(u)$ determined by the fluid concentration u, no matter how u may evolve in time. If u tends to a limit u_∞ for $t \rightarrow \infty$, then, U will tend to the limit $\Phi(u_\infty)$, and, for $\gamma \rightarrow \infty$, one arrives at equilibrium $U = \Phi(u)$. Such reversibility may not be realistic. If no desorption is possible, a mathematical description leads to

$$b(u, U) = \gamma(\Phi(u) - U)_+,$$

where the subscript $+$ has the meaning $z_+ = z$ for $z > 0$ and $z_+ = 0$ for $z \leq 0$. In such a situation, the adsorbed concentration U at a certain site can only increase in time, but never decrease.

A more sophisticated model introduces two thresholds: If $\Phi(u) - U$ exceeds a certain value, then, adsorption takes place; if $U - \Phi(u)$ exceeds another prescribed value, then desorption occurs. This means that adsorption and desorption differ in the sense that there is a *hysteresis* effect. In such a case, the process can no longer be modeled using a single-valued function, as in Eq. (2.5). Boundary conditions of the hysteresis type have been studied theoretically in [HS95b]. One expects that after homogenization, a model of the *Preisach* type should result (see [HJ93a]). Rigorous proofs have not yet been given for nonlinear adsorption processes of the kind mentioned in this section.

Two-Scale Convergence

A remark about to the mathematics needed for adsorption problems is in order here. In most homogenization problems, the ε-dependent functions, such as u^ε, are defined in a domain Ω or certain subdomains, e.g., in $\mathfrak{P}^\varepsilon$. Here, we have a different situation, in that the adsorbed concentration U^ε is defined on ε-dependent surfaces Γ^ε. In other words, their domains of definition are varying sets of codimension one. The difficulty is how to establish limits of such functions appropriately. In principle, there are two possibilities: (a) One, first, extends the functions u^ε or U^ε to the whole domain Ω and, then, studies limits in spaces of functions defined on Ω. This method, e.g., has been used for derving the Darcy law from the Stokes law and for variants thereof (see Chapters 3 and 4). Or (b), one uses the notion of *two-scale convergence*. Here the underlying idea is that the *two-scale limit U* of U^ε is no longer a function only of the global variable x, but also of the local variable y. This means that the limit $U(x, y)$ is defined for $x \in \Omega$ and $y \in \Gamma$. In [ADH95] the tools described in Appendix A.3 have been adapted to this case (see also [Rad92]).

6.2.3   Chromatography

An important process in chemical engineering is known as *chromatography*. Certain chemical species are dissolved in a fluid flowing through a porous medium. These chemical substances are transported by convection and diffusion and they penetrate into the porous matrix, where they may be absorbed by the solid material. If one assumes that the species under consideration may diffuse within the interior of the solid matrix, one comes to models that were studied theoretically in [Vog82] and experimentally in [SS88]. A nonlinear problem of this type is investigated in [CJ].

The equations that we start from here are again the Stokes equations describing fluid flow in the pore space, as in (2.1). As before, diffusion, convection, and reaction are modeled by Eq. (2.2). It is reasonable to assume that the diffusion of the substances in the interior of the porous matrix is small compared to the pore volume. Therefore,

$$\partial_t U^\varepsilon(t, x) = \varepsilon^2 \nabla \cdot (a_{\mathfrak{G}}^\varepsilon(x) \nabla U^\varepsilon(t, x)) + R_{\mathfrak{G}}(U^\varepsilon(t, x)), \ x \in \mathfrak{G}^\varepsilon \qquad (2.12)$$

for the concentration in the interior of the solid matrix. In the model we have in mind, we do not allow the particle surfaces to carry any of the substances, in contrast to the adsorption model studied before. This leads to the continuity equation,

$$a_{\mathfrak{P}}^\varepsilon(x)\vec{v} \cdot \nabla u^\varepsilon(t, x) = \varepsilon^2 a_{\mathfrak{G}}^\varepsilon(x)\vec{v} \cdot \nabla U^\varepsilon(t, x), \ x \in \Gamma^\varepsilon. \qquad (2.13)$$

This equation, which describes continuity of mass, has to be coupled with another equation that specifies either a relation between the concentrations u^ε and U^ε on Γ^ε or the actual normal mass flow taking place at the interface Γ. One reasonable choice is to assume continuity of the concentrations

$$u^\varepsilon(x) = U^\varepsilon(x), \ x \in \Gamma^\varepsilon; \qquad (2.14)$$

but the reader should keep in mind that this is a choice the modeler has to make. The microsystem consisting of equations (2.1), (2.2), (2.12), (2.13), and (2.14) has the following macrosystem as its homogenized limit.

$$
\begin{cases}
\vec{v}(x) = -\dfrac{1}{\mu}K\nabla p(x), & x \in \Omega, \\[4pt]
\nabla \cdot \vec{v}(x) = 0, & x \in \Omega, \\[4pt]
|\mathfrak{P}|\partial_t u(t, x) + Q(t, x) \\
\quad = \nabla \cdot (A(x)\nabla u(t, x) - \vec{v}(x)u(t, x)) \\
\quad + |\mathfrak{P}|R_\mathfrak{P}(u(t, x)), & x \in \Omega, \\[4pt]
Q(t, x) = \int_\Gamma a_\mathfrak{S}(y)\vec{v} \cdot \nabla_y U(t, x, y)\, d\Gamma(y), & x \in \Omega, \\[4pt]
\partial_t U(t, x, y) \\
\quad = \nabla_y \cdot (a_\mathfrak{S}(x, y)\nabla_y U(t, x, y)) \\
\quad + R_\mathfrak{S}(x, y, U(t, x, y)), & x \in \Omega, \ y \in \mathfrak{S}, \\[4pt]
U(t, x, y) = u(t, x), & x \in \Omega, \ y \in \Gamma.
\end{cases}
\tag{2.15}
$$

This macromodel is similar to the one derived in sections 1.5 and 3.4.

Convolution

In a manner similar to the adsorption problem one can also give a convolution formulation for the macromodel derived here. What we need is the time-dependent solution r of the following cell problem:

$$
\begin{cases}
\partial_t r(t, x, y) \\
\quad = \nabla_y \cdot (a_\mathfrak{S}(x, y)\nabla_y r(t, x, y)), & t > 0, \ x \in \Omega, \ y \in \mathfrak{S}, \\[4pt]
r(t, x, y) = 0, & t > 0, \ x \in \Omega, \ y \in \Gamma, \\[4pt]
r(t, x, y) = 1, & t = 0, \ x \in \Omega, \ y \in \mathfrak{S}.
\end{cases}
\tag{2.16}
$$

Under the assumption that the initial values of U do not depend on y and that there is no reaction taking place in $\mathfrak{S}$, i.e., $R_\mathfrak{S} = 0$, we arrive at a formula that is slightly different from the one obtained before, namely,

$$
U(t, x, y) = u(0, x)r(t, x, y) - \int_0^t \partial_t r(t - \tau, x, y)u(\tau, x)\, d\tau.
$$

Now, the sink term can be expressed as

$$
\begin{aligned}
Q(t, x) &= \int_\Gamma a_\mathfrak{S}(x, y)\vec{v} \cdot \nabla_y U(t, x, y)\, d\Gamma(y) \\
&= \int_\mathfrak{S} \nabla_y \cdot (a_\mathfrak{S}(x, y)\nabla_y U(t, x, y))\, dy \\
&= \partial_t \int_\mathfrak{S} U(t, x, y)\, dy.
\end{aligned}
$$

Therefore, assuming vanishing initital conditions,

$$
Q(t, x) = -\partial_t(\partial_t \rho(t, x) * u(t, x)),
$$

where the kernel ρ is given by

$$
\rho(\tau, x) = \int_\mathfrak{S} r(\tau, x, y)\, dy.
$$

Formula 2.16 yields a recipe for calculating the kernel ρ.

Proposition 2.1 *Let $x \in \Omega$ be fixed and ω_k be eigenfunctions with eigenvalues λ_k of the boundary-value problem,*

$$\begin{cases} -\nabla_y \cdot (a_{\mathfrak{S}}(x, y)\nabla_y \omega_k(y)) = \lambda_k \omega_k(y), & y \in \mathfrak{S}, \\ \omega_k(y) = 0, & y \in \Gamma, \end{cases} \qquad (2.17)$$

such that the relation

$$1 = \sum_{k=1}^{\infty} a_k \omega_k(y)$$

holds for $y \in \mathfrak{S}$. Then, the solution r of the problem 2.16 is given by

$$r(t, x, y) = \sum_{k=1}^{\infty} a_k e^{-\lambda_k t} \omega_k(y)$$

and the kernel ρ is given by

$$\rho(t, x) = \sum_{k=1}^{\infty} a_k e^{-\lambda_k t} \int_{\mathfrak{S}} \omega_k(y)\, dy.$$

Examples

Proposition 2.2 *If $\mathfrak{B}$ is an interval in $\mathbb{R}$ of length R and $a_{\mathfrak{S}}$ is a constant, then, the kernel ρ is given by*

$$\rho(\tau) = \frac{8R}{\pi^2} \sum_{k=0}^{\infty} \frac{1}{(2k+1)^2} e^{-(\frac{(2k+1)\pi}{R})^2 a_{\mathfrak{S}} \tau}.$$

Proof For $\mathfrak{B} = [0, R]$, eigenfunctions are

$$\omega_k(y) = \sin(k\pi \frac{y}{R})$$

with eigenvalues $\lambda_k = a_{\mathfrak{S}}(\frac{k\pi}{R})^2$,

$$1 = \frac{4}{\pi} \sum_{k=0}^{\infty} \frac{1}{2k+1} \sin(2k+1)\pi \frac{y}{R}$$

for $0 < y < R$, and

$$\int_0^R \omega_{2k+1}(y)\, dy = \frac{2R}{\pi(2k+1)}.$$

Q.E.D.

Proposition 2.3 *If $\mathfrak{B}$ is a disk in $\mathbb{R}^2$ of radius R and $a_{\mathfrak{S}}$ is a constant, then, the kernel ρ is given by*

$$\rho(\tau) = 4\pi R^2 \sum_{k=1}^{\infty} \frac{1}{j_k^2} e^{-(\frac{j_k}{R})^2 a_{\mathfrak{S}} \tau}.$$

Proof If $|y|$ denotes the distance from the origin, then, eigenfunctions are given by

$$\omega_k(y) = J_0(j_k \frac{r}{R})$$

(with the Bessel function of zero order J_0 and its kth zero J_k) with eigenvalues $\lambda_k = (\frac{j_k}{R})^2 a_\mathfrak{S}$,

$$1 = -2 \sum_{k=1}^{\infty} \frac{1}{j_k J_0'(j_k)} J_0(j_k \frac{r}{R}),$$

for $0 \le |y| < R$, and

$$\int_B \omega_k(y) \, dy = -\frac{2\pi R^2 J_0'(j_k)}{j_k}.$$

Q.E.D.

Proposition 2.4 *If $\mathfrak{B}$ is a ball in $\mathbb{R}^3$ of radius R and $a_\mathfrak{S}$ is a constant, then, the kernel ρ is given by*

$$\rho(\tau) = \frac{8R^3}{\pi} \sum_{k=1}^{\infty} \frac{1}{k^2} e^{-(\frac{k\pi}{R})^2 a_\mathfrak{S} \tau}.$$

Proof If $|y|$ denotes the distance from the origin, then, eigenfunctions are given by

$$\omega_k(y) = \frac{\sin k\pi \frac{|y|}{R}}{k\pi \frac{|y|}{R}}$$

with eigenvalues $\lambda_k = (\frac{k\pi}{R})^2 a_\mathfrak{S}$,

$$1 = 2 \sum_{k=1}^{\infty} (-1)^{k+1} \frac{\sin k\pi \frac{|y|}{R}}{k\pi \frac{|y|}{R}},$$

for $0 \le |y| < R$, and

$$\int_B \omega_k(y) \, dy = \frac{(-1)^{k+1} 4R^3}{k^2 \pi}.$$

Q.E.D.

In Fig. 6.1 the derivatives σ of the normalized kernels $\frac{\rho}{|\mathfrak{S}|}$ from the last three propositions are shown for $a_\mathfrak{S} = 1$. The numbers at the curves indicate the dimension of the set $\mathfrak{S}$. The reader should notice that these kernels differ only very slightly quantitatively, but not qualitatively. Therefore, one may say that the shape of $\mathfrak{S}$ does not strongly influence the form of the kernel ρ.

Robin Transmission Conditions

If we replace the transmission condition (2.14) by a *Robin* type condition of the form

$$\varepsilon d^\varepsilon(x) a_\mathfrak{S}^\varepsilon(x) \vec{v} \cdot \nabla U^\varepsilon(t, x) = c_\mathfrak{P} u^\varepsilon(t, x) - U^\varepsilon(t, x), \quad t > 0, \ x \in \Gamma^\varepsilon, \qquad (2.18)$$

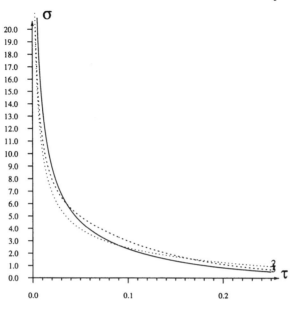

FIGURE 6.1. Normalized kernels.

similar to section 1.5, with a constant $c_{\mathfrak{P}} \neq 0$, the boundary condition on Γ in (2.15) has to be replaced by

$$d(y)a_{\mathfrak{S}}(y)\vec{v} \cdot \nabla_y U(t, x, y) = c_{\mathfrak{P}}u(t, x) - U(t, x, y), \quad t > 0, \ x \in \Omega, \ y \in \Gamma. \tag{2.19}$$

To derive a convolution formula, we now have to use the time-dependent solution r of the cell problem

$$\begin{cases} \partial_t r(t, x, y) = \nabla_y \cdot (a_{\mathfrak{S}}(x, y)\nabla_y r(t, x, y)), & t > 0, \ x \in \Omega, \ y \in \mathfrak{S}, \\ d(y)a_{\mathfrak{S}}(y)\vec{v} \cdot \nabla_y r(t, x, y) = -r(t, x, y), & t > 0, \ x \in \Omega, \ y \in \Gamma, \\ r(t, x, y) = c_{\mathfrak{P}}, & t = 0, \ x \in \Omega, \ y \in \mathfrak{S}. \end{cases} \tag{2.20}$$

In this situation, as before,

$$U(t, x, y) = u(0, x)r(t, x, y) - \int_0^t \partial_t r(t - \tau, x, y)u(\tau, x) \, d\tau,$$

and, therefore, assuming vanishing initial values,

$$Q(t, x) = -\partial_t(\partial_t \rho(t, x) * u(t, x))$$

with

$$\rho(\tau, x) = \int_{\mathfrak{S}} r(\tau, x, y) \, dy.$$

Obviously, by letting $d \equiv 0$ and $c_{\mathfrak{P}} = 1$, one gets back the formulas for the previous case of Dirichlet transmission conditions. An approach similar to Proposition 2.1 is possible here as well.

6.2.4   Semipermeable Membranes

In the previous sections, we have considered only linear transmission conditions. Formally speaking, the Robin condition (2.18),

$$\varepsilon a_{\mathbb{G}}^{\varepsilon}(x)\vec{v} \cdot \nabla U^{\varepsilon}(t, x) = g^{\varepsilon}(x, u^{\varepsilon}(t, x), U^{\varepsilon}(t, x)), \ \ t > 0, \ x \in \Gamma^{\varepsilon},$$

with

$$g(x, y, u, U) = c_{\mathfrak{P}}(x, y)u - c_{\mathbb{G}}(x, y)U,$$

can be replaced by one with a nonlinear function of the form

$$g^{\varepsilon}(x, u, U) = g(x, \frac{x}{\varepsilon}, u, U),$$

where g is Y-periodic with respect to the variable y. One expects that, in the homogenized limit, the transmission condition becomes a boundary condition of the form

$$a_{\mathbb{G}}(y)\vec{v} \cdot \nabla_y U(t, x, y) = g(x, y, u(t, x), U(t, x, y)), \ \ t > 0, \ x \in \Omega, \ y \in \Gamma.$$

This is true, indeed, if, e.g., the function g has certain monotonicity properties. In the paper [HJM94], this has been proven for an even more general case, namely, when the term g is a monotone graph in terms of the difference of u and U for fixed x and y. A situation where this is relevant occurs when the fluid-solid interface consists of semipermeable membranes.

Let us explore the situation that we want to explain by the following family of approximating problems. We assume that the function g depends only on the difference $s = u - U$ and not on x or y, i.e., $g(x, y, u, U) = g(s) = g(u - U)$. For $\delta > 0$,

$$g_{\delta}(s) = \begin{cases} \frac{s}{\delta}, & s \geq 0, \\ 0, & s < 0. \end{cases}$$

For such a function g_{δ}, instead of g, we have no normal flux on Γ^{ε} as long as $u - U < 0$, i.e., if the interior concentration is larger than the exterior concentration. But if $u - U > 0$, the normal flux is very large. Indeed, it is larger, the smaller the parameter δ becomes. The graph of this function as a function of s is the set $M_{\delta} \subset \mathbb{R}^2$ defined by

$$M_{\delta} = \{(s, q) : \ q = g_{\delta}(s)\}.$$

In the limit $\delta \searrow 0$, these sets become

$$M = \{(s, q) : \ s \leq 0 \text{ and } q \geq 0 \text{ and } sq = 0\}$$

(see Fig. 6.2; there values $\delta = 1, 10$, and 100 are chosen). Using this limit set M, we formulate our transmission condition for the micromodel with a semipermeable membrane as

$$(u^{\varepsilon}(t, x) - U^{\varepsilon}(t, x), \varepsilon a_{\mathbb{G}}^{\varepsilon}(x)\vec{v} \cdot \nabla U^{\varepsilon}(t, x)) \in M.$$

The interpretation of such a condition is the following. There are basically two different possibilities.

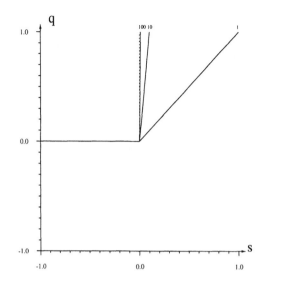

FIGURE 6.2. Family of monotone graphs.

- $u^\varepsilon < U^\varepsilon$ on Γ^ε. Then, no normal mass flux may take place.

- $u^\varepsilon = U^\varepsilon$ on Γ^ε. Then, a positive mass flux may occur on Γ^ε toward $\mathfrak{S}^\varepsilon$.

No other combination is allowed in this situation. A *unilateral* transmission condition of this form is called a *Signorini* condition. The paper [HJM94] has shown that the homogenized limit of a Signorini condition of this type is expressed as

$$(u(t, x) - U(t, x, y), a_{\mathfrak{S}}(y)\vec{v} \cdot \nabla_y U(t, x, y)) \in M.$$

The unilateral condition presented here is not the only one that one may consider. If the orientation of the semipermeable membrane is reversed, then, one uses the set,

$$M = \{(s, q) : \ s \geq 0 \text{ and } q \leq 0 \text{ and } sq = 0\},$$

instead. In this case, the substance is allowed to flow only from the solid part $\mathfrak{S}^\varepsilon$ to the fluid part $\mathfrak{P}^\varepsilon$. Another possibility is to consider membranes that tolerate a normal flux only if the difference of the concentrations on both sides of the membrane exceeds certain thresholds.

It should be mentioned that the mathematics of the homogenization proof of these nonlinear problems is based on their formulation as *variational inequalities*.

A problem related to this type of problem was treated in [FK92].

6.3   Upscaling from the Meso- to the Macroscale

6.3.1   Mobile and Immobile Water

Hydrologists very often assume that soil water does not move at the same rate in the
different domains of a porous medium. Because most soils possess pores of highly
varying size because the pores, in general, have a tortuous geometric structure,
as a crude approximation, one may conceive that a destinction between mobile
and immobile water is reasonable. Here we are going to explain how models
for miscible displacement on a mesoscale may be set up, and how one can use
homogenization to derive corresponding models on the macroscale.

We denote the part of the soil governed by *mobile water* by $\mathfrak{F}^{\varepsilon}$ and the part with
immmobile water by $\mathfrak{B}^{\varepsilon}$. Because we are starting from models on a mesoscale,
we assume that the mobile water flow is governed by Darcy's law. For one-phase
saturated flow, we, then, have equations of the form,

$$\vec{v}^{\varepsilon}(x) = -\frac{k^{\varepsilon}(x)}{\mu}\nabla p^{\varepsilon}(x), \ \nabla \cdot \vec{v}^{\varepsilon}(x) = 0, \ x \in \mathfrak{F}^{\varepsilon},$$

where $\vec{v}^{\varepsilon}$ is the Darcy velocity and p^{ε} is the pressure in the water. We are ignoring
gravitational effects here, but these may be added in a standard fashion. If u^{ε} is
the concentration of a chemical substance dissolved in the water,

$$\partial_t(\vartheta_{\mathfrak{F}}^{\varepsilon}(x)u^{\varepsilon}(t, x)) = \nabla \cdot (a_{\mathfrak{F}}^{\varepsilon}(x)\nabla u^{\varepsilon}(t, x) - \vec{v}^{\varepsilon}(x)u^{\varepsilon}(t, x)), \ x \in \mathfrak{F}^{\varepsilon},$$

describes diffusion and convection in the domain $\mathfrak{F}^{\varepsilon}$, where $\vartheta_{\mathfrak{F}}^{\varepsilon}$ is the saturation
of the mobile water, defined as the volume fraction of water per bulk volume. We
are going to assume that the saturation $\vartheta_{\mathfrak{F}}^{\varepsilon}$ does not vary in time.

Uniform Concentration

Different kinds of assumptions are possible for the regime of immobile water.
The simplest is not to consider any spatial variations in the interior of the immobile regime. Thus, we assume that the domain $\mathfrak{B}^{\varepsilon}$ of immobile water consists of
small disconnected subdomains in each of which the concentration of the chemical
substance under consideration U^{ε} is constant.

Linearizing the exchange of the substance between mobile and immobile water,
the boundary condition for the transport in the mobile water is given by

$$a_{\mathfrak{F}}^{\varepsilon}(x)\vec{v} \cdot \nabla u^{\varepsilon}(t, x) = \varepsilon(\alpha^{\varepsilon}(x)u^{\varepsilon}(t, x) - \gamma^{\varepsilon}(x)U^{\varepsilon}(t, x)), \ x \in \Gamma^{\varepsilon},$$

where $\vec{v}$ is the normal on the interface Γ^{ε} between the mobile domain $\mathfrak{F}^{\varepsilon}$ and the
immobile domain $\mathfrak{B}^{\varepsilon}$ directed into $\mathfrak{F}^{\varepsilon}$. Allowing no spatial variation of $U^{\varepsilon}(t, x)$
in each connected component of $\mathfrak{B}^{\varepsilon}$, we get a kinetic equation of the form,

$$\partial_t U^{\varepsilon}(t, x) = \int_{\Gamma^{\varepsilon}(x)} (\alpha^{\varepsilon}(x)u^{\varepsilon}(t, x) - \gamma^{\varepsilon}(x)U^{\varepsilon}(t, x)) \, d\Gamma(y), \ x \in \mathfrak{B}^{\varepsilon},$$

where $\Gamma^\varepsilon(x)$ is boundary of that component of $\mathfrak{B}^\varepsilon$ to which x belongs. Using the mathematical techniques described in sections 1.4 and 1.5, one gets the following homogenized system. For the water flow,

$$\vec{v}(x) = -\frac{K(x)}{\mu}\nabla p(x), \ \nabla \cdot \vec{v}(x) = 0, \ x \in \Omega,$$

where the tensor K is obtained from k according to formula (4.6) from Chapter 1. The transport of the chemical is governed by

$$\partial_t(\Theta_{\mathfrak{F}}(x)u(t, x)) + Q(t, x) = \nabla \cdot (A(x)\nabla u(t, x) - \vec{v}(x)u(t, x)), \ x \in \Omega,$$

where $\Theta_{\mathfrak{F}}(x) = |\mathfrak{F}|\vartheta_{\mathfrak{F}}(x)$. The homogenized concentration in the immobile water satisfies

$$\partial_t U(t, x) = \int_\Gamma (\alpha(y)u(t, x) - \gamma(y)U(t, x)) \, d\Gamma(y) = \bar{\alpha}u(t, x) - \bar{\gamma}U(t, x), \ x \in \Omega,$$

with

$$\bar{\alpha} = \int_\Gamma \alpha(y) \, d\Gamma(y)$$

and

$$\bar{\gamma} = \int_\Gamma \gamma(y) \, d\Gamma(y).$$

The sink term is $Q(t, x) = \partial_t U(t, x)$. This model is similar to the *first-order kinetic model* already encountered in section 6.2.2. There we saw how to use time convolution to rewrite the equations as integro-differential equations. Miscible displacement in mobile-immobile systems was studied in [GW76]. An approach leading to a parallel flow model (see section 9.2) was considered in [GG93b] and [GG93c].

Diffusion within Immobile Water

Now we are going to modify the model within the domain of immobile water. Instead of assuming a uniform concentration, we now allow spatial variations there, and assume that the solute undergoes diffusion. Convection does not take place because we assumed that the water is at rest. Obviously, the previous model can be looked at as a limit case where the diffusion within the immobile regime is so large that spatial equilibrium is instantaneously attained.

We get an equation of the form,

$$\partial_t(\vartheta_{\mathfrak{B}}^\varepsilon(x)U^\varepsilon(t, x)) = \varepsilon^2\nabla \cdot (a_{\mathfrak{B}}^\varepsilon(x)\nabla U^\varepsilon(t, x)), \ x \in \mathfrak{B}^\varepsilon. \quad (3.1)$$

Conservation of mass across the interface Γ^ε gives

$$a_{\mathfrak{F}}^\varepsilon(x)\vec{v} \cdot \nabla u^\varepsilon(t, x) = \varepsilon^2 a_{\mathfrak{B}}^\varepsilon(x)\vec{v} \cdot \nabla U^\varepsilon(t, x), \ x \in \Gamma^\varepsilon. \quad (3.2)$$

One natural assumption at the interface Γ^ε is continuity of concentrations, i.e.,

$$u^\varepsilon(t, x) = U^\varepsilon(t, x), \ x \in \Gamma^\varepsilon.$$

FIGURE 6.3. Structured soil (courtesy H. Gerke).

This model is practically the same as the one studied in section 1.5. Thus, it is clear that we get the following macromodel.

$$\begin{cases} \partial_t(\Theta_{\mathfrak{F}}(x)u(t,x)) + Q(t,x) \\ \quad = \nabla \cdot (A(x)\nabla u(t,x) - \bar{v}(x)u(t,x)), & x \in \Omega, \\ Q(t,x) = \int_\Gamma \bar{v} \cdot a_{\mathfrak{B}}(y)\nabla U(t,x,y)\, d\Gamma(y), & x \in \Omega, \\ \partial_t(\vartheta_{\mathfrak{B}}(y)U(t,x,y)) = \nabla_y \cdot (a_{\mathfrak{B}}(y)\nabla_y U(t,x,y)), & x \in \Omega, \; y \in \mathfrak{B}, \\ u(t,x) = U(t,x,y), & x \in \Omega, \; y \in \Gamma. \end{cases}$$

This model is similar to the model that we have already encountered in section 6.2.3; there we saw how to use time convolution to rewrite the equations as integro-differential equations. This type of problem was studied in [Hor91b] which gives numerical results for one-dimensional infiltration experiments that lead to *break-through curves*. There comparisons are also given with this system and the standard approach used in soil science.

6.3.2 Fractured Media

Fractured or *aggregated* media are considered to consist of blocks with large storage volume but low permeability which are separated by fractures or fissures with low storage volume but high permeability. Models for miscible displacement in such media are very similar to those for mobile and immobile water. The difference is that, in the interior of the blocks, convection takes place (see Fig. 6.3).

First, we will describe the micromodel. We have to distinguish between two water velocities and pressures, namely, in the system of fractures $\mathfrak{F}^\varepsilon$ and in the system of blocks $\mathfrak{B}^\varepsilon$.

$$\begin{cases} \vec{v}_{\mathfrak{F}}^\varepsilon(x) = -\frac{k_{\mathfrak{F}}^\varepsilon(x)}{\mu}\nabla p_{\mathfrak{F}}^\varepsilon(x), \; \nabla \cdot \vec{v}_{\mathfrak{F}}^\varepsilon(x) = 0, & x \in \mathfrak{F}^\varepsilon, \\ \vec{v}_{\mathfrak{B}}^\varepsilon(x) = -\varepsilon^2\frac{k_{\mathfrak{B}}^\varepsilon(x)}{\mu}\nabla p_{\mathfrak{B}}^\varepsilon(x), \; \nabla \cdot \vec{v}_{\mathfrak{B}}^\varepsilon(x) = 0, & x \in \mathfrak{B}^\varepsilon, \\ k_{\mathfrak{F}}^\varepsilon(x)\vec{v} \cdot \nabla p_{\mathfrak{F}}^\varepsilon(x) = \varepsilon^2 k_{\mathfrak{B}}^\varepsilon(x)\vec{v} \cdot \nabla p_{\mathfrak{B}}^\varepsilon(x), & x \in \Gamma^\varepsilon, \\ p_{\mathfrak{F}}^\varepsilon(x) = p_{\mathfrak{B}}^\varepsilon(x), & x \in \Gamma^\varepsilon. \end{cases}$$

The system for the concentrations has the following form:

$$\begin{cases} \partial_t(\vartheta_{\mathfrak{F}}^\varepsilon(x)u^\varepsilon(t, x)) = \nabla \cdot (a_{\mathfrak{F}}^\varepsilon(x)\nabla u^\varepsilon(t, x) - \vec{v}_{\mathfrak{F}}^\varepsilon(x)u^\varepsilon(t, x)), & x \in \mathfrak{F}^\varepsilon, \\ \partial_t(\vartheta_{\mathfrak{B}}^\varepsilon(x)U^\varepsilon(t, x)) = \nabla \cdot (\varepsilon^2 a_{\mathfrak{B}}^\varepsilon(x)\nabla U^\varepsilon(t, x) - \varepsilon \vec{v}_{\mathfrak{B}}^\varepsilon(x)U^\varepsilon(t, x)), & x \in \mathfrak{B}^\varepsilon, \\ a_{\mathfrak{F}}^\varepsilon(x)\vec{v} \cdot \nabla u^\varepsilon(t, x) = \vec{v} \cdot (\varepsilon^2 a_{\mathfrak{B}}^\varepsilon(x)\nabla U^\varepsilon(t, x) - \varepsilon \vec{v}_{\mathfrak{B}}^\varepsilon(x)U^\varepsilon(t, x)), & x \in \Gamma^\varepsilon, \\ u^\varepsilon(t, x) = U^\varepsilon(t, x), & x \in \Gamma^\varepsilon. \end{cases}$$

Now it is clear how to homogenize this system with the following result. We get for the water flow similar to the result in Chapter 3:

$$\begin{cases} \vec{v}_{\mathfrak{F}}(x) = -\frac{K(x)}{\mu}\nabla p_{\mathfrak{F}}(x), \; B(x) = -\nabla \cdot \vec{v}_{\mathfrak{F}}(x), & x \in \Omega, \\ B(x) = -\int_\Gamma \vec{v} \cdot \vec{v}_{\mathfrak{B}}(x, y) \, d\Gamma(y), & x \in \Omega, \\ \vec{v}_{\mathfrak{B}}(x, y) = -\frac{k_{\mathfrak{B}}}{\mu}\nabla_y p_{\mathfrak{B}}(x, y), \; \nabla_y \cdot \vec{v}_{\mathfrak{B}}(x, y) = 0, & x \in \Omega, \; y \in \mathfrak{B}, \\ p_{\mathfrak{F}}(x) = p_{\mathfrak{B}}(x, y), & x \in \Omega, \; y \in \Gamma. \end{cases}$$

For the solute, the result is the following system:

$$\begin{cases} \partial_t(\Theta_{\mathfrak{F}}(x)u(t, x)) + Q(t, x) \\ \quad = \nabla \cdot (A(x)\nabla u(t, x) - \vec{v}_{\mathfrak{F}}(x)u(t, x)), & x \in \Omega, \\ Q(t, x) \\ \quad = \int_\Gamma \vec{v} \cdot (a_{\mathfrak{B}}(y)\nabla_y U(t, x, y) - \vec{v}_{\mathfrak{B}}(x)U(t, x, y)) \, d\Gamma(y), & x \in \Omega, \\ \partial_t(\vartheta_{\mathfrak{B}}(y)U(t, x, y)) \\ \quad = \nabla_y \cdot (a_{\mathfrak{B}}(y)\nabla_y U(t, x, y) - \vec{v}_{\mathfrak{B}}(x, y)U(t, x, y)), & x \in \Omega, \; y \in \mathfrak{B}, \\ u(t, x) = U(t, x, y), & x \in \Omega, \; y \in \Gamma. \end{cases}$$

Models of this kind are called *double-porosity* or *dual-porosity models* (see section 3.4 and Chapters 9 and 10). The papers [Hor91a] and [GG93a] deal with a generalization of such models to unsaturated flow. A miscible displacement problem in two-phase flow is investigated in [Mik89].

6.4   Discussion

The field of miscible displacement is one of lively research. Phenomena, such as adsorption, absorption, chromatography, etc., are of great importance in soil physics and chemistry, in chemical engineering, and in the environmental sciences. Until now, all kinds of competing models have been used and applied to practical problems. There is absolutely no doubt that the final question which model is best can be answered only by experimental tests, either in the laboratory or in the field. The value of homogenization in this area is that it allows one to clearly distinguish between the type of model one starts with on the microscale and those that can be derived on the macroscale. Homogenization also sheds light on which meaning variables have on both scales. In this sense, homogenization points out the assumptions on which a specific macromodel is based.

The price paid for homogenization making the "derivation" of macromodels mathematically rigorous is that often, there are many choices possible for scaling micromodels. For example, the factor ε in Eqs. (2.5) or (2.18) is based on the scaling of the total area of surfaces Γ^ε which is like $\frac{1}{\varepsilon}$. Different scalings of the interfacial or transmission conditions on the microscale would lead to different macromodels, in all cases not reasonable ones. The same observation can be made for the factor ε^2 used in Eqs. (3.1) and (3.2) which is chosen to account for the large contrast between diffusivities in the fractures and blocks. Dropping this factor would lead to an averaged, standard diffusion-convection equation. Therefore, it has to be emphasized that homogenization can only make the proof of a statement *if ... then ...* rigorous. The validity of the *if ...* part is not a matter of mathematical investigation, but of the experimental sciences. One cannot overemphasize this last statement, because we are far from understanding all relevant physical, chemical, and biological processes taking place in porous media. On the pore scale, in particular, the interplay between solid and fluid mechanics and electrical and chemical forces has still to be investigated in more detail. Here, homogenization is a powerful tool which supports the modeling process wherever spatial upscaling is necessary.

7

Thermal Flow

Horia Ene

7.1 Introduction

Porous materials, such as sand and crushed rock underground, are saturated with water which, under a local pressure gradient, migrates and transports energy through the material. The transport of heat in the presence of exterior forces, if the fluid is in motion, is called convection. Further examples of convection through porous media may be formed in man-made systems, such as fiber and granular insulations, the cores of nuclear reactors, or chemical industries.

The convective motion arising when a Darcy-type fluid is in contact with a source of heat is also of great importance for practical problems. An example is the problem of storage tanks, where the decrease of convective and radiative transfer of heat may be obtained as a function of the insulating characteristics of the saturated porous materials used.

This chapter deals with thermal flow in porous media. We assume that we have single-phase flow. The motion of such a fluid is described by Darcy's law (see Chapters 1 and 3). The study of single-phase flow reveals the basic mechanism governing the phenomenon. For multiphase flow, we refer readers to the book [EP87].

The basic equations will be presented in section 7.2 of this chapter. Of course, the fundamental equations were obtained by using the homogenization method (see [EP87], [ESP82], [MP92], [MP94], and [Tor87]).

The Rayleigh number plays an important role in the study of thermal flow in porous media. Consequently, we present the problem at low values of the Rayleigh number, for bounded and unbounded domains, in sections 7.3 and 7.4. The problem of forced convection is presented in section 7.5, and the case of high values of the Rayleigh number is treated in section 7.6. In all cases, the examples presented are representative for the methods used or for the physical significance of different parameters.

7.2 Basic Equations

Instead of explaining the homogenization method for the energy equation in a porous medium, we will simply present the results. The details are given in [EP87], [ESP81], and [ESP82].

At the microscale (the pore level), there are two equations: the thermo-convective equation in the fluid part and the conduction equation in the solid part. To obtain a one-temperature equation at the macroscale, it is necessary to use the same hypothesis. First of all, we assume that the thermal conductivities of the fluid and of the solid are of the same order of magnitude. Secondly, we need the hypothesis of a Péclet number of order 1, which simply means that convection is important.

For other cases, e.g., the model with two-temperature equations at the macroscale, we refer to [Ene96].

Using the same type of asymptotic expansions as in Chapter 1, we obtain the macroscopic equation for temperature in the form,

$$(\rho c)^{\sim}\frac{\partial T}{\partial t} + \rho_f c_f v_k \frac{\partial T}{\partial x_k} = \frac{\partial}{\partial x_k}(\chi_{kj}\frac{\partial T}{\partial x_j}) + \mu(K^{-1})_{kj}v_j v_k, \tag{2.1}$$

where $\vec{v}$ is the velocity vector given by Darcy's law, K_{ij} is the permeability tensor, χ_{kj} is the macroscopic thermal conductivity of the solid-fluid mixture (with an expression similar to the homogenized coefficients of Chapter 1), ρ_f is the density of the fluid, c_f the specific heat capacity of the fluid, μ the dynamic viscosity, and $(\rho c)^{\sim} = n(\rho_f c_f) + (1 - n)(\rho_s c_s)$, with porosity n .

The second term in the right-hand side of Eq. (2.1) is viscous dissipation. In the case of natural convection in porous media, viscous dissipation is negligibly small [ESP81], and, consequently,

$$(\rho c)^{\sim}\frac{\partial T}{\partial t} + \rho_f c_f v_k \frac{\partial T}{\partial x_k} = \frac{\partial}{\partial x_k}\left(\chi_{kj}\frac{\partial T}{\partial x_j}\right). \tag{2.2}$$

The complete system of equations, in vectorial form, can be written as

$$\nabla \cdot \vec{v} = 0, \tag{2.3}$$

$$\vec{v} = -\frac{k}{\mu}(\nabla p - \rho\vec{g}), \tag{2.4}$$

$$(\rho c)^{\sim}\frac{\partial T}{\partial t} + \rho_f c_f \vec{v} \cdot \nabla T = \chi\Delta T, \tag{2.5}$$

and

$$\rho = \rho_0[1 - \alpha(T - T_0)]. \tag{2.6}$$

Eq. (2.6) is the Boussinesq approximation proven by the homogenization method in [EP87] and [ESP81]. This equation says that the density of the Darcy-type fluid is constant everywhere except in the buoyancy force term $\rho\vec{g}$ causing the thermal flow.

In dimensionless form [EP87], the system takes the form,

$$\nabla \cdot \vec{v} = 0, \tag{2.7}$$

$$\vec{v} + \nabla p = \mathbf{Ra}T\vec{e}, \tag{2.8}$$

and

$$\frac{\partial T}{\partial t} + \vec{v} \cdot \nabla T = \Delta T, \tag{2.9}$$

where $\vec{e} = (0, 0, 1)$ and the Rayleigh number $\mathbf{Ra}$ will be defined later.

7.3   Natural Convection in a Bounded Domain

Supposing that the solid part of the porous medium is fixed, we can agree that the velocity of the fluid is far lower than the acoustic velocity and, thus, that motion has little effect on pressure. Therefore, we neglect variations of the thermodynamic quantities resulting from pressure changes. We also assume that the temperature differences are small enough to use the Boussinesq approximation: density in the gravitational force varies with the temperature. We adopt the Darcy–Boussinesq system in dimensionless form:

$$\nabla \cdot \vec{v} = 0 \quad \text{in } \Omega, \tag{3.1}$$

$$\vec{v} + \nabla p = \mathbf{Ra} T \vec{e} \quad \text{in } \Omega, \tag{3.2}$$

and

$$\frac{\partial T}{\partial t} + \vec{v} \cdot \nabla T = \Delta T \quad \text{in } \Omega, \tag{3.3}$$

where $\vec{e} = (0, 0, 1)$, $\vec{v}$ is Darcy's velocity, p is the pressure, T is the temperature, and the Rayleigh number

$$\mathbf{Ra} = \frac{k \rho_f c_f g L \alpha (T_1 - T_2)}{\nu \chi_m}. \tag{3.4}$$

Here, k is the permeability of the medium, $T_1 > T_2$ are two temperatures, α is the thermal expansion coefficient, χ_m is the thermal conductivity of the porous medium, ν is the kinematic viscosity of the fluid, L is a characteristic length of Ω, and c_f is the specific heat of the fluid.

The boundary and initial conditions are

$$\vec{v} \cdot \vec{n} = 0 \quad \text{on } \partial\Omega \times (0, \theta), \tag{3.5}$$

$$T = T_1 \quad \text{on } \partial_1\Omega \times (0, \theta), \tag{3.6}$$

$$T = T_2 \quad \text{on } \partial_2\Omega \times (0, \theta), \tag{3.7}$$

and

$$T(0) = T^0 \quad \text{in } \Omega, \tag{3.8}$$

Here, $\theta > 0$ is the time interval, and the problem may be considered in the cylinder $\Omega \times (0, \theta)$.

To study the problem, it is important to note some results obtained in [EP87].

First of all, uniqueness of the steady solutions holds for sufficiently low Rayleigh numbers.

Secondly, *this unique solution is analytic and regular with respect to the Rayleigh number.*

As a conclusion from these general results, it is possible to use the perturbation method for investigating the convection in a domain with two different uniform temperatures at the boundary $\partial\Omega = \partial_1\Omega \cup \partial_2\Omega$.

Concerning the nonsteady solution, it is important to note that *the solution exists and is unique* [EP87]. We also have global stability in the mean: *any perturbation of the initial value decreases exponentially with time, in the sense used by Serrin* [Ser59].

As an example of the general method described below, we choose the problem of natural convection in a fluid-saturated porous medium contained between two impervious concentric spheres. The spheres have radii R_1 and R_2 ($R_1 < R_2$) and are centred at O, the origin of the coordinate system $0xyz$. The negative $0z$ direction is the direction in which the gravitational force acts. If we set $q = R_2/R_1 > 1$, the domain Ω will be $r \in (1, q)$. The problem (3.1)–(3.7) in the steady-state case is

$$\nabla \cdot \vec{v} = 0, \ r \in (1, q),$$

$$\vec{v} + \nabla p = \mathbf{Ra}T\vec{e}, \ r \in (1, q),$$

$$\vec{v} \cdot \nabla T = \Delta T, \ r \in (1, q), \tag{3.9}$$

$$\vec{v} \cdot \vec{n} = 0, \ r = 1, \ r = q,$$

$$T = 1, \ r = 1,$$

and

$$T = 0, \ r = q,$$

the dimensionless temperature being $\frac{T-T_2}{T_1-T_2}$.

Changing to spherical polar coordinates (r, θ, φ), problem (3.9) becomes

$$\frac{\partial}{\partial r}(r^2 \sin\theta\, v_r) + \frac{\partial}{\partial\theta}(r \sin\theta\, v_\theta) + \frac{\partial}{\partial\varphi}(r v_\varphi) = 0,$$

$$v_r + \frac{\partial p}{\partial r} = \mathbf{Ra}T \cos\theta,$$

$$v_\theta + \frac{1}{r}\frac{\partial p}{\partial\theta} = -\mathbf{Ra}T \sin\theta,$$

$$v_\varphi + \frac{1}{r \sin\theta}\frac{\partial p}{\partial\varphi} = 0,$$

and

$$\frac{1}{r^2 \sin\theta}\left(\frac{\partial}{\partial r}\left(r^2 \sin\theta\frac{\partial T}{\partial r}\right) + \frac{\partial}{\partial\theta}\left(\sin\theta\frac{\partial T}{\partial\theta}\right) + \frac{\partial}{\partial\varphi}\left(\frac{1}{\sin\theta}\frac{\partial T}{\partial\varphi}\right)\right)$$
$$= v_r\frac{\partial T}{\partial v} + \frac{v_\theta}{r}\frac{\partial T}{\partial\theta} + \frac{v_\varphi}{r \sin\theta}\frac{\partial T}{\partial\varphi}, \tag{3.10}$$

with the boundary conditions

$$v_r = 0, \quad r = 1, q,$$

$$T = 1, \quad r = 1,$$

and

$$T = 0, \quad r = q.$$

We look for an asymptotic expansion of the form,

$$(\vec{v}, p, T) = \sum_{m \geq 0} (\vec{v}_m, p_m, T_m) \frac{\mathbf{Ra}^m}{m!}, \qquad (3.11)$$

where the **Ra**-independent functions $(\vec{v}_m, p_m, T_m)$ are assumed to satisfy the set of equations obtained by equating the coefficients of the different powers of **Ra** which occur in the formal expansion of the system (3.10). We see that (3.11) is the Taylor series at **Ra** $= 0$. Due to the fact that we have unique solutions, they are independent of φ in the present case. Then, it is possible to introduce the stream function Ψ given by

$$r^2 \sin \theta \, v_r = \frac{\partial \Psi}{\partial \theta}, \quad r \sin \theta \, v_\theta = -\frac{\partial \Psi}{\partial r}.$$

Eliminating p, system (3.10) becomes

$$\frac{1}{\sin \theta} \frac{\partial^2 \Psi}{\partial r^2} + \frac{1}{r^2} \frac{\partial}{\partial \theta} \left(\frac{1}{\sin \theta} \frac{\partial \Psi}{\partial \theta} \right) = \mathbf{Ra}(\cos \theta \frac{\partial T}{\partial \theta} + r \sin \theta \frac{\partial T}{\partial r}),$$

$$\sin \theta \frac{\partial}{\partial r} \left(r^2 \frac{\partial T}{\partial r} \right) + \frac{\partial}{\partial \theta} (\sin \theta \frac{\partial T}{\partial \theta}) = \frac{\partial T}{\partial r} \frac{\partial \Psi}{\partial \theta} - \frac{\partial T}{\partial \theta} \frac{\partial \Psi}{\partial r},$$

$$T = 1 \text{ and } \Psi = 0, \ r = 1,$$

$$T = 0 \text{ and } \Psi = 0, \ r = q.$$

Taking (Ψ, T) in the form

$$(\Psi, T) = \sum_{m \geq 0} (\Psi_m, T_m) \mathbf{Ra}^m,$$

successively,

$$\Psi_0 = 0, \qquad T_0 = \frac{1}{q - 1} (\frac{q}{r} - 1),$$

$$\Psi_1 = \frac{q}{2(q^3 - 1)} \left[-\frac{q^2}{r} + (q^2 + q + 1)r - (q + 1)r^2 \right] \sin^2 \theta,$$

and

$$T_1 = \frac{q^2 \cos \theta}{4(q^3 - 1)^2} \cdot \left[3qr - (2q^3 + 4q^2 + 4q + 2) + \right.$$

$$\left. + \frac{2q^4 + 4q^3 + 6q^2 + 4q + 2}{r} - \frac{3q(q^3 + q^2 + q + 1)}{r^2} \right.$$

$$\left. + \frac{q^2(q^2 + q + 1)}{r^3} \right].$$

7.4   Natural Convection in a Horizontal Porous Layer

We will try to explain the convective motion in a horizontal porous layer bounded by
two isothermal planes maintaining a temperature gradient opposite to the direction
of gravity. The Rayleigh number regime near the critical point will be investigated.
The finite-amplitude steady-state solutions will be obtained more precisely by a
general method due to Malkus and Veronis [MV58] and Gor'kov [Gor58]. This
method is a successive approximation, in which the small parameter is a measure
of the deviation from the static state. We use the Darcy–Boussinesq system in the
dimensionless form (3.1)–(3.3). Now, the domain Ω is $z \in (0, 1)$.

The boundary conditions are

$$w = 0, \ z = 0, z = 1, \tag{4.1}$$

$$T = 1, \ z = 0, \tag{4.2}$$

and

$$T = 0, \ z = 1, \tag{4.3}$$

where $\vec{v} = (u, v, w)$.

Any solution of (3.1)–(3.3), (4.1)–(4.3) with $\frac{\partial}{\partial t} = 0$ is called a steady-state
solution. Obviously, the static solution

$$\vec{v}_0 = 0$$

which corresponds to the pure-conduction pattern,

$$T_0 = 1 - z,$$

$$p_0 = \mathbf{Ra}(z - \frac{1}{2}z^2) + \text{constant},$$

is a steady-state solution. It is possible to prove that this solution is unique for
sufficiently low Rayleigh numbers.

Introducing

$$h = p - p_0$$

and

$$S = T - T_0,$$

the system becomes

$$\nabla \cdot \vec{v} = 0, \ z \in (0, 1), \tag{4.4}$$

$$\vec{v} + \nabla h = \mathbf{Ra} S \vec{e}, \ z \in (0, 1), \tag{4.5}$$

$$\vec{v} \cdot \nabla S = w + \Delta S, \ z \in (0, 1), \tag{4.6}$$

and

$$w = S = 0, \ z = 0, z = 1. \tag{4.7}$$

Note that S represents the difference between a steady-state solution occurring
after the critical point and the static-state / pure-conduction solution. We assume

that the amplitude of S is a small parameter, denoted by ε. Because we want to study the case of Rayleigh numbers close to the critical value, the solutions and the Rayleigh number may be approximated by the expansions,

$$\mathbf{Ra} = \mathbf{Ra}_0 + \varepsilon \mathbf{Ra}_1 + \varepsilon^2 \mathbf{Ra}_2 + \cdots, \tag{4.8}$$

$$\vec{v} = \varepsilon \vec{v}_1 + \varepsilon^2 \vec{v}_2 + \cdots, \tag{4.9}$$

$$h = \varepsilon h_1 + \varepsilon^2 h_2 + \cdots, \tag{4.10}$$

and

$$S = \varepsilon S_1 + \varepsilon^2 S_2 + \cdots. \tag{4.11}$$

The small parameter ε, defined as the amplitude of the eigenfunction S, is precisely

$$\varepsilon^2 = \sup_{(x,y) \in Y} \int_0^1 S^2 dz.$$

An important remark: because $\mathbf{Ra}$ is an externally given parameter, Eq. (4.8) implicitly defines ε.

The characteristic domain is $D = Y \times [0, 1]$, $Y \subseteq \mathbb{R}^2$. Usually the convective motion occurs in a periodic cell pattern, where a cell is defined by a closed boundary with vertical walls between $z = 0$ and $z = 1$, the bottom and top layers.

For any $n \geq 1$, the coefficient $\mathbf{Ra}_n$, $\vec{v}_n$, h_n, and S_n are formed by substituting (4.8)–(4.11) in (4.4)–(4.7), and by solving the characteristic-value problems obtained by collecting the terms with the same power of ε.

The first-order problem is given by

$$\nabla \cdot \vec{v}_1 = 0, \; z \in (0, 1),$$

$$\vec{v}_1 + \nabla h_1 = \mathbf{Ra}_0 S_1 \vec{e}, \; z \in (0, 1),$$

$$\Delta S_1 + w_1 = 0, \; z \in (0, 1),$$

and

$$S_1 = w_1 = 0, \; z = 0, \; z = 1.$$

By straightforward elimination,

$$\Delta^2 S_1 + \mathbf{Ra}_0 \Delta_{xy} S_1 = 0, \; \Delta_{xy} = \frac{\partial^2}{\partial x^2} + \frac{\partial^2}{\partial y^2}, \tag{4.12}$$

$$S_1 = \Delta S_1 = 0, \; z = 0, \; z = 1.$$

The problem now consists of finding the first eigenvalue of (4.12). Because of physical considerations, we must seek eigenfunctions of the form,

$$S_1(x, y, z) = F_1(x, y) G_1(z).$$

One finds that (4.12) is separable, if F_1 satisfies the equation for a vibrating membrane,

$$\Delta_{xy} F_1 + n^2 F_1 = 0. \tag{4.13}$$

Thus, the horizontal section Y of a cell in the convection pattern can have any shape compatible with (4.13).

The separation constant n corresponds to the principal mode of vibration of a membrane of that shape, and it will be determined explicitly.

If $D = \frac{d}{dz}$, the problem for G reduces to

$$D^4 G_1 - 2n^2 D^2 G_1 + (n^4 - \mathbf{Ra}_0 n^2) G_1 = 0, \quad z \in (0, 1),$$

$$G_1 = D^2 G_1 = 0, \quad z = 0, \ z = 1. \tag{4.14}$$

Setting

$$\alpha^2 = n^2 - n \sqrt{\mathbf{Ra}_0} \tag{4.15}$$

and

$$\beta^2 = n^2 + n \sqrt{\mathbf{Ra}_0} \tag{4.16}$$

we can write G_1 in the form

$$G_1(z) = C_1 e^{\alpha z} + C_2 e^{-\alpha z} + C_3 e^{\beta z} + C_4 e^{-\beta z}.$$

To satisfy the boundary conditions (4.14),

$$e^\alpha C_1 + e^{-\alpha} C_2 + e^\beta C_3 + e^{-\beta} C_4 = 0,$$

$$C_1 + C_2 + C_3 + C_4 = 0,$$

$$-e^\alpha C_1 - e^{-\alpha} C_2 + e^\beta C_3 + e^{-\beta} C_4 = 0,$$

and

$$-C_1 - C_2 + C_3 + C_4 = 0.$$

As β is real and nonzero,

$$C_3 = C_4 = 0.$$

Because we are looking for nontrivial values of G_1 and $C_1 = -C_2$, it is necessary that $\alpha \neq 0$ and $e^\alpha - e^{-\alpha} = 0$, that is,

$$\alpha = ik\pi, \quad k \in \mathbb{Z} \backslash \{0\}.$$

We obtain $\mathbf{Ra}_0$ from (4.15) and (4.16) as

$$\mathbf{Ra}_0 = \left(n + \frac{k^2 \pi^2}{n} \right)^2, \quad k \in \mathbb{Z} \backslash \{0\}.$$

The smallest value of $\mathbf{Ra}_0$ is obtained for $k^2 = 1$ and $n = \pi$. Then, at the first order,

$$\mathbf{Ra}_0 = 4\pi^2,$$

$$S_1 = F_1(x, y) \sin \pi z,$$

$$h_1 = -2\pi F_1(x, y) \cos \pi z,$$

$$u_1 = 2\pi \frac{\partial F_1(x, y)}{\partial x} \cos \pi z,$$

$$v_1 = 2\pi \frac{\partial F_1(x_1 y)}{\partial y} \cos \pi z,$$

$$w_1 = 2\pi^2 F_1(x, y) \sin \pi z,$$

where F_1 satisfies

$$\Delta_{xy} F_1 + \pi^2 F_1 = 0,$$

$$\int_Y F_1^2(x, y) dx dy = 2, \tag{4.17}$$

and

$$\frac{\partial F_1}{\partial n} = 0 \text{ on } \partial Y.$$

For the nth order,

$$\Delta^2 S_n + 4\pi^2 \Delta_{xy} S_n = \sum_{k=1}^{n-1} (\Delta(v_k \nabla S_{n-k}) - \mathbf{Ra}_k \Delta_{xy} S_{n-k}),$$

$$S_n = \Delta S_n - \sum_{k=1}^{n-1} v_k \nabla S_{n-k} = 0, \ z = 0, \ z = 1.$$

For details, see [EP87]. In the second-order problem,

$$\mathbf{Ra}_1 = 0,$$

and

$$S_2 = \pi \sin 2\pi z F_2(x, y) + C_1 S_1,$$

where the constant C_1 is given by

$$\int_D S_2 S_1 dx = 0$$

and $F_2(x, y)$ is the Y-periodic solution of

$$(\Delta_{xy}^2 - 4\pi^2 \Delta_{xy} + 16\pi^4) F_2(x, y) = \Delta_{xy} G_1 - 4\pi^2 G_1.$$

If we introduce the Nusselt number by

$$\mathbf{Nu} = \frac{\int_Y \frac{\partial T}{\partial z} |_{z=1} dx dy}{\int_Y \frac{\partial T_0}{\partial z} |_{z=1} dx dy},$$

we finally obtain

$$\mathbf{Nu} = 1 + 2\pi^2 \varepsilon^2 + O(\varepsilon^3).$$

For the third-order approximation,

$$\mathbf{Ra} = 4\pi^2 + \mathbf{Ra}_2 \varepsilon^2 + O(\varepsilon^3),$$

where $\mathbf{Ra_2}$ can be computed from

$$\mathbf{Ra_2} = -\pi^2 \int_Y G_1(x, y)F_2(x, y)dxdy.$$

Note that many experimental papers have confirmed $\mathbf{Ra_0} = 4\pi^2$ as the critical Rayleigh number. The Y-periodic pattern (4.17) is not well specified. Because there are no preferred points or directions in the horizontal sections, we assume that the entire horizontal will be covered by regular polygons, i.e., equilateral triangles, squares, or regular hexagons.

For the square pattern of convection,

$$F_1(x, y) = \cos\frac{\pi x}{\sqrt{2}} \cos\frac{\pi y}{\sqrt{2}}, \quad Y = [-\sqrt{2}, \sqrt{2}] \times [-\sqrt{2}, \sqrt{2}],$$

$$G_1(x, y) = \frac{\pi^2}{4}\left(2 + \cos\frac{\pi x}{2} + \cos\frac{\pi y}{\sqrt{2}}\right),$$

$$F_2(x, y) = -\frac{1}{56}\left(7 + 3\cos(\pi x\sqrt{2} + 3)\cos\frac{\pi y}{\sqrt{2}}\right),$$

$$\mathbf{Ra_2} = \frac{17\pi^4}{28},$$

$$\varepsilon \simeq \frac{2\sqrt{119}}{17\pi^2}\sqrt{\mathbf{Ra} - 4\pi^2},$$

and

$$\mathbf{Nu} \simeq 1 + \frac{56}{17\pi^2}(\mathbf{Ra} - 4\pi^2).$$

The regular hexagonal pattern is given by

$$F_1(x, y) = 2\cos\frac{\pi\sqrt{3}}{2}x \cos\frac{\pi y}{2} + \cos\pi y.$$

7.5   Mixed Convection in a Horizontal Porous Layer

In the problem studied in section 7.4, the existence of a transverse flow was neglected. More precisely, the interaction between a longitudinal flow and the vertical temperature gradient represents a problem of interest from a practical point of view in packed bed reactors, thermal oil recovery, and groundwater motion.

Using the same method as in section 7.4, we consider the system (3.1)–(3.3) and the boundary conditions:

$$\vec{v} \cdot \vec{e} = 0, \quad T = 1 \text{ for } z = 0, \tag{5.1}$$

$$\vec{v} \cdot \vec{e} = 0, \quad T = 0 \text{ for } z = 1,$$

and

$$\vec{v} \text{ and } T \text{ are bounded as } (x, y) \to \infty.$$

If V_0 is the velocity of the translational flow in physical variables, we can also introduce the Péclet number:

$$\mathbf{Pe} = \frac{\rho_f c_f L V_0}{\chi_m}.$$

Now supposing that the temperature varies slowly on each stream line, we consider the uniform translational flow (which corresponds to the pure-conduction pattern) as the basic solution of (3.1)–(3.3), (5.1):

$$\vec{v}_b = (\mathbf{Pe}, 0, 0),$$

$$T_b = 1 - z,$$

and the corresponding p_b. Introducing

$$\vec{U} = \vec{v} - \vec{v}_b,$$

$$h = p - p_b,$$

and

$$S = T - T_b,$$

the problem becomes

$$\nabla \cdot \vec{U} = 0, \ z \in (0, 1),$$

$$\vec{U} + \nabla h = \mathbf{Ra} S \vec{e}, \ z \in (0, 1),$$

$$\mathbf{Pe} \frac{\partial S}{\partial x} + \vec{U} \cdot \nabla S = w + \Delta S, \ z \in (0, 1), \qquad (5.2)$$

$$w = S = 0, \ z = 0, \ z = 1,$$

and

$$\vec{U} \text{ and } S \text{ are bounded as } (x, y) \to \infty.$$

For $\mathbf{Ra}$, $\vec{U}$, h, and S we search expansions of the form (4.8)–(4.11), with $\vec{U}$ instead of $\vec{v}$.

The first-order problem is expressed as

$$F_1(x, y)[(D^2 - n^2)^2 \ G_1(z) - \mathbf{Ra}_0 n^2 G_1(z)] = $$
$$= \mathbf{Pe} \frac{\partial F_1(x,y)}{\partial x}(D^2 - n^2)G_1(z). \qquad (5.3)$$

It follows that $(D^2 - n^2)G_1(z)$ is not identically zero, because, otherwise, $S_1 = 0$. Then, (5.3) yields

$$\frac{\partial F_1(x, y)}{\partial x} = \lambda F_1(x, y),$$

for some real λ. As $F_1(x, y)$ must be bounded for $(x, y) \to \infty$, it follows that $\lambda = 0$. Thus,

$$F_1(x, y) = A \sin ny, \quad A \neq 0,$$

and (5.3) becomes

$$(D^2 - n^2)^2 G_1 - \mathbf{Ra}_0 n^2 G_1 = 0 \; z \in (0, 1).$$

It follows that the first-order solution is given by (see Ene and Polişevski [EP90])

$$\mathbf{Ra}_0 = 4\pi^2,$$

$$\vec{U}_1 = 2\pi^2 \sqrt{2}(0, -\sin \pi y \cos \pi z, \cos \pi y \sin \pi z),$$

$$h_1 = -2\pi \sqrt{2} \cos \pi y \cos \pi z,$$

and

$$S_1 = \sqrt{2} \cos \pi y \sin \pi z.$$

As previously shown, it can be proven by induction that $\frac{\partial S_k}{\partial x} = 0$ for any $k \geq 1$. Then, the system (5.2) reduces to the classical system of natural convection in a horizontal layer.

Then, the total velocity in the x direction remains unchanged, but the components in the y and z direction are periodic. *Hence, the only steady convective solutions are the helical motions with axes parallel to the $0x$ direction.* Another conclusion is that *the critical Rayleigh number required for the onset of free convection is independent of the magnitude of the horizontal current.*

Note that the increase in the mean heat transfer, resulting from convection, can be computed, as in section 7.4, in terms of the Nusselt number:

$$\mathbf{Nu} = 1 + \varepsilon^2 \pi^2 + O(\varepsilon^3)$$

On the other hand, for the third-order approximation,

$$\mathbf{Ra} = 4\pi^2 + 2\pi^4 \varepsilon^2 + O(\varepsilon^3).$$

If we try to study the stability of this convective motion, we obtain the result that *any infinitesimal disturbance leads to stability* (see [EP87] and [EP90]).

7.6   Thermal Boundary Layer Approximation

All problems studied so far have concerned small Rayleigh numbers. Now, we will focus on the case of high values of the Rayleigh number, named thermal boundary layer approximation. From a practical point of view this case is of considerable importance in geothermal energy for power generation and other technologies, such as porous insulation, gas-cooled electric machinery, and nuclear reactors.

The thermal boundary-layer approximation is based on the assumption that convection takes place in a thin layer around the heating surface. The analogy with the classical boundary-layer theory for a freely flowing viscous fluid is based on the fact that the motion takes place at a high Rayleigh number. Then, the Rayleigh number, instead of Reynolds number, is the parameter that indicates this possible approach to convection problems in porous media. Consequently, the system of equations can be simplified by a method similar to that proposed by Prandtl for the classical boundary layer theory in a free viscous fluid.

The idea of considering the thermal boundary layer approximation for large Rayleigh numbers appeared in Wooding's paper [Woo63] for the first time. The methods used to study this problem are similar to those used in the theory of a viscous boundary layer in a freely flowing fluid. However, it is important to note that this similarity in methods and equations is formal, because the physical significance of different terms is completely different.

We take the classical Darcy–Boussinesq system (3.1)–(3.3) written in a rectangular Cartesian coordinate system with the origin fixed at the leading edge of the vertical surface, so that the x axis is directed upwards along the wall and the y axis is normal to it:

$$\frac{\partial u}{\partial x} + \frac{\partial v}{\partial y} = 0, \tag{6.1}$$

$$u + \frac{\partial p}{\partial x} = \mathbf{Ra}T, \tag{6.2}$$

$$v + \frac{\partial p}{\partial y} = 0, \tag{6.3}$$

and

$$\frac{\partial T}{\partial t} + u\frac{\partial T}{\partial x} + v\frac{\partial T}{\partial y} = \frac{\partial^2 T}{\partial x^2} + \frac{\partial^2 T}{\partial y^2}. \tag{6.4}$$

Eliminating p from (6.2) and (6.3) by cross-differentiation,

$$\frac{\partial u}{\partial y} - \frac{\partial v}{\partial x} = \mathbf{Ra}\frac{\partial T}{\partial y}. \tag{6.5}$$

If δ is the boundary-layer thickness, then, $x \sim O(1)$, $y \sim O(\delta)$, $u \sim O(1)$, $v \sim O(\delta)$, $T \sim O(1)$, and we find $\frac{\partial}{\partial x} \sim O(1)$, $\frac{\partial}{\partial y} \sim O(\delta^{-1})$, and $\frac{\partial^2}{\partial y^2} \sim O(\delta^{-2})$. Consequently, from (6.1), (6.5), and (6.4),

$$\frac{\partial u}{\partial x} + \frac{\partial v}{\partial y} = 0, \tag{6.6}$$

$$\frac{\partial u}{\partial y} = \mathbf{Ra}\frac{\partial T}{\partial y}, \tag{6.7}$$

and

$$\frac{\partial T}{\partial t} + u\frac{\partial T}{\partial x} + v\frac{\partial T}{\partial y} = \frac{\partial^2 T}{\partial y^2}. \tag{6.8}$$

In this case, note that the definition of the Rayleigh number is a little bit different from the definition (3.4):

$$\mathbf{Ra} = \frac{k\rho_f c_f g L\alpha(T_w - T_\infty)}{\nu \chi_m},$$

where T_w is the wall temperature and T_∞ is the reference temperature at great distance from the wall.

Using the new variables $x^* = x$, $y^* = y\mathbf{Ra}^{1/2}$, $t^* = t\mathbf{Ra}$, $u^* = uR_a^{-1}$, $v^* = vR_a^{-1}$, and $\theta = T$, we can rewrite the system (6.6)–(6.8) in the form

$$\frac{\partial u}{\partial x} + \frac{\partial v}{\partial y} = 0, \tag{6.9}$$

$$\frac{\partial u}{\partial y} = \frac{\partial \theta}{\partial y}, \tag{6.10}$$

and

$$\frac{\partial \theta}{\partial t} + u\frac{\partial \theta}{\partial x} + v\frac{\partial \theta}{\partial y} = \frac{\partial^2 \theta}{\partial y^2}, \tag{6.11}$$

where the stars were omitted.

Concerning the initial and boundary conditions which must be imposed, it is important to note that, in all cases, the temperature and the vertical component of the velocity v must vanish at $y = 0$ (on the wall). As a consequence of these conditions, we observe from (6.10) that the vertical velocity and the temperature distributions are of the same shape. Then, the vertical velocity at the wall varies similarly to the prescribed wall temperature. In particular, the vertical velocity is constant along an isothermal wall.

The initial and other boundary conditions for the temperature field may be prescribed for each specific problem in an adequate form. For example, it is possible to take into consideration a step increase in the wall temperature, a step increase in the heat flux on the wall, or other types of boundary conditions.

Another consequence of the above remark is an integral form for the temperature equation. We integrate the equation (6.11) with respect to y, using (6.9), the divergence formula, and the fact that $u = \theta$. Then, (Cheng [Che78])

$$\frac{\partial}{\partial t}\int_0^\infty \theta\, dy + \frac{\partial}{\partial x}\int_0^\infty \theta^2\, dy = -\left(\frac{\partial \theta}{\partial y}\right)_{y=0}. \tag{6.12}$$

If we introduce $\lambda = \left(\frac{\partial \theta}{\partial y}\right)^{-1}$ in (6.12), and substitute the variable θ for the variable y,

$$\frac{\partial}{\partial t}\int_0^1 \lambda\theta\, d\theta + \frac{\partial}{\partial x}\int_0^1 \lambda\theta^2\, d\theta = \frac{1}{\lambda_0},$$

where λ_0 is the value of λ at $y = 0$ and has the clear physical meaning,

$$\frac{\mathbf{Nu}}{\mathbf{Ra}^{1/2}} = \frac{1}{\lambda_0}.$$

As an example we take the problem of a step increase in the wall temperature. We introduce the new variables and the stream function

$$\eta = \frac{y}{x^{1/2}},$$

$$\xi = \frac{t}{x},$$

and

$$\Psi = x^{1/2} f(\xi, \eta),$$

and (6.10) and (6.11) become

$$\frac{\partial f}{\partial y} = \theta,$$ (6.13)

and

$$\frac{\partial^2 \theta}{\partial y^2} + \left(\frac{1}{2} f - \xi \frac{\partial f}{\partial \xi} \right) \frac{\partial \theta}{\partial y} = (1 - \xi\theta) \frac{\partial \theta}{\partial \xi}.$$ (6.14)

The initial and boundary conditions are

$$f(0, \eta) = 0,$$

$$\theta(0, \eta) = 0,$$

$$\theta(\xi, 0) = 1,$$

$$f(\xi, 0) = 0,$$

and

$$\theta \to 0 \text{ as } \eta \to \infty.$$

If we introduce

$$f = 2\xi^{1/2} F(\xi), \theta = G(\xi), \zeta = \frac{\eta}{2\xi^{1/2}}$$

for a small value of time, we have an exact solution:

$$\theta = \text{erfc } \zeta,$$

$$f = 2\xi^{1/2} \left[\zeta \text{erfc } \zeta - \frac{e^{-\zeta^2} - 1}{\sqrt{\pi}} \right],$$

$$\text{erfc } \zeta = 1 - \text{erf } \zeta,$$

and

$$\text{erf } \zeta = \frac{2}{\sqrt{\pi}} \int_0^\zeta e^{-s^2} ds.$$

For large values of time it is possible to integrate Eq. (6.14) numerically.

In a steady-state situation, it is sufficient to take $\xi = 0$, and, instead of (6.13) and (6.14),

$$f' = \theta,$$ (6.15)

and

$$\theta'' + \frac{1}{2}\theta' f = 0.$$

From (6.15), we obtain only one equation,

$$f''' + \frac{1}{2}ff'' = 0,$$

which, formally, is the *Blasius* equation.

For other examples, the reader is referred to the book [EP87] and Ene [Ene91].

7.7   Conclusion

Two methods were presented which permit us to analytically solve the problem of natural convection in porous media at low values of the Rayleigh number. They permit us to construct the solutions of the problem in bounded or unbounded domains and the solution for forced convection.

In fact, both methods represent perturbation methods, the small parameter being the Rayleigh number itself or the measure of the deviation from the static state. It is important to note that, in the case of a horizontal porous layer, the method permits describing any geometric shape of the horizontal section of a cell.

From a practical point of view, note the conclusion of the problem of forced convection: The critical Rayleigh number required for the onset of free convection is independent of the magnitude of the horizontal current. For example, in the problem of storage tanks, the decrease of convective transfer of heat may be obtained only accounting for the small value of the Rayleigh number.

The case of the high value of the Rayleigh number is an example of heat transfer from a surface embedded in a porous medium, as in the problem of geothermal energy. Here the methods used are very different, and they are due to the analogy with classical boundary-layer theory for freely flowing viscous fluids. In any case note that this similarity in methods and equations is formal, because the physical significance of different terms is completely different. For example, in a porous medium, the velocity is tangential to the impervious wall and has its maximum value there.

8

Poroelastic Media

Jean-Louis Auriault

In this chapter, we will investigate the acoustic macroscopic behavior of saturated porous media. The macroscopic quasi-static behavior will be deduced classically by neglecting inertial terms. The material of the porous matrix is elastic and the saturating fluid is viscous, Newtonian, and incompressible. The local structure is assumed periodic with a period Y of characteristic length l. The characteristic length L at the macroscopic scale can be associated with the wave length λ. For simplicity, we will use $L = \lambda/2\pi$. The two characteristic lengths introduce two dimensionless space variables:

$$y = \frac{X}{l}$$

and

$$x = \frac{X}{L}$$

where X is the physical space variable. The solid part of the period is denoted $\mathfrak{S}$, the pores are $\mathfrak{P}$, and the interface is Γ (Fig. 8.1). In the first section, we will investigate the acoustics of an empty porous medium that represents one of the basic problems to be used in all the following investigations. In the second section, we will present some estimates that permit us to point out other macroscopic behavior of the saturated porous medium, in relation to the value of a dimensionless number as a function of the small parameter ε of scale separation:

$$\varepsilon = \frac{l}{L} \ll 1$$

and

$$x = \varepsilon y.$$

The following sections are devoted to these different types of macroscopic behavior: fluid acoustics in a rigid matrix, diphasic behavior, monophasic elastic behavior, and monophasic viscoelastic behavior, respectively. The acoustics of a fluid in a rigid porous medium is also a basic problem. Finally, in the last section, we will investigate the acoustics of double-porosity media.

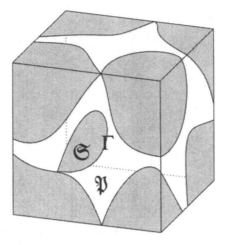

FIGURE 8.1. Period of the porous medium.

8.1   Acoustics of an Empty Porous Medium

Consider a porous medium filled with a very compressible and low viscosity fluid, excited so that the fluid can be ignored. The macroscopic behavior will be referred to as model I.

8.1.1   Local Description and Estimates

The medium satisfies the Navier equation in the solid part. The classical continuity equations apply at the interface Γ, i.e., the continuity of the normal stress and of the displacement:

$$\frac{\partial \sigma_{\mathfrak{S}ij}}{\partial X_j} = \rho_{\mathfrak{S}} \frac{\partial^2 u_{\mathfrak{S}i}}{\partial t^2}, \tag{1.1}$$

and

$$\sigma_{\mathfrak{S}ij} = a_{ijkl}\, e_{kl}(\mathbf{u}_{\mathfrak{S}}) \qquad \text{in } \mathfrak{S}. \tag{1.2}$$

Due to the pore emptiness, the normal stress is zero-valued on Γ:

$$\sigma_{\mathfrak{S}ij} v_j = 0 \qquad \text{on } \Gamma. \tag{1.3}$$

a is an elastic tensor. It is, a priori, a periodic function of **y**, with possible discontinuities to which the classical continuity of the displacement and of the normal stress applies. The tensor **a** verifies

$$a_{ijkl} = a_{jikl} = a_{ijlk} = a_{klij}$$

and the ellipticity condition

$$a_{ijkl}\, e_{ij}\, e_{kl} \geq \gamma\, e_{ij}\, e_{ij}, \qquad \gamma > 0.$$

We assume that all the components of **a** are of the same order of magnitude a with respect to the powers of ε. The tensor **e** is the small deformation

$$e_{ij}(\mathbf{u}) = \frac{1}{2}\left(\frac{\partial u_i}{\partial X_j} + \frac{\partial u_j}{\partial X_i}\right).$$

Equation (1.1) introduces the dimensionless number P

$$P = \frac{\left|\rho_\mathfrak{S}\dfrac{\partial^2 u_{\mathfrak{S}i}}{\partial t^2}\right|}{\left|\dfrac{\partial \sigma_{\mathfrak{S}ij}}{\partial X_j}\right|}.$$

We consider a wave of pulsation ω, and we use l as the characteristic length to make the equations dimensionless. We follow the methodology defined in [Aur91b] which gives

$$P_l = O\left(\frac{\rho\omega^2 l^2}{a}\right) = O(\varepsilon^2),$$

where we have used the relation

$$\frac{\rho\omega^2}{a} = O\left(\frac{(2\pi)^2}{\lambda^2}\right) = O\left(\frac{1}{L^2}\right).$$

Finally, the dimensionless form of the local monochromatic description is written as follows. Because no confusion is possible, we use the same notations for dimensional and dimensionless quantities, with the exception of the space variable. The subscript y shows operators with respect to the space variable y.

$$\frac{\partial \sigma_{\mathfrak{S}ij}}{\partial y_j} = -\varepsilon^2 \rho_\mathfrak{S}\omega^2 u_{\mathfrak{S}i}, \tag{1.4}$$

$$\sigma_{\mathfrak{S}ij} = a_{ijkl}\, e_{ykl}(\mathbf{u}_\mathfrak{S}) \qquad \text{in } \mathfrak{S}, \tag{1.5}$$

and

$$\sigma_{\mathfrak{S}ij}\, v_j = 0 \qquad \text{on } \Gamma. \tag{1.6}$$

8.1.2 Macroscopic Description

We look for the unknown $\mathbf{u}_s$ in the form,

$$u_{\mathfrak{S}i}(x, y) = u_i^{(0)}(x, y) + \varepsilon u_i^{(1)}(x, y) + \varepsilon^2 u_i^{(2)}(x, y) + \dots$$

with $x = \varepsilon y$. The $\mathbf{u}^{(i)}$'s are Y-periodic in y. We introduce the above expansion in the set (1.4–1.6), and we use the following differentiation rule (see Chapter 1):

$$\frac{d}{dy} = \frac{\partial}{\partial y} + \varepsilon \frac{\partial}{\partial x}.$$

Identifying the same powers of ε gives successive boundary-value problems. The first order shows a boundary-value problem for $\mathbf{u}^{(0)}$

$$\frac{\partial}{\partial y_j}(a_{ijkh}e_{ykh}(\mathbf{u}^{(0)})) = 0 \qquad \text{in } \mathfrak{S}, \qquad (1.7)$$

and

$$a_{ijkh}e_{ykh}(\mathbf{u}^{(0)})v_j = 0 \qquad \text{on } \Gamma, \qquad (1.8)$$

$\mathbf{u}^{(0)}$ Y periodic in y. Clearly, the solution is an arbitrary function of x, independent of y

$$\mathbf{u}^{(0)} = \mathbf{u}^{(0)}(x).$$

The following orders define $\mathbf{u}^{(1)}$:

$$\frac{\partial}{\partial y_j}(a_{ijkh}(e_{ykh}(\mathbf{u}^{(1)}) + e_{xkh}(\mathbf{u}^{(0)}))) = 0 \qquad \text{in } \mathfrak{S}, \qquad (1.9)$$

and

$$a_{ijkh}(e_{ykh}(\mathbf{u}^{(1)}) + e_{xkh}(\mathbf{u}^{(0)}))v_j = 0 \qquad \text{on } \Gamma, \qquad (1.10)$$

$\mathbf{u}^{(1)}$ Y periodic in y. It appears as a linear vectorial function of $e_{xkh}(\mathbf{u}^{(0)})$, to an arbitrary vector, independent of y:

$$u_i^{(1)} = \xi_i^{lm} e_{xlm}(\mathbf{u}^{(0)}) + \bar{u}_i^{(1)}(x) \qquad (1.11)$$

where $\bar{u}_i^{(1)}(x)$ is arbitrary. $\xi(y)$ is taken of zero average, that insures its uniqueness,

$$\int_{\mathfrak{S}} \xi dy = 0.$$

Finally, the boundary-value problem for $\mathbf{u}^{(2)}$ is obtained from the next orders. It can be written in the form,

$$\frac{\partial \sigma_{\mathfrak{S}ij}^{(1)}}{\partial y_j} + \frac{\partial \sigma_{\mathfrak{S}ij}^{(0)}}{\partial x_j} = -\rho_{\mathfrak{S}}\omega^2 u_i^{(0)} \qquad \text{in } \mathfrak{S}, \qquad (1.12)$$

and

$$\sigma_{\mathfrak{S}ij}^{(1)}v_j = 0 \qquad \text{on } \Gamma, \qquad (1.13)$$

where $\sigma_{\mathfrak{S}}^{(0)}$ and $\sigma_{\mathfrak{S}}^{(1)}$ are the zero- and first-order terms, respectively, in the asymptotic expansion of $\sigma_{\mathfrak{S}}$:

$$\sigma_{\mathfrak{S}ij}^{(0)} = a_{ijkh}(e_{ykh}(\mathbf{u}^{(1)}) + e_{xkh}(\mathbf{u}^{(0)}))$$

$$\sigma_{\mathfrak{S}ij}^{(1)} = a_{ijkh}(e_{ykh}(\mathbf{u}^{(2)}) + e_{xkh}(\mathbf{u}^{(1)})).$$

Eq. (1.12) is a balance equation for the periodic quantity $\sigma^{(1)}$, with the source terms $-\partial\,\sigma_{\mathfrak{G}ij}^{(0)}/\partial x_j$ and $\rho_{\mathfrak{G}}\omega^2 u_i^{(0)}$. Therefore, it introduces a compatibility condition for the existence of $\mathbf{u}^{(2)}$. Let us classically define the total stress σ^T by

$$\sigma^T = \left\{ \begin{array}{ll} \sigma_{\mathfrak{G}} & \text{in } \mathfrak{G} \\ 0 & \text{in } \mathfrak{P} \end{array} \right. .$$

The compatibility condition is obtained by integrating (1.12) on $\mathfrak{G}$. By using the relation (1.13) and the periodicity of $\sigma^{(1)}$,

$$\langle \frac{\partial\,\sigma_{\mathfrak{G}ij}^{(0)}}{\partial x_j}\rangle_{\mathfrak{G}} = \frac{\partial\,\langle\sigma_{\mathfrak{G}ij}^{(0)}\rangle_{\mathfrak{G}}}{\partial x_j} = -\rho_{\mathfrak{G}}^{eff}\,\omega^2 u_i^{(0)},$$

and

$$\rho_{\mathfrak{G}}^{eff} = \langle\rho_{\mathfrak{G}}\rangle_{\mathfrak{G}},$$

where we have introduced the volume average symbol

$$\langle\bullet\rangle_{\mathfrak{G}} = |Y|^{-1}\int_{\mathfrak{G}} \bullet\, dY.$$

With the definition of the total stress,

$$\langle\sigma_{ij}^T\rangle_{\mathfrak{G}} = a_{ijkh}^{eff}\, e_{xkh}(\mathbf{u}^{(0)}) \tag{1.14}$$

with

$$a_{ijkh}^{eff} = \langle a_{ijkh} + a_{ijlm}e_{ylm}(\boldsymbol{\xi}^{kh})\rangle_{\mathfrak{G}}. \tag{1.15}$$

Finally, by returning to dimensional quantities, one obtains the macroscopic description - model I - at the first order of approximation:

$$\frac{\partial\,\langle\sigma_{ij}^T\rangle_{\mathfrak{G}}}{\partial X_j} = \rho_{\mathfrak{G}}^{eff}\frac{\partial^2\,u_{\mathfrak{G}i}}{\partial t^2} + O_r(\varepsilon), \tag{1.16}$$

and

$$\langle\sigma_{ij}^T\rangle_{\mathfrak{G}} = a_{ijkh}^{eff}\, e_{kh}(\mathbf{u}_{\mathfrak{G}}) + O_r(\varepsilon). \tag{1.17}$$

It is possible to show [ASP77] that $\mathbf{a}^{eff}$ is an elastic tensor. Therefore, the set (1.16, 1.17) is a classical Navier description. The symbol $O_r(\varepsilon)$ shows that the macroscopic description is an approximation of relative order ε.

8.2   A Priori Estimates for a Saturated Porous Medium

To anticipate the different structures of the macroscopic description of the acoustics of saturated porous media, consider the volume balance of such media. Without lack of generality, we assume in this subsection that the material constituting the porous matrix is incompressible. Therefore, at the macroscopic scale,

$$\frac{\partial \theta}{\partial t} = -\frac{\partial v_{ri}}{\partial X_i} \tag{2.1}$$

where v_r is the fluid velocity relative to the matrix and θ is the volume change of the matrix. Eq. (2.1) introduces the Strouhal number **Sh**

$$\mathbf{Sh} = \frac{\left|\dfrac{\partial \theta}{\partial t}\right|}{\left|\dfrac{\partial v_{ri}}{\partial X_i}\right|}.$$

Clearly $\mathbf{Sh} = O(1)$ corresponds to diphasic behavior with well-identified macroscopic displacements of the matrix and the fluid. For $\mathbf{Sh} \leq O(\varepsilon)$ the matrix displacement becomes negligible when compared to the fluid displacement. The matrix can be considered rigid as a first approximation. The opposite $\mathbf{Sh} \geq O(\varepsilon^{-1})$ has a relative velocity v_r small compared to the matrix velocity. The macroscopic behavior is monophasic:

$$\mathbf{Sh} = O(\frac{u\omega}{v_r}),$$

where u is the displacement in the solid. It is related to the solid stress σ by

$$u = O(\frac{\sigma L}{a}).$$

On the other hand, v_r is given by a Darcy-like relation,

$$\mathbf{v}_{ri} = -K_{ij}\frac{\partial p}{\partial X_j},$$

where the filtration tensor is of the order

$$K = O(\frac{l^2}{\mu}),$$

where μ is the fluid viscosity. From the above relations,

$$v_r = O(\frac{l^2 p}{\mu L}).$$

The Strouhal number becomes

$$\mathbf{Sh} = O(\frac{L^2}{l^2}\frac{\sigma}{p}\frac{\mu\omega}{a}).$$

We will assume that the stress in the matrix of the same order as the fluid pressure on the macroscopic scale. Therefore, the Strouhal number simplifies to

$$\mathbf{Sh} = O(\varepsilon^{-2}C), \qquad\qquad C = \frac{\mu\omega}{a}, \tag{2.2}$$

where C is the contrast of property number. From the meaning of $\mathbf{Sh}$, four cases of interest will be investigated in the four following sections:

- Model II, $\mathbf{C} = O(\varepsilon^3)$: acoustics of a fluid in a rigid porous matrix

- Model III, $\mathbf{C} = O(\varepsilon^2)$: diphasic macroscopic behavior

- Model IV, $\mathbf{C} = O(\varepsilon)$: monophasic elastic macroscopic behavior

- Model V, $\mathbf{C} = O(1)$: monophasic viscoelastic macroscopic behavior

8.3   Local Description of a Saturated Porous Medium

Let us now introduce the local description. When the matrix is not rigid (models III, IV, and V), the medium satisfies the Navier equation in the solid part and the Navier–Stokes equation in the fluid part. The classical continuity equations apply at the interface Γ, i.e., the continuity of the normal stress and the displacement

$$\frac{\partial \sigma_{\mathfrak{S}ij}}{\partial X_j} = \rho_{\mathfrak{S}} \frac{\partial^2 u_{\mathfrak{S}i}}{\partial t^2}, \tag{3.1}$$

$$\sigma_{\mathfrak{S}ij} = a_{ijkl}\, e_{kl}(\mathbf{u}_{\mathfrak{S}}) \qquad \text{in } \mathfrak{S}, \tag{3.2}$$

$$\frac{\partial \sigma_{\mathfrak{P}ij}}{\partial X_j} = \rho_{\mathfrak{P}} \left(\frac{\partial v_i}{\partial t} + (v_j \frac{\partial}{\partial X_j}) v_i \right), \tag{3.3}$$

$$\sigma_{\mathfrak{P}ij} = 2\mu D_{ij} - p\, I_{ij}, \tag{3.4}$$

$$\frac{\partial v_i}{\partial X_i} = 0 \qquad \text{in } \mathfrak{P}, \tag{3.5}$$

$$[u_i] = 0, \tag{3.6}$$

and

$$[\sigma_{ij}]\, v_j = 0 \qquad \text{on } \Gamma. \tag{3.7}$$

We assume that the amplitude of the perturbation is sufficiently small to neglect the nonlinear terms in (3.3). Moreover we consider movements at constant pulsation. The set (3.1–3.7) introduces several dimensionless numbers. For simplicity, we restrict ourselves to the following features:

- The displacements and densities are of the same order of magnitude in the solid and the fluid parts, respectively,

$$|\mathbf{u}_{\mathfrak{S}}| = O(|\mathbf{u}_{\mathfrak{P}}|),$$

and

$$\rho_{\mathfrak{S}} = O(\rho_{\mathfrak{P}}).$$

- The pressure in the fluid and the stress in the solid are of the same order of magnitude at the macroscopic level, i.e., when estimated with the character-istic length L,

$$\Sigma = \left| \frac{\sigma_{\mathfrak{S}}}{p} \right|, \quad \Sigma_L = \frac{au}{pL} = O(1), \quad \text{and} \quad \Sigma_l = \frac{au}{pl} = O(\varepsilon^{-1}). \tag{3.8}$$

As shown in the first section, the acoustic behavior of the matrix is governed by the ratio,

$$P_l = \frac{|\rho_{\mathfrak{S}}\omega^2 u_{\mathfrak{S}i}|}{|\frac{\partial \sigma_{\mathfrak{S}ij}}{\partial X_j}|} = O(\frac{\rho\omega^2 l^2}{a}) = O(\varepsilon^2).$$

Therefore, for the fluid, with (3.8),

$$T_l = \frac{|\rho_{\mathfrak{P}}\frac{\partial v_i}{\partial t}|}{|\frac{\partial p}{\partial X_i}|} = O(\frac{\rho\omega^2 ul}{p}) = O(\varepsilon).$$

With the above estimates, there is only one free dimensionless number, that concerns the fluid:

$$Q_l = \frac{|\mu\frac{\partial^2 v_i}{\partial X_j \partial X_j}|}{|\frac{\partial p}{\partial X_i}|} = O(\frac{\mu\omega u}{pl}) = O(\varepsilon^{-1}\frac{\mu\omega}{a}).$$

The value of Q_l is directly related to the contrast of property number $C = \mu\omega/a$. Let $Q_l = O(\varepsilon^{-r})$. Following the discussion in the preceding section leads to the evaluation of Q_l for models III, IV, and V. They correspond to increasing the viscosity at constant elastic properties and constant pulsation:

- Model III, $r = -1$: diphasic macroscopic behavior

- Model IV, $r = 0$: monophasic macroscopic elastic behavior

- Model V, $r = 1$: monophasic macroscopic viscoelastic behavior

Finally, using l to make the local equations dimensionless,

$$\frac{\partial \sigma_{\mathfrak{S}ij}}{\partial y_j} = -\varepsilon^2 \rho_{\mathfrak{S}} \omega^2 u_{\mathfrak{S}i}, \tag{3.9}$$

$$\sigma_{\mathfrak{S}ij} = a_{ijkl} e_{ykl}(\mathbf{u}_{\mathfrak{S}}) \qquad \text{in } \mathfrak{S}, \tag{3.10}$$

$$\frac{\partial \sigma_{\mathfrak{P}ij}}{\partial y_j} = \varepsilon\rho_{\mathfrak{P}} i\omega v_i, \tag{3.11}$$

$$\sigma_{\mathfrak{P}ij} = 2\mu\varepsilon^{-r} \mathcal{D}_{yij}(\mathbf{v}) - p\, I_{ij}, \tag{3.12}$$

$$\frac{\partial v_i}{\partial y_i} = 0 \qquad \text{in } \mathfrak{P}, \tag{3.13}$$

$$[u_i] = 0, \tag{3.14}$$

and

$$(\sigma_{\mathfrak{S}ij} - \varepsilon\sigma_{\mathfrak{P}ij}) v_j = 0 \qquad \text{on } \Gamma. \tag{3.15}$$

When the matrix can be considered rigid (model II), the fluid is in motion relative to the matrix. Therefore the dimensionless equations of the fluid are those of model III, with the continuity condition (3.14) reducing to the adherence condition,

$$\frac{\partial \, \sigma_{\mathfrak{P}ij}}{\partial y_j} = \varepsilon \rho_{\mathfrak{P}} i \omega v_i,$$
(3.16)

$$\sigma_{\mathfrak{P}ij} = 2\mu \varepsilon D_{yij}(\mathbf{v}) - p \, I_{ij} \qquad \text{in } \mathfrak{P},$$
(3.17)

and

$$v_i = 0 \qquad \text{on } \Gamma.$$
(3.18)

8.4   Acoustics of a Fluid in a Rigid Porous Medium

The acoustics of an incompressible fluid was first investigated in [Lev77] and [Aur80]. Into (3.16–3.18), we introduce asymptotic expansions for p and $\mathbf{v}$ in the form

$$p(x, y) = p^{(0)}(x, y) + \varepsilon p^{(1)}(x, y) + \varepsilon^2 p^{(2)}(x, y) + \dots ,$$

and

$$v_i(x, y) = v_i^{(0)}(x, y) + \varepsilon v_i^{(1)}(x, y) + \varepsilon^2 v_i^{(2)}(x, y) + \dots ,$$

with $x = \varepsilon y$. The $p^{(i)}$'s and the $\mathbf{v}^{(i)}$'s are Y-periodic in y. Successively,

$$\frac{\partial p^{(0)}}{\partial y_i} = 0$$

i.e.,

$$p^{(0)} = p^{(0)}(x),$$

and

$$\mu \frac{\partial^2 \, v_i^{(0)}}{\partial y_j \partial y_j} - \frac{\partial p^{(0)}}{\partial x_i} - \frac{\partial p^{(1)}}{\partial y_i} = \rho_{\mathfrak{P}} i \omega v_i^{(0)},$$

$$\frac{\partial \, v_i^{(0)}}{\partial y_i} = 0 \qquad \text{in } \mathfrak{P},$$

and

$$v_i^{(0)} = 0 \qquad \text{on } \Gamma.$$

$v_i^{(0)}$ and $p^{(1)}$ are Y-periodic in y. The linearity of the above system enables us to write $\mathbf{v}^{(0)}$ in the form,

$$v_i = - \, k_{ij} \frac{\partial p^{(0)}}{\partial x_j},$$
(4.1)

where the tensor $k(y)$ is complex-valued and dependent on ω. Finally, the volume balance at the next order,

$$\frac{\partial \, v_i^{(1)}}{\partial y_i} = - \frac{\partial \, v_i^{(0)}}{\partial x_i},$$

imposes the compatibility condition,

$$\frac{\partial \langle v_i^{(0)} \rangle_{\mathfrak{P}}}{\partial x_i} = 0,$$

$$\langle v_i^{(0)} \rangle_{\mathfrak{P}} = -K_{ij}^{eff} \frac{\partial p^{(0)}}{\partial x_j}, \qquad (4.2)$$

and

$$K_{ij}^{eff} = \langle k_{ij} \rangle_{\mathfrak{P}},$$

where $\langle \bullet \rangle_{\mathfrak{P}}$ represents the volume average, on $\mathfrak{P}$

$$\langle \bullet \rangle_{\mathfrak{P}} = |Y|^{-1} \int_{\mathfrak{P}} \bullet dY.$$

Relationship (4.2) represents a monochromatic seepage law with a complex-valued and ω-dependent filtration tensor $\mathbf{K}^{eff}$. Returning to physical quantities, model II is expressed by

$$\frac{\partial \langle v_i \rangle}{\partial X_i} = O_r(\varepsilon),$$

$$\langle v_i \rangle = -K_{ij}^{eff} \frac{\partial p}{\partial X_j} + O_r(\varepsilon),$$

and

$$K_{ij} = \langle k_{ij} \rangle.$$

For nonmonochromatic acoustics, model II shows memory effects.

8.5   Diphasic Macroscopic Behavior

More details, especially concerning the properties of effective coefficients, can be found in [ASP77], [Aur80], and [Aur91a]. The local description is now given by the set (3.9–3.15), where Eq. (3.12) is written with $r = -1$. We look for the unknowns $\mathbf{u}_{\mathfrak{S}}$, $\mathbf{v}$ and p in the form,

$$u_{\mathfrak{S}i}(x, y) = u_i^{(0)}(x, y) + \varepsilon u_i^{(1)}(x, y) + \varepsilon^2 u_i^{(2)}(x, y) + \dots,$$

$$v_i(x, y) = v_i^{(0)}(x, y) + \varepsilon v_i^{(1)}(x, y) + \varepsilon^2 v_i^{(2)}(x, y) + \dots,$$

and

$$p(x, y) = p^{(0)}(x, y) + \varepsilon p^{(1)}(x, y) + \varepsilon^2 p^{(2)}(x, y) + \dots,$$

with $x = \varepsilon y$. $\mathbf{u}^{(i)}$, $\mathbf{v}^{(i)}$, and $p^{(i)}$ are Y-periodic in y. We introduce the above expansions into the set (3.9–3.15). These give successive boundary-value problems. $\mathbf{u}^{(0)}$ and $p^{(0)}$ are given by the problem similar to those in sections 8.1 and 8.4, respectively. Therefore,

$$\mathbf{u}^{(0)} = \mathbf{u}^{(0)}(x),$$

and

$$p^{(0)} = p^{(0)}(x).$$

The boundary-value problem for $\mathbf{u}^{(1)}$ resembles the problem in section 8.1:

$$\frac{\partial}{\partial y_j}(a_{ijkh}(e_{ykh}(\mathbf{u}^{(1)}) + e_{xkh}(\mathbf{u}^{(0)}))) = 0 \qquad \text{in } \mathfrak{S},$$

$$a_{ijkh}(e_{ykh}(\mathbf{u}^{(1)}) + e_{xkh}(\mathbf{u}^{(0)}))v_j = -p^{(0)}v_i \qquad \text{on } \Gamma,$$

$\mathbf{u}^{(1)}$ Y periodic in y. When $p^{(0)} = 0$, this set is similar to the set (1.9, 1.10) for empty pores. The linearity shows $\mathbf{u}^{(1)}$ in the form,

$$u_i^{(1)} = \xi_i^{lm}e_{xlm}(\mathbf{u}^{(0)}) - \eta_i p^{(0)} + \bar{u}_i^{(1)}(x),$$

where $\xi(y)$ is as introduced in section 8.1 and $\eta(y)$ is a new vector field of zero average,

$$\int_{\mathfrak{S}} \eta \, dY = 0.$$

The problem for $p^{(1)}$ and $\mathbf{v}^{(0)}$ can be put in the form already encountered in section 8.4:

$$\mu \frac{\partial^2 v_i^{(0)}}{\partial y_j \partial y_j} - \frac{\partial p^{(0)}}{\partial x_i} - \frac{\partial p^{(1)}}{\partial y_i} = \rho_{\mathfrak{P}} i\omega v_i^{(0)},$$

$$\frac{\partial v_i^{(0)}}{\partial y_i} = 0 \qquad \text{in } \mathfrak{P},$$

and

$$v_i^{(0)} = i\omega u_i^{(0)} \qquad \text{on } \Gamma.$$

$v_i^{(0)}$ and $p^{(1)}$ are Y-periodic in y. After setting $w_i = v_i^{(0)} - \dot{u}_i^{(0)}$,

$$\mu \frac{\partial^2 w_i}{\partial y_j \partial y_j} - \frac{\partial p^{(0)}}{\partial x_i} - \frac{\partial p^{(1)}}{\partial y_i} = \rho_{\mathfrak{P}} i\omega w_i - \rho_{\mathfrak{P}} \omega^2 u_i^{(0)},$$

$$\frac{\partial w_i}{\partial y_i} = 0 \qquad \text{in } \mathfrak{P},$$

and

$$w_i = 0 \qquad \text{on } \Gamma.$$

w_i and $p^{(1)}$ are Y-periodic in y. This is the boundary-value problem of section 8.4 with $\frac{\partial p^{(0)}}{\partial x_j}$ replaced by $\frac{\partial p^{(0)}}{\partial x_j} - i\omega^2 u_i^{(0)}$. Therefore,

$$w_i = v_i^{(0)} - i\omega u_i^{(0)} = -k_{ij}(\frac{\partial p^{(0)}}{\partial x_j} - i\omega^2 u_i^{(0)}), \tag{5.1}$$

where the tensor field $\mathbf{k}$ is the complex-valued permeability tensor defined in section 8.4.

Contrary to the preceding boundary-value problems, the following need compatibility conditions for the existence of solutions. Two compatibility conditions are obtained. First,

$$\frac{\partial \, \sigma_{\mathfrak{G}ij}^{(2)}}{\partial y_j} + \frac{\partial \, \sigma_{\mathfrak{G}ij}^{(1)}}{\partial x_j} = -\rho_{\mathfrak{G}}\omega^2 u_i^{(0)} \qquad \text{in } \mathfrak{G},$$

$$\frac{\partial \, \sigma_{\mathfrak{P}ij}^{(1)}}{\partial y_j} + \frac{\partial \, p_{ij}^{(0)}}{\partial x_j} = \rho_{\mathfrak{P}}i\omega v_i^{(0)} \qquad \text{in } \mathfrak{P},$$

and

$$\sigma_{\mathfrak{G}ij}^{(2)} v_j = \sigma_{\mathfrak{P}ij}^{(1)} v_j \qquad \text{on } \Gamma.$$

Let the total stress σ^T now be defined by

$$\sigma^T = \begin{cases} \sigma_{\mathfrak{G}}, & \text{in } \mathfrak{G}, \\ \sigma_{\mathfrak{P}}, & \text{in } \mathfrak{P}. \end{cases}$$

Integrating the first equation over $\mathfrak{G}$, the second over $\mathfrak{P}$, using the divergence theorem together with the periodicity condition, and taking into account the last relation,

$$\frac{\partial \, \langle \sigma_{ij}^{T(0)} \rangle}{\partial x_j} = -\rho_{\mathfrak{G}}^{eff} \omega^2 u_i^{(0)} + \rho_{\mathfrak{P}}i\omega \langle v_i^{(0)} \rangle_{\mathfrak{P}}, \qquad (5.2)$$

with

$$\sigma_{ij}^{T(0)} = a_{ijkh}^{eff} \, e_{xkh}(\mathbf{u}^{(0)}) - \alpha_{ij}^{eff} \, p^{(0)},$$

$$\langle \bullet \rangle = |Y|^{-1} \int_Y \bullet \, dY.$$

The tensor $\mathbf{a}^{eff}$ was defined in section 8.1. α^{eff} is a new effective tensor,

$$\alpha_{ij}^{eff} = \Phi I_{ij} + \langle a_{ijlm} \, e_{ylm}(\eta) \rangle_{\mathfrak{G}} = \alpha_{ji}^{eff},$$

where Φ is the porosity. When the medium is isotropic, α is shown [ASP77] to verify

$$\Phi < \alpha < 1.$$

Secondly, the incompressibility is written at the second order in the form,

$$\frac{\partial \, v_i^{(1)}}{\partial y_i} = -\frac{\partial \, v_i^{(0)}}{\partial x_i}.$$

By integrating over $\mathfrak{P}$, using the divergence theorem, the periodicity, the displacement continuity on Γ and the divergence theorem, and the periodicity one, once again, obtains the second compatibility relation in the form,

$$\frac{\partial \, (\langle v_i^{(0)} \rangle_P - \Phi i\omega u_i^{(0)})}{\partial x_i} = -\alpha_{ij}^{eff} i\omega e_{xij}(\mathbf{u}^{(0)}) - \beta^{eff} i\omega p^{(0)}. \qquad (5.3)$$

β^{eff} is a positive scalar effective coefficient [ASP77] defined by

$$\beta^{eff} = \langle \frac{\partial \eta_p}{\partial y_p} \rangle_\mathfrak{S}.$$

The compatibility relations (5.1) and (5.2) represent the first-order macroscopic behavior of a acoustics of the saturated porous medium. Returning to physical quantities, the macroscopic model III is given as follows:

$$\frac{\partial \langle \sigma_{ij}^T \rangle}{\partial X_j} = \rho_\mathfrak{S}^{eff} \frac{\partial^2 u_{\mathfrak{S}i}}{\partial t^2} + \rho_\mathfrak{P} \frac{\partial \langle v_i \rangle_\mathfrak{P}}{\partial t} + O_r(\varepsilon),$$

$$\sigma_{ij}^T = a_{ijkh}^{eff} \, e_{Xkh}(u_\mathfrak{S}) - \alpha_{ij}^{eff} p + O_r(\varepsilon),$$

$$\frac{\partial (\langle v_i \rangle_\mathfrak{P} - \Phi \frac{\partial u_{\mathfrak{S}i}}{\partial t})}{\partial X_i} = -\alpha_{ij}^{eff} \frac{\partial e_{Xij}(u_\mathfrak{S})}{\partial t} - \beta^{eff} \frac{\partial p}{\partial t} + O_r(\varepsilon),$$

and

$$\langle v_i \rangle_\mathfrak{P} - \Phi \frac{\partial u_{\mathfrak{S}i}}{\partial t} = -K_{ij}^{eff} (\frac{\partial p}{\partial X_j} - \rho_\mathfrak{P} \omega^2 u_{\mathfrak{S}i}) + O_r(\varepsilon),$$

where a^{eff}, α^{eff}, and β^{eff} are elastic effective coefficients. The permeability K^{eff} is complex-valued and ω-dependent. The model introduces two macroscopic displacement fields and, therefore, the description is diphasic. It can be shown to have a structure similar to Biot's model ([Aur91a] and [Bio56]).

8.6   Monophasic Elastic Macroscopic Behavior

Let us increase the viscosity while keeping a and ω constant. Eq. (3.12) is now written with $p = 0$:

$$\sigma_{\mathfrak{P}ij} = 2\mu \mathcal{D}_{yij}(v) - p \, I_{ij} \qquad \text{in } \mathfrak{P}.$$

It appears that $u^{(0)}$ is obtained in the same way as in the preceding section:

$$u^{(0)} = u^{(0)}(x).$$

Concerning the fluid, the first boundary-value problem is changed into

$$\mu \frac{\partial^2 v_i^{(0)}}{\partial y_j \partial y_j} - \frac{\partial p^{(0)}}{\partial y_i} = 0,$$

$$\frac{\partial v_i^{(0)}}{\partial y_i} = 0 \qquad \text{in } \mathfrak{P},$$

and

$$v_i^{(0)} = i \omega u_i^{(0)} \qquad \text{on } \Gamma.$$

Obviously, the solution is given as

$$v_i^{(0)} = i\omega u_i^{(0)},$$

and

$$p^{(0)} = p^{(0)}(x).$$

Taking these results into account, the set of equations for $\mathbf{u}^{(1)}$ is the same as above. Therefore,

$$u_i^{(1)} = \xi_i^{lm} e_{xlm}(\mathbf{u}^{(0)}) - \eta_i \, p^{(0)} + \overline{u}_i^{(1)}(x).$$

The compatibility relation for the total momentum balance also appears to be unchanged:

$$\frac{\partial \langle \sigma_{ij}^{T(0)} \rangle}{\partial x_j} = -\rho_{\mathbb{G}}^{eff} \omega^2 u_i^{(0)} + \rho_{\mathbb{P}} i\omega \langle v_i^{(0)} \rangle_{\mathbb{P}}, \tag{6.1}$$

with, as in the preceding section,

$$\sigma_{ij}^{T(0)} = a_{ijkh}^{eff} \, e_{xkh}(\mathbf{u}^{(0)}) - \alpha_{ij}^{eff} \, p^{(0)}. \tag{6.2}$$

In contrast, using $v_i^{(0)} = i\omega u_i^{(0)}$, the compatibility condition (5.2), after integration, becomes

$$\alpha_{ij}^{eff} e_{xij}(\mathbf{u}^{(0)}) + \beta^{eff} p^{(0)} = 0. \tag{6.3}$$

Eliminating $p^{(0)}$ between (6.1) and (6.2) yields the following first-order macroscopic description:

$$\frac{\partial \langle \sigma_{ij}^{T(0)} \rangle}{\partial x_j} = -(\rho_{\mathbb{G}}^{eff} + \Phi \rho_{\mathbb{P}})\omega^2 u_i^{(0)},$$

and

$$\sigma_{ij}^{T(0)} = a_{ijkh}^{*eff} \, e_{xkh}(\mathbf{u}^{(0)}),$$

where $\mathbf{a}^{*eff}$ is an elastic tensor defined by

$$a_{ijkh}^{*eff} = a_{ijkh}^{eff} + \alpha_{ij}^{eff} \alpha_{kl}^{eff} \beta^{eff-1}.$$

The macroscopic model takes the form,

$$\frac{\partial \langle \sigma_{ij}^{T} \rangle}{\partial X_j} = (\rho_{\mathbb{G}}^{eff} + \Phi \rho_{\mathbb{P}}) \frac{\partial^2 u_{\mathbb{G}i}}{\partial t^2} + O_r(\varepsilon)$$

$$\sigma_{ij}^{T} = a_{ijkh}^{*eff} \, e_{Xkh}(\mathbf{u}_{\mathbb{G}}) + O_r(\varepsilon).$$

The porous medium behaves macroscopically like a monophasic elastic medium with the elastic tensor $\mathbf{a}^{*eff}$ and the density $\rho_{\mathbb{G}}^{eff} + \Phi \rho_{\mathbb{P}}$.

8.7   Monophasic Viscoelastic Macroscopic Behavior

Again increasing the viscosity, there is no more contrast between the mechanical properties of the constituents, $C = O(1)$. Eq. (3.12) is now written with $p = 1$:

$$\sigma_{\mathfrak{P}ij} = 2\mu\varepsilon^{-1}\mathcal{D}_{yij}(\mathbf{v}) - p\,I_{ij}.$$

It is convenient to introduce the displacement $\mathbf{u}$ defined by $\mathbf{u}_{\mathfrak{S}}$ in $\mathfrak{S}$ and $\mathbf{u}_{\mathfrak{P}}$ in $\mathfrak{P}$. We look for $\mathbf{u}$ in the form,

$$u_i(x, y) = u_i^{(0)}(x, y) + \varepsilon u_i^{(1)}(x, y) + \varepsilon^2 u_i^{(2)}(x, y) + \dots .$$

The first boundary-value problem concerns $\mathbf{u}^{(0)}$:

$$\frac{\partial}{\partial y_j}(a_{ijkl}\,e_{ykl}(\mathbf{u}^{(0)})) = 0 \qquad\qquad \text{in } \mathfrak{S},$$

$$\mu\frac{\partial^2(i\omega u_i^{(0)})}{\partial y_j \partial y_j} = 0,$$

$$\frac{\partial\,u_i^{(0)}}{\partial y_i} = 0 \qquad\qquad \text{in } \mathfrak{P},$$

$$a_{ijkl}\,e_{ykl}(\mathbf{u}^{(0)})\,v_j = 2\mu\,e_{yij}(i\omega\mathbf{u}^{(0)})\,v_j,$$

and

$$[u_i^{(0)}] = 0 \qquad\qquad \text{on } \Gamma,$$

where $\mathbf{u}^{(0)}$ is Y-periodic with respect to y. It follows that

$$\mathbf{u}^{(0)} = \mathbf{u}^{(0)}(x).$$

The following orders show $\mathbf{u}^{(1)}$ and $p^{(0)}$ as the solutions of the set:

$$\frac{\partial}{\partial y_j}[a_{ijkl}\,(e_{ykl}(\mathbf{u}^{(1)}) + e_{xkl}(\mathbf{u}^{(0)}))] = 0 \qquad\qquad \text{in } \mathfrak{S},$$

$$\mu\frac{\partial^2(i\omega u_i^{(1)})}{\partial y_j \partial y_j} - \frac{\partial p^{(0)}}{\partial y_i} = 0,$$

$$\frac{\partial\,u_i^{(1)}}{\partial y_i} + \frac{\partial\,u_i^{(0)}}{\partial x_i} = 0 \qquad\qquad \text{in } \mathfrak{P},$$

$$a_{ijkl}\,(e_{ykl}(\mathbf{u}^{(1)}) + e_{xkl}(\mathbf{u}^{(0)}))\,v_j$$
$$= [-p^{(0)}I_{ij} + 2\mu i\omega(e_{ykl}(\mathbf{u}^{(1)}) + e_{xkl}(\mathbf{u}^{(0)}))]\,v_j,$$

and

$$[u_i^{(1)}] = 0 \qquad\qquad \text{on } \Gamma.$$

Because of the linearity of the equations, the solutions can be written in the form,

$$u_i^{(1)} = \chi_i^{lm} e_{xlm}(\mathbf{u}^{(0)}) + \bar{u}_i^{(1)}(x),$$

and

$$p^{(0)} = \zeta_{lm} e_{xlm}(\mathbf{u}^{(0)}).$$

Here $\bar{u}_i^{(1)}(x)$ is an arbitrary vector, independent of y. The tensors $\chi(y)$ and $\zeta(y)$ are complex-valued, and they depend on y and ω. Due to the stress continuity on Γ at the order ε^0, there is no additive constant in the expression of $p^{(0)}$.

The only compatibility relation is obtained from

$$\frac{\partial \, \sigma_{\mathfrak{G}ij}^{(2)}}{\partial y_j} + \frac{\partial \, \sigma_{\mathfrak{G}ij}^{(1)}}{\partial x_j} = -\rho_{\mathfrak{G}} \omega^2 u_i^{(0)} \qquad \text{in } \mathfrak{G},$$

$$\frac{\partial \, \sigma_{\mathfrak{P}ij}^{(1)}}{\partial y_j} + \frac{\partial \, p_{ij}^{(0)}}{\partial x_j} = -\rho_{\mathfrak{P}} \omega^2 u_i^{(0)} \qquad \text{in } \mathfrak{P},$$

and

$$\sigma_{\mathfrak{G}ij}^{(2)} \nu_j = \sigma_{\mathfrak{P}ij}^{(1)} \nu_j \qquad \text{on } \Gamma.$$

By integrating the first two equations over their respective domains of definition,

$$\frac{\partial \, \langle \sigma_{ij}^{T(0)} \rangle}{\partial x_j} = -(\rho_{\mathfrak{G}}^{eff} + \Phi \rho_{\mathfrak{P}}) \omega^2 u_i^{(0)},$$

and

$$\sigma_{ij}^{T(0)} = a_{ijkh}^{**eff} \, e_{xkh}(\mathbf{u}^{(0)}),$$

where $\mathbf{a}^{**eff}$ is given by

$$a_{ijkh}^{**eff} = \langle a_{ijkh} + a_{ijlm} e_{ylm}(\chi^{kh}) \rangle_{\mathfrak{G}} + \langle -\zeta_{ij} \zeta_{kh} + 2i\omega\mu e_{yij}(\chi^{kh}) + 2i\omega\mu I_{ij} I_{kh} \rangle_{\mathfrak{P}}.$$

The monochromatic macroscopic model takes the form,

$$\frac{\partial \, \langle \sigma_{ij}^T \rangle}{\partial X_j} = -(\rho_{\mathfrak{G}}^{eff} + \Phi \rho_F) \omega^2 u_{\mathfrak{G}i} + O_r(\varepsilon),$$

and

$$\sigma_{ij}^T = a_{ijkh}^{**eff} \, e_{Xkh}(\mathbf{u}_{\mathfrak{G}}) + O_r(\varepsilon).$$

The fourth-order tensor $\mathbf{a}^{**eff}$ is complex-valued and ω-dependent. The model V describes a monophasic viscoelastic medium at constant pulsation.

8.8   Acoustics of Double-Porosity Media

Now we will consider porous saturated media with double porosity, where the porous system is composed of two connected subsystems, the pores and the fractures, with very different characteristic lengths. Denoting these two lengths l and l', respectively, the porous media exhibit a small dimensionless parameter:

$$\frac{l}{l'} = \eta \ll 1. \qquad (8.1)$$

The medium undergoes an acoustic excitation of wave length $\lambda \gg l'$. It introduces a third characteristic length $l'' = \lambda/2\pi$. Therefore, a second small dimensionless parameter appears:

$$\frac{l'}{l''} = \varepsilon \ll 1. \qquad (8.2)$$

For simplicity, the two porous subsystems are assumed to be periodic. At the fracture scale, the periodic cell Y' comprises the fracture domain $\mathfrak{F}$ and the domain $\mathfrak{SP}'$ occupied by the microporous saturated matrix. Γ' is their common interface. At the micropore scale, the periodic cell Y comprises the pore domain $\mathfrak{P}$ and the solid domain $\mathfrak{S}$. Γ is their common interface. We denote Φ_P and $\Phi_{\mathfrak{F}}$ as the porosities of the micropores and the fractures, respectively. We assume again an elastic porous matrix and a Newtonian incompressible fluid, with macroscopic stresses in the two media of the same order of magnitude.

   The study is restricted to situations where the macroscopic description is two-phase. Therefore, the fractured medium satisfies the conditions for model III (section 8.5). In this case, the relative value of the two scale separations strongly influences the macroscopic description. The introduction of double porosity in models I, II, IV, and V does not change the structure of the macroscopic description. Details of the calculus can be found in [AB92], [AB93], and [AB94]. Partial homogenization can be found in [ADH90].

8.8.1   A Priori Estimate

The fractured media satisfy the conditions of validity of model III. Therefore,

$$C = \frac{\mu\omega}{a} = O(\varepsilon^2).$$

The fractured medium is excited at the characteristic pulsation $\omega_{\mathfrak{F}}$ for diphasic behavior defined by

$$\omega_{\mathfrak{F}} = O(\varepsilon^2 \frac{a}{\mu}).$$

The characteristic pulsation $\omega_{\mathfrak{P}}$ for the diphasic behavior of the microporous medium is introduced similarly:

$$\omega_{\mathfrak{P}} = O(\eta^2 \frac{a}{\mu}).$$

The behavior of the double-porosity medium is characterized by the dimensionless number,

$$\tau = \frac{\omega_{\mathfrak{F}}}{\omega_{\mathfrak{P}}} = O(\varepsilon^2 \eta^{-2}).$$

Three different macroscopic models can be anticipated:

- Model III-1: small pore-fracture scale separation

$$\varepsilon = O(\eta^2).$$

The two characteristic pulsation times are related by

$$\omega_{\mathfrak{P}} \gg \omega_{\mathfrak{F}}.$$

As the double-porosity medium is excited at the pulsation ω_F, the microporous medium is subjected to a pulsation very much smaller than its characteristic pulsation $\omega_{\mathfrak{P}}$. The flow in the microporous matrix is instantaneous when compared to the flow in the fractured medium. The microporous matrix plays the role of a fluid reservoir with instantaneous exchanges with the fractures.

- Model III-2: equal scale separations

$$\varepsilon = O(\eta).$$

The two characteristic pulsations are of the same order of magnitude:

$$\omega_{\mathfrak{P}} = O(\omega_{\mathfrak{F}}).$$

The flow in the microporous matrix strongly influences the flow in the fractured medium. The coupling between the two porosity systems is important.

- Model III-3: large pore-fracture scale separation

$$\varepsilon = O(\sqrt{\eta}).$$

The two characteristic pulsation times are related by

$$\omega_{\mathfrak{P}} \ll \omega_{\mathfrak{F}}.$$

The flow in the microporous matrix is negligible when compared to the flow in the fractured medium. It does not influence the flow in the fractured medium. The microporous medium behaves as a monophasic, elastic medium, model IV, $C = O(\eta)$.

8.8.2   Double-Porosity Macroscopic Models

Again, we assume that the amplitudes are small and the nonlinearities are neglected. The local description in the solid $\mathfrak{S}$, the fractures $\mathfrak{F}$, and the micropores $\mathfrak{P}$ is as follows:

$$\frac{\partial \sigma_{\alpha ij}}{\partial X_j} = -\rho_\alpha \omega^2 u_{\alpha i} \qquad in\ \alpha = \mathfrak{S}, \mathfrak{P}, \mathfrak{F},$$

with

$$\sigma_{\mathfrak{S}ij} = a_{ijkh}\, e_{Xkh}(\mathbf{u}_\mathfrak{S}) \qquad in\ \mathfrak{S},$$

$$\frac{\partial v_{\alpha i}}{\partial X_i} = 0, \qquad \sigma_{\alpha ij} = -p_\alpha I_{ij} + 2\mu \mathcal{D}_{Xij}(\mathbf{v}_\alpha) \qquad in\ \alpha = \mathfrak{P}, \mathfrak{F},$$

$$[\sigma_{ij} v_j] = 0, \qquad [u_i] = 0, \qquad on\ \Gamma,$$

$$(\langle \sigma_{\mathfrak{S}ij}\rangle_Y + \langle \sigma_{\mathfrak{P}ij}\rangle_Y) v_i' = \sigma_{\mathfrak{F}ij}' v_i', \qquad p_\mathfrak{F} = p_\mathfrak{P},$$

and

$$v_{\mathfrak{F}i} = \langle v_{\mathfrak{S}i}\rangle_\mathfrak{S} + \langle v_{\mathfrak{P}i}\rangle_\mathfrak{P} \qquad on\ \Gamma'.$$

The three characteristic lengths l, l' and l'' introduce three dimensionless space variables x, x' and x''

$$x = \frac{X}{l},$$

$$x' = \frac{X}{l'},$$

and

$$x'' = \frac{X}{l''},$$

where X is the physical space variable. Each quantity φ depends, *a priori*, on these three dimensionless space variables:

$$\varphi = \varphi(x, x', x'').$$

The method of developing macroscopic models consists of introducing adequate three-scale asymptotic expansions in the above set. For details, see [AB92], [AB93], and [AB94]. For simplicity, the density $\rho_\mathfrak{S}$ of the solid is assumed homogeneous. The following three macroscopic models are obtained:

Model III-1: small pore-fracture scale separation

$$\varepsilon = O(\eta^2).$$

The macroscopic behavior at constant pulsation ω is given by

$$\frac{\partial \langle \sigma_{ij}^T\rangle}{\partial X_j} =$$

$$-((1 - \Phi_\mathfrak{P})\rho_\mathfrak{S} + \phi_\mathfrak{P}\, \rho_\mathfrak{P}) (1 - \Phi_\mathfrak{F})\, \omega^2 u_{\mathfrak{S}i} + i\omega \rho_\mathfrak{P} \langle v_{\mathfrak{F}i}\rangle_\mathfrak{F} + O_r(\varepsilon),$$

$$\langle \sigma_{ij}^T \rangle = a_{ijkh}^{eff} \, e_{kh}(\mathbf{u}_\mathfrak{S}) - \gamma_{ij}^{eff} \, p + O_r(\varepsilon),$$

$$\frac{\partial \left(\langle v_{\mathfrak{F}i} \rangle_\mathfrak{F} - \Phi_\mathfrak{F} \, i\omega \, u_{\mathfrak{S}i} \right)}{\partial X_i} = - \alpha_{ij}^{eff} \, i\omega \, e_{Xij}(\mathbf{u}_\mathfrak{S}) - \beta \, i\omega \, p + O_r(\varepsilon),$$

and

$$\langle v_i \rangle_\mathfrak{F} - \Phi_\mathfrak{F} \, i\omega \, u_{\mathfrak{S}i} = - K_{\mathfrak{F}ij}^{eff} \left(\frac{\partial p}{\partial X_j} - \omega^2 \rho_\mathfrak{W} \, u_{\mathfrak{S}i} \right) + O_r(\varepsilon).$$

$\mathbf{a}^{eff}$, γ^{eff}, α^{eff}, and β^{eff} are classical, quasi-static, effective elastic tensors. p is the pressure in the fractures and in the pores. $\mathbf{K}_\mathfrak{F}^{eff}$ is the acoustic permeability of the fracture system, already introduced for models II and III. It is complex-valued and ω-dependent. We obtain a Biot-like description, but with broken symmetry, because, in general,

$$\gamma^{eff} \neq \alpha^{eff}.$$

Model III-2: equal scale separations

$$\varepsilon = O(\eta).$$

The macroscopic description is given by the above set for model III-2 with the following modifications:

- $\mathbf{a}^{eff}$, γ^{eff}, α^{eff}, $\gamma^{eff} \neq \alpha^{eff}$, and β^{eff} are now complex-valued and ω-dependent tensors.

- $p = p_\mathfrak{F} \neq p_\mathfrak{W}$ is the pressure in the fractures.

Model III-3: large pore-fracture scale separation

$$\varepsilon = O(\sqrt{\eta}).$$

The macroscopic description is again given by the set for model III-1. However, $\mathbf{a}^{eff}$, γ^{eff}, α^{eff}, $\gamma^{eff} = \alpha^{eff}$, and β^{eff} are the classical effective elastic tensors of model III. The description is similar to the description of a single-porosity medium.

8.9   Conclusion

In the above presentation, we have restricted our investigations to elastic matrices and incompressible Newtonian fluids. Several extensions are available. Concerning simple porous media, the case of a medium saturated by two incompressible and immiscible fluids is studied in [ALB89] and [Aur91a], whereas very compressible fluids are introduced in [ASBH90] for the quasi-statics. In [AR93b], the results of section 8.8 are extended to the case of a very compressible fluid. Finally, it is possible to show that model III-2, where full coupling exists between the two porous systems, is not reducible to a two-pressure field model, as, for example, the model of Barrenblatt, [BZK60]. The problem was analyzed in [AR93a] for double-conductivity media.

9

Microstructure Models of Porous Media

Ralph E. Showalter[1]

9.1   Introduction

Every attempt to exactly model laminar flow through highly inhomogeneous media, e.g., fissured or layered media, leads to very singular problems of partial differential equations with rapidly oscillating coefficients. Various methods of averaging will yield corresponding types of *double-porosity* models, and we shall describe some of these.

As a first approximation to flow in a region Ω which consists of such a composite of two finely interspersed materials, one can consider *averaged solutions*, one for each material and both defined at every point $x \in \Omega$. This leads to a pair of partial differential equations, one identified with each of the two components and a coupling term that describes the flow across the interface between these components. The values of the two dependent variables in this system (the solutions) at each point x have been obtained by averaging in the respective media over a generic neighborhood, which is located at $x \in \Omega$ and is sufficiently large to contain a representative sample of each component. Because the two components are treated symmetrically in the resulting system of two parabolic partial differential equations, such a double-porosity model is said to be of *parallel flow* type. Although appropriate for many situations, this symmetric treatment of the two components can be a real limitation. For a fissured medium, for example, such a representation is particularly restrictive, because the porous and permeable cells within the structure have flow properties radically different from those of the surrounding highly developed system of fissures. Moreover the geometry of the individual cells and the corresponding interface are lost in the averaging process leading to such models. For layered media, similar remarks apply, and these could be supplemented by nonisotropic considerations. In general, essential limitations of the parallel flow models are the suppression of the geometry of the cells and their corresponding interfaces on which coupling occurs and the lack of any distinction between the space and time scales of the two components of the medium.

---

[1]**Acknowledgements:** The author gratefully acknowledges the National Science Foundation for the support of research reported in this chapter.

To overcome these deficiencies, we consider the class of double-porosity models which we call *distributed microstructure* models. These are known in many cases to be the limit (by homogenization) as the scale of the inhomogeneity tends to zero, and they provide a way to represent a continuous distribution of cells with prescribed geometry. At each point $x \in \Omega$, there is given a representative model cell $\mathfrak{B}_x$. The flow within each such cell is described independently by an initial-boundary-value problem, and the solution of this local problem on $\mathfrak{B}_x$ is coupled on the boundary Γ_x to the value near x of the solution of the flow problem in the global region Ω. Thus, we have a continuum of partial differential equations to describe the local flow on the microscale, one at each point $x \in \Omega$, and these are each coupled to a single partial differential equation in Ω for the macroscale flow. Such *continuous* models represent a good approximation of the real, but discrete (and very singular), case of a finite but very large number of cells. This concept was introduced in the early 1930s (see section 9.3), and it has arisen repeatedly in a variety of applications which we will mention below.

We shall illustrate these two classes of double-porosity models with examples of single-phase flow in various types of fissured media. The class of parallel models is briefly treated in section 9.2 for motivation and orientation. The distributed microstructure models are introduced in section 9.3 by a simple and classical example, and we discuss this class in much more detail. We use both types to describe, first, the example of a *totally fissured medium* in which the individual cells are completely isolated from each other by the fissure system. In this situation, the cells act only as storage sites, and there is no direct diffusion from cell to cell within the matrix. Then, we introduce corresponding models for the more general example of a *partially fissured medium* in which there is some fluid flow directly through the cell structure. This flow through the cells is driven indirectly by the pressure gradient in the fissure system, and it contributes an additional component to the velocity field in the fissure system which we call the *secondary flux*. We will also display the form of the functional differential equations by which these problems can be represented; here, the family of local problems is replaced by convolution terms. We shall describe some typical results of the theory of such systems in section 9.4. This is illustrated in the simplest case as an application of continuous direct sums of Hilbert spaces which arise rather naturally as the energy or state spaces for the corresponding (stationary) variational or (temporal) dynamic problems. In section 9.5, we will indicate in some detail how the distributed microstructure models arise, by way of homogenization, from exact but singular models, partial differential equations with highly periodic coefficients or geometry, and, then, we will close with some additional remarks.

9.2 Parallel Flow Models

The classical example of a *parallel flow model* for single-phase flow in a composite medium is the parabolic system,

$$\frac{\partial}{\partial t}(au_1) - \nabla \cdot (A\nabla u_1) + \frac{1}{\delta}(u_1 - u_2) = f_1, \qquad (2.1.a)$$

and

$$\frac{\partial}{\partial t}(bu_2) - \nabla \cdot (B\nabla u_2) + \frac{1}{\delta}(u_2 - u_1) = f_2, \qquad (2.1.b)$$

discussed in [BZK60] for which u_1 represents the density of fluid in the first material and u_2 the density in the second. The coefficients $a(x)$ and $A(x)$ are the porosity and permeability of the first material, respectively, whereas $b(x)$ and $B(x)$ are the corresponding properties of the second material. The first equation quantifies the rate of flow in the first component of the composite, and the second equation quantifies the corresponding flow rate in the second. Both of these equations are to be understood macroscopically; that is, they were obtained by averaging over a generic neighborhood sufficiently large to contain contributions from each component. The third term in each equation is an attempt to quantify the exchange of fluid between the two components. See [Rub48] for a corresponding system which describes heat conduction in such a composite medium.

9.2.1 Totally Fissured Media

A *fissured medium* consists of a matrix of porous and permeable material cells through which is intertwined a highly developed system of fissures. The bulk of the flow occurs in the highly permeable fissure system, and most of the storage of fluid is in the matrix of cells which accounts for almost all of the total volume. One approach to constructing a model of such a medium is to regard the fissure system as the first component and the cell matrix as the second component of a general composite by adjusting the coefficients in (2.1) appropriately. These fissured media characteristics are modeled by choosing very small values for the coefficients $a(x)$ and $B(x)$ in (2.1). Because one component is essentially responsible for storage and the other for transport, the distributed exchange of fluid between the two components is of fundamental importance. The parameter δ represents the resistance of the medium to this exchange. (When $\delta = \infty$, no exchange flow is possible, and the system is completely decoupled.) An alternative interpretation is that $1/\delta$ represents the degree of fissuring in the medium. (When the degree of fissuring is infinite, the exchange flow encounters no resistance and $u_1 = u_2$.)

To specialize the system (2.1) to a totally fissured medium in which the individual cells are isolated from each other, one sets $B = 0$, because there is no direct flow through the matrix of cells. Only an indirect exchange occurs by way of the fissures. Thus, the condition $B = 0$ corresponds to a *totally fissured* medium in which each cell of the matrix is isolated from adjacent cells by the fissure system. The resulting system of parabolic-ordinary differential equations,

$$\frac{\partial}{\partial t}(au_1) - \nabla \cdot (A\nabla u_1) + \frac{1}{\delta}(u_1 - u_2) = f,　\qquad (2.2.a)$$

and

$$\frac{\partial}{\partial t}(bu_2) + \frac{1}{\delta}(u_2 - u_1) = 0,　\qquad (2.2.b)$$

is called the *first-order kinetic* model, because the cell storage is regarded as an added kinetic storage perturbation of the global fissure system.

Eq. (2.2.b) models the delay inherent in the flow between the fissures and blocks. It is precisely this delay that led to the introduction of such models by Barenblatt, Zheltov, and Kochina [BZK60] and Warren and Root [WR63], three decades ago, to better match observed reservoir behavior. See [Anz26], [Dea63], [CS64], [LN73], [GW76], [CHP87], [DK91], [Kna89], and [SW90] for additional applications and mathematical developments of such models.

If we further specialize this model by setting $a = 0$ to realize that the relative volume of the fissures is zero, we obtain the *pseudoparabolic*, partial differential equation

$$\frac{\partial}{\partial t}b\left(u_1 - \delta \nabla \cdot (A\nabla u_1)\right) - \nabla \cdot (A\nabla u_1) = f + \delta\frac{\partial f}{\partial t}.　\qquad (2.3)$$

See [CS76] for the development of such equations. Their solutions are determined by a *group* of operators on appropriate spaces, and their dynamics is regularity-*preserving*.

9.2.2   Partially Fissured Media

Next we present a parallel flow model for a *partially fissured* medium, a fissured medium in which there are substantial flow paths directly joining the cells in addition to the predominant connection with the surrounding fissure system. Thus, the cells are not completely isolated from one another by the fissure system, and the matrix is somewhat connected. This model allows for a *secondary flux* that arises from this bridging between cells. The model above is based on assuming that the exchange flow between the components has a spatially distributed density proportional to the pressure differences, that is, the fluid stored in the cell system at a point in space is determined solely by the history of the *values* of the pressures of the components at that point. To induce flow within the cell matrix as a response to local effects, however, it is necessary to apply a pressure *gradient* from the fissure system. Thus, we shall model the flux exchange as a response to both the value and the *gradient* of the pressure. Equivalently, we assume that the local cell structure at a point is responsive to the best linear approximation of the pressure in a neighborhood of that point. Furthermore, if the geometry of the cell matrix is symmetric with respect to the coordinate system, then, the response of the cell to the value and to the gradient of the pressure is additive, the terms representing even and odd responses, respectively, to even and odd input. Thus, we are led to model the resulting storage and transport responses within the cell matrix as two independent processes whose effects are additive.

In this situation, one must account for the effect of the gradient of the fissure flow on the local flow within the cells, for it is this fissure pressure gradient which necessarily provides the driving force for this transport within the matrix. To implement this in a model of the parallel flow type, we introduce, into the first-order kinetic model (2.2), a *secondary flux* $\vec{u}_3$ through the cell system. This flux is assumed to respond to the fissure pressure gradient with a delay analogous to that of the cell storage in response to the value of fissure pressure in (2.2). This leads to the *second-order kinetic model*,

$$\frac{\partial}{\partial t}(au_1) - \nabla \cdot (A\nabla u_1) + \frac{1}{\delta}(u_1 - u_2) + \nabla \cdot C^*\vec{u}_3 = f, \qquad (2.4.a)$$

$$\frac{\partial}{\partial t}(bu_2) + \frac{1}{\delta}(u_2 - u_1) = 0, \qquad (2.4.b)$$

and

$$\frac{\partial}{\partial t}(c\vec{u}_3) + \frac{1}{\beta}(\vec{u}_3 + C\nabla u_1) = 0. \qquad (2.4.c)$$

We assume that the responses of the cell structure to the value and to the gradient of fissure pressure are additive at a point, an assumption valid for cell structures which are symmetric with respect to coordinate directions. The third and fourth terms in (2.4.a) give the distributed mass flow rate into the cell matrix from the fissure system at a point. According to (2.4.b), the first of these goes toward the storage of fluid in the cells. Fluid from the fissure system enters the cell system at a point of higher pressure, it flows through the cell matrix to a point of lower pressure, and, then, it exits from the cell back into the fissure system according to the second of these exchange terms in (2.4.a). The secondary flux $\vec{u}_3$ follows the fissure pressure gradient according to (2.4.c). The matrix C^*C arises from the bridging of the cells, and it distinguishes the *partially fissured* model (2.4) from the *totally fissured* model (2.2). Such a system may also be appropriate to describe totally fissured media composed of *larger* cells for which the more accurate approximation of u is necessary. See [RS95] for a discussion and development of such models with multiple nonlinearities.

9.3   Distributed Microstructure Models

We first describe an elementary example in which a distributed microstructure model arises very naturally. For historical reasons, it will be presented in the framework of a heat conduction model, but, clearly, it has a meaningful analog for the analogous problem of diffusion and absorption of a dissolved chemical in a fluid flowing through a porous medium. The problem concerns the following situation. A system of pipes is used to absorb heat from circulating hot water and, then, later, to return this heat to colder water, thereby, serving as an energy storage

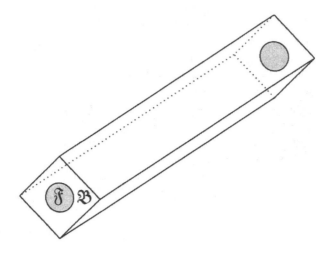

FIGURE 9.1. Heat recuperator.

device. This system has two sources of singularity, a geometric one due to the high ratio between the pipe length and the cross section and a material one due to the very different conductive properties of the two materials involved. We seek a model in which these two singularities are properly balanced to obtain a good description of the exchange process.

We begin by examining the heat exchange in a single pipe in such a system. The heat transport in the water is primarily convective due to the unit velocity of the water. The transport in the pipe walls is purely conductive and relatively very slow. The pipe cross-sections are very small, and these must be properly scaled to accurately model the heat exchange between them and the water. We introduce a small parameter $\varepsilon > 0$ to quantify this. A reference cell $Y = \{y = (y_1, y_2) \in \mathbb{R}^2 : |y_1| + |y_2| < 2^{3/2}\}$ defines the structure of the cross-sections, and we write the parts of this double component domain as $\overline{Y} = \overline{\mathfrak{F}} \cup \overline{\mathfrak{B}}$, with $\mathfrak{F} = \{y \in \mathbb{R}^2 : \|y\| < 1\}$ representing the internal flow region and $\mathfrak{B} = \{y \in \mathbb{R}^2 : 1 < \|y\|$ and $|y_1| + |y_2| < 2^{3/2}\}$ the walls of the pipe. The boundary of $\mathfrak{F}$ is the unit circle $\Gamma_{\mathfrak{F}}$, and that of Y is the square $\Gamma_{\mathfrak{B}}$. The boundary of $\mathfrak{B}$ is given by $\Gamma_{\mathfrak{F}} \cup \Gamma_{\mathfrak{B}}$. We denote the unit vector normal to $\Gamma_{\mathfrak{F}} \cup \Gamma_{\mathfrak{B}}$ by $\vec{\nu}$ which points in the direction *out* of $\mathfrak{B}$. We shall represent the very small cross-sections by $\mathfrak{F}^\varepsilon = \varepsilon\,\mathfrak{F} = \{z = (z_1, z_2) = \varepsilon(y_1, y_2) \in \mathbb{R}^2 : (y_1, y_2) \in \mathfrak{F}\}$ and $\mathfrak{B}^\varepsilon = \varepsilon\,\mathfrak{B} = \{z = \varepsilon y \in \mathbb{R}^2 : y \in \mathfrak{B}\}$. Similarly, we denote the corresponding boundaries by $\Gamma_{\mathfrak{F}}^\varepsilon$ and $\Gamma_{\mathfrak{B}}^\varepsilon$ (see Fig. 9.1).

Let the interval $\Omega = (0, 1)$ denote the axis of the very narrow pipe in which water is flowing at unit velocity through the cross section $\mathfrak{F}^\varepsilon$ and for which $\mathfrak{B}^\varepsilon$ is the pipe wall, a material of relatively *very low* conductivity. This situation is described by the initial-boundary-value problem,

$$\frac{\partial U^\varepsilon}{\partial t} + \frac{\partial U^\varepsilon}{\partial x} = \frac{\partial^2 U^\varepsilon}{\partial x^2} + \nabla_z \cdot \nabla_z U^\varepsilon(t, x, z), \qquad z \in \mathfrak{F}^\varepsilon, \ x \in \Omega,$$

$$\frac{\partial U^\varepsilon}{\partial t} = \varepsilon^2 \left(\nabla_z \cdot \nabla_z U^\varepsilon(t, x, z) + \frac{\partial^2 U^\varepsilon}{\partial x^2} \right), \qquad z \in \mathfrak{B}^\varepsilon, \ x \in \Omega,$$

$$\nabla_z U^\varepsilon \cdot \vec{v}|_{\mathfrak{F}^\varepsilon} = \varepsilon^2 \nabla_z U^\varepsilon \cdot \vec{v}|_{\mathfrak{B}^\varepsilon}, \qquad z \in \Gamma_{\mathfrak{F}}^\varepsilon, \ x \in \Omega,$$

and

$$\nabla_z U^\varepsilon \cdot \vec{v} = 0, \quad z \in \Gamma_{\mathfrak{B}}^\varepsilon, \ x \in \Omega,$$

which is the *exact ε-model*. The ε^2-scaled conductivity permits *very high* gradients in $\Omega \times \mathfrak{B}^\varepsilon$. However, only those components in the cross-sectional direction are responsible for the exchange flux, and, to see that it is balanced with the other transport terms, we rescale with $z = \varepsilon y$ and $\nabla_z = \frac{1}{\varepsilon} \nabla_y$ to get

$$\frac{\partial U^\varepsilon}{\partial t} + \frac{\partial U^\varepsilon}{\partial x} = \frac{\partial^2 U^\varepsilon}{\partial x^2} + \frac{1}{\varepsilon^2} \nabla_y \cdot \nabla_y U^\varepsilon(t, x, y), \quad y \in \mathfrak{F}, \ x \in \Omega, \qquad (3.1.a)$$

$$\frac{\partial U^\varepsilon}{\partial t} = \nabla_y \cdot \nabla_y U^\varepsilon(t, x, y) + \varepsilon^2 \frac{\partial^2 U^\varepsilon}{\partial x^2}, \quad y \in \mathfrak{B}, \ x \in \Omega, \qquad (3.1.b)$$

$$\nabla_y U^\varepsilon \cdot \vec{v}|_{\mathfrak{F}} = \varepsilon^2 \nabla_y U^\varepsilon \cdot \vec{v}|_{\mathfrak{B}}, \quad y \in \Gamma_{\mathfrak{F}}, \ x \in \Omega, \qquad (3.1.c)$$

and

$$\nabla_y U^\varepsilon \cdot \vec{v} = 0, \quad y \in \Gamma_{\mathfrak{B}}, \ x \in \Omega. \qquad (3.1.d)$$

Now it is easy to recognize the limit as $\varepsilon \to 0$. In $\mathfrak{F}$, we get $U(t, x, y) = u(t, x)$ in the limit, i.e., it is independent of $y \in \mathfrak{F}$. Average (3.1.a) over $\mathfrak{F}$ and use (3.1.c) with Gauss' theorem to get an integral over $\Gamma_{\mathfrak{F}}$. This gives the following limiting form of the problem:

$$\frac{\partial u}{\partial t} + \frac{\partial u}{\partial x} = \frac{\partial^2 u}{\partial x^2} - \frac{1}{|\mathfrak{F}|} \int_{\Gamma_{\mathfrak{F}}} \nabla_y U \cdot \vec{v} \, d\Gamma, \quad x \in \Omega, \qquad (3.2.a)$$

$$\frac{\partial U}{\partial t} = \nabla_y \cdot \nabla_y U(t, x, y), \quad y \in \mathfrak{B}, \ x \in \Omega, \qquad (3.2.b)$$

$$U(t, x, y) = u(t, x), \quad s \in \Gamma_{\mathfrak{F}}, \ x \in \Omega, \qquad (3.2.c)$$

and

$$\nabla_y U^\varepsilon \cdot \vec{v} = 0, \quad s \in \Gamma_{\mathfrak{B}}, \ x \in \Omega. \qquad (3.2.d)$$

The integral term in (3.2.a) is the total heat lost to the pipe wall at the cross-section $x \in \Omega$, computed from the normal component of the heat gradient in the wall. It is comparable to the transport terms representing convection and conduction along the pipe. This balances the heat gained in the water with its moderate gradients against that exchanged with the pipe where the gradients are very high.

This problem approximates the heat transport and the exchange between the water and the pipe walls. It is essentially the *heat recuperator* problem introduced by Lowan [Low34], and it is our first example of a *distributed microstructure*

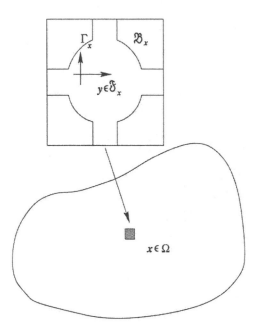

FIGURE 9.2. Domain and microcell.

model. The *global* domain is the interval Ω. For each point $x \in \Omega$, a *local cell* $\mathfrak{B}$ is identified, which is located at or identified with that point. In the heat exchange with the global medium, this family of cells acts as a distributed source, and the global medium is coupled to the cell only at that point on its boundary $\Gamma_{\mathfrak{F}}$. It provides a well-posed problem which is a good approximation to the apparently singular ε-problem. We see that the apparently singular term in (3.1.a) arose in rescaling from the singular geometry in the original problem, but it is balanced by the corresponding term in (3.1.b) without the small coefficient because of the flux condition (3.1.c). In particular, the choice of ε^2 as a coefficient in (3.1.b) exactly balanced the competing singularities introduced by the geometry and the materials.

Remark To describe the complete system of parallel pipes which makes up a heat recuperator, we need a periodic array of such cells εY to cover a fixed region in $\mathbb{R}^2$. Because the domain, then, has an ε-dependent structure not eliminated by a simple scale change as above, we need a more sophisticated technique to describe the limit. This is provided by *homogenization*. (See section 9.5.)

The introduction of *distributed microstructure models* for diffusion in porous media represents an attempt to recognize the geometry and the multiple scales in these problems to better quantify the exchange of fluid across the intricate interface between the components. We are given a domain Ω which represents the global region of the model. At each point $x \in \Omega$, there is specified a cell $\mathfrak{B}_x$, a magnified or scaled representation of the microstructure that is present near x (see Fig. 9.2). One partial differential equation is specified to describe the global flow in the region Ω, and a separate partial differential equation is specified in each cell $\mathfrak{B}_x$

to describe the flow internal to that cell. Any coupling between these equations will occur on the boundary of $\mathfrak{B}_x$, denoted by Γ_x. It is the collection $\{\Gamma_x : x \in \Omega\}$ which provides the interface on which this exchange takes place. Now we will use this concept to model some examples of single-phase flow in a fissured medium. These include analogs of the parallel flow models which were given above.

9.3.1   Totally Fissured Media

The global flow in the fissure system is described in the macroscale x by

$$\frac{\partial}{\partial t}\left(a(x)u(t,x)\right) - \nabla \cdot A(x)\nabla u + Q(t,x) = f(t,x), \qquad x \in \Omega, \qquad (3.3.\text{a})$$

where $Q(t,x)$ is the exchange term representing the flow into the cell $\mathfrak{B}_x$. The flow within each local cell $\mathfrak{B}_x$ is described by

$$\frac{\partial}{\partial t}\left(b(x,y)U(t,x,y)\right) - \nabla_y \cdot (B(x,y)\nabla_y U) = F(t,x,y), \qquad y \in \mathfrak{B}_x. \quad (3.3.\text{b})$$

The subscript y on the gradient indicates that the gradient is with respect to the *local* variable y. A gradient operator without any subscript will mean that the gradient is taken with respect to the *global* variable x. Because of the smallness of the cells, the fissure pressure is assumed to be well approximated by the "constant" value $u(t,x)$ at every point of the cell boundary, so that the effect of the fissures on the cell pressure is given by the interface condition,

$$B(x,y)\nabla_y U(t,x,y) \cdot \vec{v} + \frac{1}{\delta}(U(t,x,y) - u(t,x)) = 0, \qquad s \in \Gamma_x, \qquad (3.3.\text{c})$$

where $\vec{v}$ is the unit outward normal on Γ_x. (When $\delta = 0$, this becomes (and converges to) the *matched* boundary condition, $u(t,x) = U(t,x,y)$ for $s \in \Gamma_x$.) Finally, the amount of fluid flux across the interface, scaled by the cell size, determines the remaining term in (3.3.a) by

$$Q(t,x) = \frac{1}{|\mathfrak{B}_x|} \int_{\Gamma_x} B(x,y)\nabla_y U \cdot \vec{v} \, d\Gamma, \qquad (3.3.\text{d})$$

where $|\mathfrak{B}_x|$ denotes the Lebesgue measure of $\mathfrak{B}_x$, and this contributes to the *cell storage*. Thus, the system (3.3) comprises a double-porosity model of *distributed microstructure* type for a totally fissured medium. It needs only to be supplemented by appropriate boundary conditions for the global pressure $u(t,x)$ and initial conditions for $u(0,x)$ and $U(0,x,y)$ to comprise a well-posed problem. See Chapter 6 and [Low34], [RW50], [Ros52], [DJW53], [Vog82], [Bar85], [GD86], [ADH90], [ADH91], [Arb89a], [Arb91], [DJW53], [DPLAS87], [FK92], [FT87], [Hor91a], [Hor91b], [Hor92b], [Hor92a], [HJ87], [HJ91], [HJM94], [HJ93b], [PS94], [SW93], [SW91b], [SW91a], [Sho91], [Sho93], and [Sho89] for applications and mathematical theory for (3.3) and various related problems. A typical development of well-posedness results is given below in section 9.4.

Finally, we remark that the system (3.3) can be rewritten as a single equation of functional-differential type. By applying Gauss' theorem to (3.3.b) we obtain, from (3.3.d),

$$\frac{\partial}{\partial t} \int_{\mathfrak{B}_x} bU \, dy = \int_{\Gamma_x} B \frac{\partial U}{\partial \vec{v}} \, d\Gamma + \int_{\mathfrak{B}_x} F \, dy \, .$$

Then, by using Green's function for the problem (3.3.b) to represent the solution $U(t, x, y)$ as an integral over Γ_x of $u(t, x)$, we substitute this in (3.3.a) to get the implicit *convolution evolution equation*,

$$\frac{\partial}{\partial t} \left\{ a(x)u(t, x) + \int_0^t k(t - \tau, x)u(\tau, x) \, d\tau \right\} - \nabla \cdot A(x)\nabla u = f(t, x) \, . \quad (3.4)$$

The convolution term represents a storage effect with memory. See [HS90] for a direct treatment of this equation and particularly [Pes92], [Pes95a], and [Pes96] where this equation forms the basis for an independent theoretical and numerical analysis. Also see [Pes95b] and [Pes94] for related work.

9.3.2   Partially Fissured Media

We shall present two *distributed microstructure* models for flow in a partially fissured medium, a porous medium composed of *two* interwoven and connected components, the first being the system of fissures and the second being the matrix of porous cells. (Note that such a construction is impossible in $\mathbb{R}^2$.) In partially fissured media, the matrix of cells is connected, so that some of the flow passes directly through the block interconnections. Whereas the primary flow will continue to be the flow from cells into fissures followed by flow within the fissures, the flow within the porous matrix has more than only a local character, as in the case of a totally fissured medium. This effect is less promiment than the bulk flow in the fractures, but it can have a noticeable effect when the interconnections between the cells are sufficiently large.

For our first model of a partially fissured medium, we are motivated by our parallel flow model above, in which the cell system responds additively to the value and the gradient of fissure pressure, that is, to the *best linear approximation* of the fissure pressure at each cell location. As before, the global fluid flow in the fissures is described by

$$\frac{\partial}{\partial t} (a(x)u(t, x)) - \nabla \cdot A(x)\nabla u + Q(t, x) = f(t, x) \, , \qquad x \in \Omega \, , \qquad (3.5.a)$$

and the local flow within the matrix in the cell at each point x is given by

$$\frac{\partial}{\partial t} (b(x, y)U(t, x, y)) - \nabla_y \cdot B(x, y)\nabla_y U = F(t, x, y) \, , \qquad y \in \mathfrak{B}_x \, . \quad (3.5.b)$$

The boundary values for each cell problem are taken from information about the solution to the global equation in the vicinity of that point. The tacit assumption of the microstructure models is that the matrix cells are so small that the global solution u may be effectively approximated over the matrix cell boundary by an appropriate approximation to u. In the usual models, as above, the approximation for this purpose is merely the constant value of the global solution $u(t, x)$. Our objective here is to refine this model to more accurately describe the flow through the matrix. Thus, we shall assume that the pressure on the matrix boundary is driven by the best linear approximation of the fissure pressure, and this leads to the boundary condition,

$$B(x, y)\nabla_y U(t, x, y) \cdot \vec{v} + \frac{1}{\delta}(U(t, x, y) - u(t, x) - \nabla u(t, x) \cdot s) = 0, \qquad s \in \Gamma_x.$$
(3.5.c)

Here Γ_x is taken to be that part of the fissure-matrix interface that is *interior* to the unit cell, Y; periodic conditions are prescribed on the remaining part of that interface $\Gamma_\mathfrak{B}$, which intersects the boundary of the unit cell. Finally, the exchange term q in (3.5.a) consists of two parts, the average amount flowing into the cell to be stored and the divergence of the secondary flux flowing through the cell structure. The total exchange is given by

$$Q(t, x) = \frac{1}{|\mathfrak{B}_x|} \int_{\Gamma_x} B(x, y)\nabla_y U \cdot \vec{v} \, d\Gamma - \frac{1}{|\mathfrak{B}_x|} \nabla \cdot \left(\int_{\Gamma_x} B(x, y)(\nabla_y U \cdot \vec{v})s \, d\Gamma \right).$$
(3.5.d)

System (3.5) comprises the first of our two distributed microstructure models for a partially fissured medium. This model was introduced in [CS94] to describe the highly anisotropic situation in layered media and developed in [CS95] for more general media. See [Arb88] for an earlier discrete version and numerical work.

When the cells $\mathfrak{B}_x$ are symmetric in coordinate directions, one can separate the effects of storage from those of the secondary flux. Specifically, the storage can, then, be expressed in terms of the value of the fissure pressure at the point over a time interval through a convolution integral obtained, as before, from a Green's function representation of the cell problem, and the secondary flux and its corresponding contribution to the global flow are expressed, likewise, in terms of the global flux. This leads, just as before, to a functional partial differential equation of the form,

$$\frac{\partial}{\partial t}\left(a(x)u(t, x) + k_1(\cdot, x) * u(t, x)\right)$$
$$-\nabla \cdot \left(A(x)\nabla u(t, x) + \frac{\partial}{\partial t}k_{12}(\cdot, x) * \nabla u(t, x) + k_2(\cdot, x) * \nabla u(t, x)\right) \qquad (3.6)$$
$$= f(t, x), \qquad x \in \Omega, \ t > 0.$$

known as *Nunziato's equation*. This equation was presented in [Nun71], without any physical or philosophical justification, as an interesting generalization of heat conduction with memory models due to Gurtin and Chen. See [Mil78] for mathematical development of these equations.

Next we describe our second microstructure model for flow in a partially fissured medium. We use *three* variables in this model. The *fast* diffusion through the fissure system is determined, as before, by a concentration $u_1(t, x)$ there, but we specify the concentration in the matrix in two components. The first of these, $U(t, x, y)$, results from the ε^2-scaled *slow* diffusion in the matrix, as before, but we account also for a *moderate* component of flow in the matrix which has concentration $u_2(t, x)$. Thus, u_2 is the (low spatial frequency) component of the density in the matrix that leads to global diffusion, and U is the (high frequency) component of the density in the matrix that gives the local storage. These three variables, $u_1(t, x)$, $u_2(t, x)$, and $U(t, x, y)$ are defined in the fissures, the matrix, and individual cells, respectively. The model system is given by

$$\Phi_1 \frac{\partial u_1}{\partial t} - \nabla \cdot (A_{\mathfrak{F}} \nabla u_1) + Q(t, x) = 0, \tag{3.7.a}$$

$$\Phi_2 \frac{\partial u_2}{\partial t} - \nabla \cdot (A_{\mathfrak{B}} \nabla u_2) - Q(t, x) = 0, \quad x \in \Omega, \tag{3.7.b}$$

$$Q(t, x) = \int_\Gamma B\nabla_y U(t, x, y) \cdot \vec{v} ds, \tag{3.7.c}$$

$$\phi_2 \frac{\partial U}{\partial t} - \nabla_y \cdot (B\nabla U) = 0, \quad s \in \mathfrak{B}_x, \tag{3.7.d}$$

where $U(t, x, y)$ and $B\nabla_y U(t, x, y) \cdot \vec{v}$ are Y-periodic on $\Gamma_{\mathfrak{B}}$, and   (3.7.e)

$$U(t, x, y) + u_2(t, x) = u_1(t, x), \quad s \in \Gamma. \tag{3.7.f}$$

Note the different spatial domains on which the problems (3.7.b) and (3.7.d) are to be solved. The equation (3.7.d) is to be solved in the individual blocks which are now artificially disconnected along $\Gamma_{\mathfrak{B}}$, whereas (3.7.b) is to be solved in the whole of Ω, which includes all of the blocks and their interfaces to form a globally connected set. In view of (3.7.d), the flux (3.7.c) can be expressed by

$$q = \int_{\mathfrak{B}} \phi_2 \frac{\partial U}{\partial t} dy.$$

With appropriate boundary conditions for u_1 and u_2 and initial conditions on each of u_1, u_2, and U, this is a well-posed problem.

Condition (3.7.f) expresses a matching of *total* pressures across the matrix-fissure interface. We note that the condition (3.7.f), in a mathematical sense, is the dual to the right sides of (3.7.a) and (3.7.b). The system is complemented by a pair of conservation equations (3.7.e) on the artificial interface $\Gamma_{\mathfrak{B}}$ where pressure and flux are *localized*. This problem is a *hybrid* between parallel and microstructure models: the introduction of two matrix components is a *parallel* construction which was built into the ε-model, and it persisted in the limit. This is one of a family of such models which were derived in [DPS] by homogenization (see section 9.5.1) from corresponding ε-models.

9.4   A Variational Formulation

We will illustrate the mathematical formulation of microstructure models as evolution equations on various Hilbert spaces with the case of a *totally fissured medium*. This provides a means of establishing that they are well-posed problems and identifies the natural *energy* and *state* spaces for these dynamical problems. Let Ω be an open, bounded domain in $\mathbb{R}^3$ and, for every $x \in \Omega$, let $\mathcal{B}_x$ be a bounded region contained in $\mathbb{R}^3$. Identify the product space $\prod_{x \in \Omega} \mathcal{B}_x \equiv \mathbf{Q}$ as a subset of $\mathbb{R}^6$; we require that $\mathbf{Q}$ be a *measurable* subset of $\mathbb{R}^6$, hence, each of the cells $\mathcal{B}_x \equiv \{y : (x, y) \in \mathbf{Q}\}$ is a measurable subset of $\mathbb{R}^3$. Here, we will formulate the Cauchy–Dirichlet problem for the *linear parabolic system*,

$$\frac{\partial}{\partial t}\big(a(x)u(t, x)\big) - \nabla \cdot A(x)\nabla u(t, x)$$
$$+ \int_{\Gamma_x} B(x, y)\nabla_y U(t, x, y) \cdot \vec{v} \, d\Gamma = f(t, x), \qquad x \in \Omega, \tag{4.1.a}$$

$$\frac{\partial}{\partial t}\big(b(x, y)U(t, x, y)\big) - \nabla_y \cdot B(x, y)\nabla_y U(t, x, y) = F(t, x, y),$$
$$x \in \Omega, \ y \in \mathcal{B}_x, \tag{4.1.b}$$

and

$$B(x, y)\nabla_y U(x, y) \cdot \vec{v} = \frac{1}{\delta}\big(U(t, x, y) - u(x)\big), \qquad x \in \Omega, \ s \in \Gamma_x, \tag{4.1.c}$$

as an evolution equation in an appropriate Hilbert space. This is just the system (3.3) in which the measure $d\Gamma$ on Γ_x is used to absorb the extra factor of $|\mathcal{B}_x|$. We shall assume $a \in L^\infty(\Omega)$, $b \in L^\infty(\mathbf{Q})$, A and B are uniformly positive definite and bounded measurable matrix functions, $\vec{v}$ is the unit outward normal on Γ_x, and $\delta > 0$. We will further assume that each boundary Γ_x is piecewise C^1 and that the measures $|\Gamma_x|$ and $|\mathcal{B}_x|$ are uniformly bounded in x. We shall use the Lebesgue space $L^2(\mathbf{Q})$ with the norm,

$$\|U\|_{L^2(\mathbf{Q})} = \left(\int_\Omega \int_{\mathcal{B}_x} |U(x, y)|^2 \, dy \, dx\right)^{1/2},$$

for $U \in L^2(\mathbf{Q})$. Note that, for a.e. $x \in \Omega$, $U(x, \cdot) \in L^2(\mathcal{B}_x)$, and this determines a function U on Ω whose value at x belongs to the vector space $L^2(\mathcal{B}_x)$. Thereby, we identify $L^2(\mathbf{Q}) = L^2(\Omega; L^2(\mathcal{B}_x))$ as a space of vector-valued functions on Ω. We denote, by $W_0^{1,2}(\Omega)$, the Sobolev space of those functions in $L^2(\Omega)$ for which each first-order derivative is in $L^2(\Omega)$ and for which the (trace) boundary values are zero. See [Ada75] for information about Sobolev spaces. Similarly, we construct the space of vector-valued functions

$$L^2(\Omega; W^{1,2}(\mathcal{B}_x)) = \Big\{U \in L^2\big(\Omega; L^2(\mathcal{B}_x)\big) : \nabla_y U \in L^2\big(\Omega; L^2(\mathcal{B}_x)\big)\Big\}.$$

For the norm on $L^2(\Omega; W^{1,2}(\mathfrak{B}_x))$, we employ the notation,

$$|U|_2 = \left(\int_\Omega \int_{\mathfrak{B}_x} |\nabla_y U(x, y)|^2 \, dy \, dx \right)^{1/2} ,$$

so $\|U\|_2^2 = |U|_2^2 + \|U\|_{L^2(Q)}^2$. Denote, by $V = W_0^{1,2}(\Omega) \times L^2(\Omega; W^{1,2}(\mathfrak{B}_x))$, the indicated product space with norm,

$$\|[u, U]\|_V = \|u\|_{W_0^{1,2}(\Omega)} + \|U\|_2 .$$

Let γ_x be the usual trace map of $W^{1,2}(\mathfrak{B}_x)$ into $L^2(\Gamma_x)$, and define $\mathcal{B} = L^2(\Omega; L^2(\Gamma_x))$ and the distributed trace $\gamma : L^2(\Omega; W^{1,2}(\mathfrak{B}_x)) \to \mathcal{B}$ by $\gamma U(x, y) = (\gamma_x U(x))(s)$. We will require that the trace maps γ_x be uniformly bounded, so γ is continuous from $L^2(\Omega; W^{1,2}(\mathfrak{B}_x))$ into $\mathcal{B}$. Define $\lambda : W_0^{1,2}(\Omega) \to \mathcal{B}$ by: $\lambda u(x, y) = u(x)1_s$, $x \in \Omega$, $s \in \Gamma_x$, where $u(x)1_s$ is the constant function on Γ_x with value $u(x)$. We will employ the notation $\tilde{u} = [u, U]$.

Define the Hilbert space $H = L^2(\Omega) \times L^2(\Omega; L^2(\mathfrak{B}_x))$ with the inner product,

$$(\tilde{u}, \tilde{\varphi})_H = \int_\Omega a(x)u(x)\varphi(x) \, dx + \int_\Omega \int_{\mathfrak{B}_x} b(x, y)U(x, y)\Phi(x, y) \, dy \, dx ,$$

for $\tilde{u} = [u, U]$, $\tilde{\varphi} = [\varphi, \Phi] \in H$.

Define $V_h = \{\tilde{u} \in V : \gamma U = \lambda u \text{ in } \mathcal{B}\}$. Because γ and λ are continuous, V_h is a closed subspace of V. Also define $V_0 \equiv \{U \in L^2(\Omega; W^{1,2}(\mathfrak{B}_x)) : \gamma U = 0\}$. It can be shown that $W_0^{1,2}(\Omega) \times V_0$ and V_h are dense in H.

We shall write the system (4.1) as an evolution equation over the spaces described above. To obtain the variational form for the system, choose $[\varphi, \Phi] \in V$, multiply (4.1.a) by φ, and integrate over Ω. Multiply (4.1.b) by Φ, and integrate over both $\mathfrak{B}_x$ and Ω. Add these equations and apply Green's theorem to obtain

$$\int_\Omega \left\{ \frac{\partial}{\partial t}(a(x)u(t, x))\varphi(x) + \int_{\mathfrak{B}_x} \frac{\partial}{\partial t}\big(b(x, y)U(t, x, y)\big)\Phi(x, y) \, dy \right\} dx$$

$$+ \int_\Omega \left\{ A(x)\nabla u(t, x) \cdot \nabla \varphi(x) + \int_{\Gamma_x} B(x, y)\nabla_y U(t, x, y) \cdot \vec{\nu} \varphi(x) \, d\Gamma \right.$$

$$+ \int_{\mathfrak{B}_x} B(x, y)\nabla_y U(t, x, y) \cdot \nabla_y \Phi(x, y) \, dy$$

$$\left. - \int_{\Gamma_x} B(x, y)\nabla_y U(t, x, y) \cdot \vec{\nu}\gamma_x \Phi(x, y) \, d\Gamma \right\} dx$$

$$= \int_\Omega f(x)\varphi(x) \, dx + \int_\Omega \int_{\mathfrak{B}_x} F(x, y)\Phi(x, y) \, dy \, dx .$$

Combining the boundary integrals and substituting for $B(x, y)\nabla_y U(t, x, y) \cdot \vec{\nu}$ yields

$$[u(t), U(t)] \in V : \int_\Omega \frac{\partial}{\partial t} a(x)u(t, x)\varphi(x) \, dx \qquad (4.2)$$

$$+ \int_\Omega \int_{\mathcal{B}_x} \frac{\partial}{\partial t} b(x, y) U(t, x, y) \Phi(x, y) \, dy \, dx$$

$$+ \int_\Omega A(x) \nabla u(t, x) \cdot \nabla \varphi(x) \, dx$$

$$+ \int_\Omega \int_{\mathcal{B}_x} B(x, y) \nabla_y U(t, x, y) \cdot \nabla_y \Phi(x, y) \, dy \, dx$$

$$+ \int_\Omega \int_{\Gamma_x} \frac{1}{\delta} \big(\gamma U(t, x, y) - \lambda u(t, x, y) \big) \big(\gamma \Phi(t, x, y) - \lambda \varphi(t, x, y) \big) \, d\Gamma \, dx$$

$$= \int_\Omega f(x) \varphi(x) \, dx + \int_\Omega \int_{\mathcal{B}_x} F(x, y) \Phi(x, y) \, dy \, dx , \quad [\varphi, \Phi] \in V .$$

A special case of the above is obtained when (4.1.c) is replaced by

$$\gamma U(t, x, y) = \lambda u(t, x, y) \qquad x \in \Omega , \ s \in \Gamma_x , \ t > 0 . \tag{4.1.c$'$}$$

This is the formal result obtained by allowing $\delta \to 0^+$, so that (4.1.c)$'$ is forced to hold, and it corresponds to

$$[u(t), U(t)] \in V_h : \int_\Omega \frac{\partial}{\partial t} a(x) u(t, x) \varphi(x) \, dx \tag{4.2$'$}$$

$$+ \int_\Omega \int_{\mathcal{B}_x} \frac{\partial}{\partial t} b(x, y) U(x, y) \Phi(x, y) \, dy \, dx$$

$$+ \int_\Omega A(x) \nabla u(t, x) \cdot \nabla \varphi(x) \, dx$$

$$+ \int_\Omega \int_{\mathcal{B}_x} B(x, y) \nabla_x U(t, x, y) \cdot \nabla_y \Phi(x, y) \, dy \, dx$$

$$= \int_\Omega f(x) \varphi(x) \, dx + \int_\Omega \int_{\mathcal{B}_x} F(x, y) \Phi(x, y) \, dy \, dx , \quad [\varphi, \Phi] \in V_h .$$

The problem (4.1) will be called the *regularized model*, and (4.1)$'$, i.e., (4.1.a), (4.1.b), and (4.1.c)$'$ will be called the *matched model*. Conversely, starting from (4.2), it is not difficult to recover (4.1).

Define the Hilbert spaces

$$
\begin{aligned}
\mathcal{H} &= L^2(0, T; H), \\
\mathcal{W}_2 &= L^2(0, T; W_0^{1,2}(\Omega)), \\
\mathcal{V}_2 &= L^2(0, T; L^2(\Omega; W^{1,2}(\mathcal{B}_x))), \\
\mathcal{V} &= \mathcal{W}_2 \times \mathcal{V}_2 , \quad \text{and} \\
\mathcal{V}_h &= \{ \bar{u} \in \mathcal{V} : \gamma U(t) = \lambda u(t) \text{ in } \mathcal{B} \text{ for almost every } t > 0 \} .
\end{aligned}
$$

Define $L : V \to V'$ by

$$Lu(\tilde{\varphi}) = \int_\Omega A(x)\nabla u \cdot \nabla\varphi\, dx + \int_\Omega \int_{\mathcal{B}_x} B(x, y)\nabla_y U \cdot \nabla_y \Phi\, dy\, dx \quad \tilde{u}, \tilde{\varphi} \in V .$$

The conditions above on A and B imply that L is bounded and coercive from V into V'. The exchange term is given by the operator

$$\mathcal{M}(\tilde{u}, \tilde{\varphi}) \equiv \int_\Omega \int_{\Gamma_x} (\gamma U - \lambda u)(\gamma \Phi - \lambda\varphi)\, d\Gamma\, dx , \qquad \tilde{u}, \tilde{\varphi} \in V .$$

This gives a continuous and linear function $\mathcal{M} : V \to V'$. Let $\tilde{f} \in V'$ be given in the form

$$\tilde{f}(\tilde{\varphi}) = \int_\Omega f\varphi\, dx + \int_\Omega \int_{\mathcal{B}_x} F\Phi\, dy\, dx , \qquad \tilde{\varphi} \in V .$$

We will use the same notation to refer to the corresponding realizations of these operators on the spaces $\mathcal{H}$, V and V_h.

Integrating (4.2) from 0 to T,

$$\tilde{u} \in V : \int_0^T \left(\frac{\partial}{\partial t}\tilde{u}, \tilde{\varphi}\right)_H dt + \int_0^T L\tilde{u}(\tilde{\varphi})\, dt + \int_0^T \frac{1}{\delta}\mathcal{M}\tilde{u}(\tilde{\varphi})\, dt \qquad (4.5)$$

$$= \int_0^T \tilde{f}(\tilde{\varphi})\, dt , \qquad \tilde{\varphi} \in V .$$

Similarly, from (4.2)',

$$\tilde{u} \in V_h : \int_0^T \left(\frac{\partial}{\partial t}\tilde{u}, \tilde{\varphi}\right)_H dt + \int_0^T L\tilde{u}(\tilde{\varphi})\, dt = \int_0^T \tilde{f}(\tilde{\varphi})\, dt , \qquad \tilde{\varphi} \in V_h . \quad (4.5)'$$

In the situation described above, (4.5) and (4.5)' have unique solutions, and the solutions $\tilde{u}_\delta$ of (4.5) converge to the solution $\tilde{u}$ of (4.5)' as $\delta \to 0$.

Theorem 4.1 *Given the spaces and operators, as above, assume that $\tilde{u}_0 = [u_0, U_0] \in H$ and $\tilde{f} = [f, F] \in V'$. Then, for every $\delta > 0$, there is a unique $\tilde{u}_\delta \in V$ which satisfies (4.5) and $\tilde{u}_\delta(0) = \tilde{u}_0$. There is also a unique $\tilde{u} \in V_h$ which satisfies (4.5)' and where $\tilde{u}(0) = \tilde{u}_0$. Furthermore, $\tilde{u}_\delta$ converges weakly to $\tilde{u}$ in V as $\delta \to 0$.*

The proof is a direct application of standard results on evolution equations in Hilbert space.

We have shown that (4.5) and (4.5)' have unique solutions and that the two models which they represent are related. We remarked previously that allowing $\delta \to 0^+$ formally transformed the regularized model into the matched model. We have substantiated that observation by showing that the solutions u_δ converge to the solution of the matched model.

Note also that the variational form (in (4.2)', for example) leads directly back to (4.1)'. This confirms that our choice of the exchange term in the physical model is the correct one.

Finally note that the models and results here could be generalized or extended in several ways. In (4.1.c), we might choose $\frac{1}{\delta}$ to be something other than a constant. If, for example, $\frac{1}{\delta}$ is assumed to be a monotone graph which is also a subgradient operator, an approach similar to that in [CS93] might be used to show the existence of a solution. As stated earlier, Dirichlet boundary conditions on $\partial \Omega$ are not necessary, so that some generalization is also possible in that respect. Finally, if additional assumptions about the differentiability of A and B and the smoothness of Γ_x and $\partial \Omega$ were made, then, it might be possible to say more about the regularity of u and U.

9.5   Remarks

We will describe, first, how microstructure models can arise, by way of homogenization, from partial differential equations with highly periodic coefficients or geometry *and* appropriately scaled coefficients. This provides the connection with the exact models described by classical equations. Then, we will close with some additional remarks and observations.

9.5.1   Homogenization

So far, we have given a direct, but only heuristic justification of the microstructure models. To employ them for simulating real phenomena, one must obtain realistic values for the coefficients, e.g., by matching with data. Here, we will briefly recall the derivation by homogenization of the distributed microstructure model of a totally fissured medium following [Arb88] and [ADH90]. Simultaneously, this provides a justification of the model and a means to compute the *effective* coefficients in the microstructure model from known coefficients in the *exact* case.

We begin with the exact microscopic model of single-phase flow in a fissured domain Ω, a bounded open subset of $\mathbb{R}^3$. We assume that the geometry and the physical parameters of the problem have ε-periodic character. This implies that the solution to the problem exhibits periodic behavior. It has also some macroscopic (nonperiodic) behavior which is seen on the scale of the whole region Ω. The basic problem is to investigate the asymptotics of the solution as $\varepsilon \to 0$ in a family of properly scaled problems posed on domains Ω^ε formed by a lattice of copies of cells εY, where the unit reference cell is the cube $Y = (0, 1)^3$. We use ε as a superscript or subscript on coefficients or variables to denote objects periodic with respect to εY; we omit this notation when $\varepsilon = 1$.

The reference cell Y defines the double-component structure of the fissured domain, and we write $\overline{Y} = \overline{\mathfrak{F}} \cup \overline{\mathfrak{B}}$, with $\mathfrak{F}$ and $\mathfrak{B}$ denoting the fissure and solid parts of the cell, respectively. That part of the fissure–solid interface which is contained in Y is given by Γ. Let $\Gamma_{\mathfrak{F}}$ and $\Gamma_{\mathfrak{B}}$ denote the respective intersections of ∂Y with $\mathfrak{F}$ and $\mathfrak{B}$. In the totally fissured case, we assume $\overline{\mathfrak{B}} \subset Y$, so the matrix interface $\Gamma_\mathfrak{B}$ is empty. We, then, refer to $\mathfrak{B}$ as a *block*. By $\vec{v}$, we denote the normal unit vector to Γ which points in the direction *out* of $\mathfrak{B}$.

The system of fissures and matrix blocks in Ω^ε are denoted by $\mathfrak{F}^\varepsilon$ and $\mathfrak{B}^\varepsilon$, respectively. The exact (but singular) ε–model consists of a pair of differential equations, one on each of the subdomains $\mathfrak{F}^\varepsilon$ and $\mathfrak{B}^\varepsilon$ for the density, which will be denoted by u^ε. These equations are coupled by standard interface conditions on Γ^ε to insure conservation of mass and momentum across the fissure–matrix interface Γ^ε. An exterior boundary condition and an initial condition must also be specified, but they do not enter into the derivation of the limit model. To preserve the magnitude of the flux crossing the interfaces contained within a fixed volume of the medium as $\varepsilon \to 0$, it is necessary to scale the permeability in the blocks by the factor ε^2. Thus, the ε–model of diffusion in a totally fissured medium has the form,

$$\phi_1 \frac{\partial u^\varepsilon}{\partial t} - \nabla \cdot (a_\mathfrak{F} \nabla u^\varepsilon(t, x)) = 0, \quad x \in \mathfrak{F}^\varepsilon, \tag{5.1.a}$$

$$\phi_2 \frac{\partial u^\varepsilon}{\partial t} - \nabla \cdot (\varepsilon^2 a_\mathfrak{G} \nabla u^\varepsilon(t, x)) = 0, \quad x \in \mathfrak{B}^\varepsilon, \tag{5.1.b}$$

$$u^\varepsilon|_{\mathfrak{F}^\varepsilon}(t, y) = u^\varepsilon|_{\mathfrak{B}^\varepsilon}(t, y), \quad s \in \Gamma^\varepsilon, \tag{5.1.c}$$

and

$$a_\mathfrak{F} \nabla u^\varepsilon|_{\mathfrak{F}^\varepsilon} \cdot \vec{v} = \varepsilon^2 a_\mathfrak{G} \nabla u^\varepsilon|_{\mathfrak{B}^\varepsilon} \cdot \vec{v}, \quad s \in \Gamma^\varepsilon. \tag{5.1.d}$$

If u^ε is expanded in powers of ε and the formal analysis of this expansion is carried out, it will be seen that the leading terms for the density $u^\varepsilon|_{\mathfrak{F}^\varepsilon}$ in the fractures and the density $u^\varepsilon|_{\mathfrak{B}^\varepsilon}$ in the matrix blocks will be a pair of functions, $u(t, x)$ and $U(t, x, y)$, $x \in \Omega$, $y \in \mathfrak{B}$, $t > 0$, respectively, which satisfy the system of equations,

$$|\mathfrak{F}| \frac{\partial u}{\partial t} - \nabla \cdot (A_\mathfrak{F} \nabla u(t, x)) + Q(t, x) = 0, \quad x \in \Omega, \tag{5.2.a}$$

$$\phi_2 \frac{\partial U}{\partial t} - \nabla_y \cdot (a_\mathfrak{G} \nabla_y U(t, x, y)) = 0, \quad y \in \mathfrak{B}_x, x \in \Omega, \tag{5.2.b}$$

$$U(t, x, y) = u(t, x), \quad s \in \Gamma, x \in \Omega, \tag{5.2.c}$$

and

$$Q(t, x) = \int_\Gamma a_\mathfrak{G} \nabla_y U \cdot \vec{v} \, dS, \quad x \in \Omega. \tag{5.2.d}$$

This is just the *matched* microstructure model (3.3). Here, $|\mathfrak{F}|$ denotes the ϕ_1–weighted volume of the reference set $\mathfrak{F}$. The *effective* permeability tensor $A_\mathfrak{F}$ is given by

$$(A_\mathfrak{F})_{ij} = \int_\mathfrak{F} a_\mathfrak{F} \left(\delta_{i,j} + \frac{\partial w_i}{\partial y_j} \right) dy,$$

with the auxiliary functions w_k, $k = 1, 2, 3$, being Y-periodic solutions of the *cell problem*

$$\Delta_y w_k = 0, \quad y \in \mathfrak{F},$$

$$\nabla_y w_k \cdot \vec{v} = -\vec{e}_k \cdot \vec{v}, \quad y \in \Gamma,$$

where $\vec{e}_k$ is the unit vector in the direction of the k axis.

Eq. (5.2.a) is the macroscopic equation to be solved in Ω for the (macroscopic) density u. The distributed source term q accounts for the flux across the boundary Γ of the block $\mathfrak{B}$; we denote it, here, by $\mathfrak{B}_x$ to emphasize that a copy of it is identified with each point $x \in \Omega$. Blocks over different points in Ω are disconnected, thus, no flow can take place directly from one such block to another. It is this feature that limits this model to flow in a totally fissured medium. If the ε^2 scaling of the permeability in the blocks had been omitted, then, the limit process would have led to a single-diffusion equation with an *effective* or averaged coefficient that fails to represent the desired delayed storage effects. Vogt [Vog82] appears to have been the first to have recognized this idea in developing a model for chromatography.

9.5.2 Further Remarks

The basic distributed microstructure model (3.3) is obtained as the limit by homogenization of a corresponding exact but highly singular partial differential equation with rapidly oscillating coefficients. This provides not merely another derivation of the model equations, but also shows the relation to the classical, but singular, formulation as a single-diffusion equation, and it provides a method for directly computing the coefficients in (3.3) which necessarily represent averaged material properties. A similar result of the convergence of a classical system to (3.5) is to be expected. Also, the convergence of the solution of the well-posed ε-problem to a solution of this system provides a proof of existence for the distributed microstructure system. An alternative direct proof is independently available, as in section 9.4. Furthermore, the *parallel flow* model systems can be recovered as limiting cases of corresponding microstructure models. For example, the first-order kinetic model (2.2) is the limiting case of (3.3), as the permeability coefficient B tends to infinity. In this limit, the cell behaves as a single point. Equivalently, the function U is independent of the local variable y, and the geometry of the cell is lost. The system (3.5) converges to (2.4), likewise, as B tends to infinity. When δ tends to zero in (3.3) or (3.5), the solution converges to that of a limiting problem with the boundary conditions (3.3.c) or (3.5.c) replaced by the corresponding "matched" conditions of Dirichlet type. These results provide useful connections between the various classes of models [CS93].

An interesting open problem is determining the coefficients in the system (3.3) from measurements of data on the boundary of the global region. It would be particularly interesting to obtain information on the cell geometry from such boundary measurements.

Systems (2.1) and (2.2) comprise parabolic and degenerate parabolic dynamical systems, respectively, in the product space $L^2(\Omega) \times L^2(\Omega)$. The functional differential equations (3.4) and (3.6) lead to dynamical systems in $L^2(\Omega)$, but these are governed by C_0 semigroups without regularizing effects, and the estimates and techniques for these are comparatively difficult. These equations lack the parabolic structure one seeks in such models. However the systems (3.3) and (3.5) do retain all of the parabolic structure and corresponding estimates and regularity of classical parabolic systems when they are posed on the spaces $L^2(\Omega) \times L^2(\Omega; L^2(\mathfrak{B}_x))$. The dynamics of these microstructure systems is given by an analytic semigroup. Thus, they are truly *parabolic* problems.

Interesting models of a *layered structure* are easily obtained from the examples above for partially fissured media. Rather than having a secondary flux in all three coordinate directions, the fissure pressure gradient induces transverse normal flow across the layers whereas the parallel flow within the layered matrix is driven by the internal pressure gradient. See [BSA78], [BS89], [PS95] and [HS95a] for alternative approaches.

All of the models and results presented here could be generalized or extended in several ways. In fact, most are special cases of what is already available in the literature. For example, the linear elliptic operators in the above examples can be replaced by quasi-linear operators of divergence type, such as p-Laplace operators, and one can include semilinear operators such as the porous medium equation. We have not even mentioned results for equations of other types, such as hyperbolic. Furthermore, we have restricted discussion to problems with the simplest geometry, and we have not mentioned those involving, e.g., flow on boundaries with concentrated capacity, or, more generally, manifolds arising from periodic cells that have a nonflat geometry.

Experience suggests that the distributed microstructure models are conceptually easy to work with, they provide accurate models which include the fine scales and geometry appropriate for many problems, and their theory can be developed straightforwardly using conventional techniques. The numerical analysis of these systems provides a natural application of parallel methods.

10

Computational Aspects of Dual-Porosity Models

Todd Arbogast

Macroscopic microstructure models of the dual-porosity type were introduced in several places earlier, such as Chapter 1, section 1.5, Chapter 3, section 3.4, and Chapter 9, section 9.3. Among other things, they model, the flow of fluids in highly fractured porous media; that is, media comprised of porous matrix rock divided into relatively small blocks by thin fractures [BZK60], [WR63], [Arb89a], [ADH90], [DA90], [ADH91], and [Arb93a]. Mathematically, dual-porosity models are a relatively complex system of partial differential equations in seven variables, (t, x, y). At first glance, it may not be apparent that there is any advantage to the macroscopic description versus the mesoscopic (i.e., the Darcy-scale description that explicitly models flow within the fractures and matrix). However, the dual-porosity model explicitly captures the length scales of the physical problem and is, thus, much easier to approximate computationally.

For example, single-phase flow is governed on the mesoscopic scale by a parabolic equation that is relatively easy to approximate. However, the grid must be extremely fine to resolve the fractures. Fractures are on the order of 100 micrometers wide, and matrix blocks are on the order of a meter wide. In three space dimensions, a uniform grid would require about a trillion grid elements to resolve the flow in a *single* matrix block. A graded grid would be more efficient, but would still require over a thousand elements per matrix block. There may be many millions of matrix blocks in an entire reservoir or aquifer, so many billions of grid elements are required to resolve the length scales of the mesoscopic system. The enormous amount of computational resources required to solve this time-dependent problem renders it intractable.

On the other hand, as we discuss in this chapter, the homogenized, dual-porosity model requires far less computational effort to solve numerically, and it approximates the mesoscopic model well.

10.1   Single-Phase Flow

In this section, we consider the problem of approximating a linear, dual-porosity model of the flow of a single, compressible fluid phase in a naturally fractured reservoir or aquifer. Further details can be found in the papers [Arb89a], [DA90], [ADH90], and [ADH91]. The numerical technique described here can be extended to nonlinear problems, as we describe in the next section on two-phase flow.

10.1.1   The Mesoscopic and Dual-Porosity Models

The single-phase model we consider was introduced briefly in Chapter 3, section 3.4. To illustrate our numerical techniques, we restrict to a linear set of equations and ignore gravity. In practice, gravitational effects are quite important, so we must include them [Arb93b]; this has been done in the computational examples to follow.

As before, Ω is the reservoir or aquifer domain, assumed to be periodic with period ε. To simplify the description of the physical scaling, in this chapter, we assume that ε is the dimensionless fracture spacing, $\varepsilon = 1$ in Ω, and, as usual, homogenization takes $\varepsilon \to 0$. The unit cell is $Y = (0, 1)^3$ containing a single porous solid matrix block $\mathcal{B}$ and the surrounding fracture set $\mathfrak{F}$ ($\mathcal{B} \cup \mathfrak{F} = Y$, $\mathfrak{F} \cap \mathcal{B} = \emptyset$, and $\partial \mathcal{B} \subset Y$). The matrix and fracture parts of Ω are $\mathcal{B}^\varepsilon$ and $\mathfrak{F}^\varepsilon$, respectively, as defined in Chapter chap:one.phase Eqs. (4.1) and (4.2).

Recall from Chapter 3 Eq. (4.3) that the porosity ϕ_ε and permeability k_ε are discontinuous over Ω, as we go from matrix to fracture: $\phi_\varepsilon(x) = \phi_\mathcal{B}$, and $k_\varepsilon(x) = \varepsilon^2 k_\mathcal{B}$ in the matrix blocks $\mathcal{B}^\varepsilon$, and $\phi_\varepsilon(x) = \phi_\mathfrak{F}$, and $k_\varepsilon(x) = k_\mathfrak{F}$ in the fractures $\mathfrak{F}^\varepsilon$. To simplify the discussion, we restrict the model to be linear in the density ρ by assuming the equation of state,

$$\rho(p) = \rho_0 \, e^{c(p-p_0)}, \tag{1.1}$$

where p is the pressure, c is the fluid (i.e., liquid) compressibility, and subscript 0 denotes a reference quantity. The mesoscopic equation is, then, simply

$$\phi_\varepsilon \frac{\partial \rho^\varepsilon}{\partial t} - \nabla \cdot \left(\frac{k_\varepsilon}{\mu c} \nabla \rho^\varepsilon \right) = f \quad \text{in } (0, T) \times \Omega, \tag{1.2}$$

plus appropriate initial and boundary conditions, where μ is the fluid viscosity and f represents the injection and extraction wells.

Homogenization yields the dual-porosity model for the fracture density $\rho_\mathfrak{F}(t, x)$ and the matrix density $\rho_\mathcal{B}(t, x, y)$. Fracture flow is described by

$$\phi^* \frac{\partial \rho_\mathfrak{F}}{\partial t} + \frac{\phi_\mathcal{B}}{|Y|} \int_\mathcal{B} \frac{\partial \rho_\mathcal{B}}{\partial t}(t, x, y) \, dy - \nabla_x \cdot \left(\frac{k^*}{\mu c} \nabla_x \rho_\mathfrak{F} \right) = f \quad \text{in } (0, T) \times \Omega, \tag{1.3}$$

where $\phi^* = (|\mathfrak{F}|/|Y|)\phi_{\mathfrak{F}}$ and k^* is the homogenized permeability tensor defined in Theorem 4.2. Matrix flow is described by

$$\begin{cases} \phi_{\mathfrak{B}} \dfrac{\partial \rho_{\mathfrak{B}}}{\partial t} - \nabla_y \cdot \left(\dfrac{k_{\mathfrak{B}}}{\mu c} \nabla_y \rho_{\mathfrak{B}} \right) = f(t, x) & \text{in } (0, T) \times \Omega \times \mathfrak{B}, \\ \rho_{\mathfrak{B}}(t, x, y) = \rho_{\mathfrak{F}}(t, x) & \text{on } (0, T) \times \Omega \times \partial \mathfrak{B}. \end{cases} \tag{1.4}$$

10.1.2 Numerical Solution

To discretize time, we select a series of time levels

$$0 = t^0 < t^1 < t^2 < \dots < t^N.$$

A backward Euler, time approximation of (1.4) results in the semidiscrete equations

$$\begin{cases} \phi_{\mathfrak{B}} \dfrac{\rho_{\mathfrak{B}}^n - \rho_{\mathfrak{B}}^{n-1}}{t^n - t^{n-1}} - \nabla_y \cdot \left(\dfrac{k_{\mathfrak{B}}}{\mu c} \nabla_y \rho_{\mathfrak{B}}^n \right) = f(t^n, x) & \text{in } \Omega \times \mathfrak{B}, \\ \rho_{\mathfrak{B}}^n(x, y) = \rho_{\mathfrak{F}}^n(x) & \text{on } \Omega \times \partial \mathfrak{B}. \end{cases} \tag{1.5}$$

This is a linear partial differential equation in y for each fixed x, and $\rho_{\mathfrak{F}}^n(x)$ is a single parameter of the system. Therefore, we can express the solution as an affine operator of $\rho_{\mathfrak{F}}^n$ by computing the derivative of the system with respect to $\rho_{\mathfrak{F}}^n$ and finding the solution for any particular choice of $\rho_{\mathfrak{F}}^n$ (such as 0) ([Arb89a] and[DA90]), that is, for fixed x, we find the solution $\tilde{\rho}_{\mathfrak{B}}^n(x, y)$ to

$$\begin{cases} \phi_{\mathfrak{B}} \dfrac{\tilde{\rho}_{\mathfrak{B}}^n}{t^n - t^{n-1}} - \nabla_y \cdot \left(\dfrac{k_{\mathfrak{B}}}{\mu c} \nabla_y \tilde{\rho}_{\mathfrak{B}}^n \right) = 0 & \text{in } \Omega \times \mathfrak{B}, \\ \tilde{\rho}_{\mathfrak{B}}^n(x, y) = 1 & \text{on } \Omega \times \partial \mathfrak{B}, \end{cases} \tag{1.6}$$

and the solution $\bar{\rho}_{\mathfrak{B}}^n(x, y)$ to

$$\begin{cases} \phi_{\mathfrak{B}} \dfrac{\bar{\rho}_{\mathfrak{B}}^n - \rho_{\mathfrak{B}}^{n-1}}{t^n - t^{n-1}} - \nabla_y \cdot \left(\dfrac{k_{\mathfrak{B}}}{\mu c} \nabla_y \bar{\rho}_{\mathfrak{B}}^n \right) = f(t^n, x) & \text{in } \Omega \times \mathfrak{B}, \\ \bar{\rho}_{\mathfrak{B}}^n(x, y) = 0 & \text{on } \Omega \times \partial \mathfrak{B}. \end{cases} \tag{1.7}$$

Then,

$$\rho_{\mathfrak{B}}^n(x, y) = \rho_{\mathfrak{F}}^n(x) \, \tilde{\rho}_{\mathfrak{B}}^n(x, y) + \bar{\rho}_{\mathfrak{B}}^n(x, y), \tag{1.8}$$

and the matrix equations have been decoupled from the fracture equations. Now (1.3) is approximated semidiscretely as

$$\phi^* \dfrac{\rho_{\mathfrak{F}}^n - \rho_{\mathfrak{F}}^{n-1}}{t^n - t^{n-1}} + \left(\dfrac{\phi_{\mathfrak{B}}}{|Y|} \int_{\mathfrak{B}} \dfrac{\tilde{\rho}_{\mathfrak{B}}^n(x, y)}{t^n - t^{n-1}} dy \right) \rho_{\mathfrak{F}}^n + \dfrac{\phi_{\mathfrak{B}}}{|Y|} \int_{\mathfrak{B}} \dfrac{\bar{\rho}_{\mathfrak{B}}^n(x, y) - \rho_{\mathfrak{B}}^{n-1}(x, y)}{t^n - t^{n-1}} dy$$

$$- \nabla_x \cdot \left(\dfrac{k^*}{\mu c} \nabla_x \rho_{\mathfrak{F}}^n \right) = f(t^n, x) \quad \text{in } \Omega.$$

$$\tag{1.9}$$

Altogether, (1.6)–(1.9) is simply a backward Euler, semidiscrete approximation of (1.3)–(1.4) written in a decoupled form.

To (1.6)–(1.9), we apply any appropriate spatial discretization (Galerkin finite elements, finite differences, mixed finite element methods, collocation, finite volume methods, etc.). The solution algorithm is, then, as follows:

1. Find $\rho_{\mathfrak{F}}^0$ and $\rho_{\mathfrak{B}}^0$ from the initial conditions.

2. For each $n = 1, 2, ..., N$,

   (a) solve for $\tilde{\rho}_{\mathfrak{B}}^n$ and $\bar{\rho}_{\mathfrak{B}}^n$ from $\rho_{\mathfrak{B}}^{n-1}$ and (1.6)–(1.7) *only* at the x points where the discrete scheme requires a value;

   (b) solve for $\rho_{\mathfrak{F}}^n$ from $\rho_{\mathfrak{F}}^{n-1}$, $\tilde{\rho}_{\mathfrak{B}}^n$, $\bar{\rho}_{\mathfrak{B}}^n$, $\rho_{\mathfrak{B}}^{n-1}$, and (1.9); and

   (c) define $\rho_{\mathfrak{B}}^n$ from $\rho_{\mathfrak{F}}^n$, $\tilde{\rho}_{\mathfrak{B}}^n$, $\bar{\rho}_{\mathfrak{B}}^n$, and (1.8).

We note that the matrix equations can be solved in parallel quite naturally. Also, if we consider the discretization of the original equations (1.3)–(1.4), then our decomposition is a simple and parallel way to form the Shur complement system that arises when the matrix unknowns are eliminated.

Recall that the mesoscopic equations would require the solution of a single coupled system on a grid with perhaps billions of grid elements. In contrast, the dual-porosity system requires solving a single large fracture system and many small, decoupled matrix problems. The fracture equations can be solved on a grid with spacing much larger than the fracture spacing, say on the order of 1 to 100 meters, so that perhaps 100,000 grid elements are needed. Depending on the spatial discretization, we need to approximate matrix flow in roughly one matrix block per grid element (*not* in every matrix block!). The two subproblems on each matrix block are small, physically symmetric, and uniformly parabolic. Thus, a relatively coarse grid can be used of, perhaps, less than 100 grid elements (in fact, good results are sometimes reported using as few as one grid element per matrix block). The total number of grid points involved in the system is on the order of 20 million, a considerable saving over the mesoscopic system. The real advantage of the dual-porosity approach, however, is that the length scales have been explicitly modeled. We have a relatively easy time solving the matrix flow (possibly even in parallel) and need only to invest substantial computational resources in the coupled fracture calculation, which is thousands of times smaller than the coupled mesoscopic problem.

10.2   Two-Phase Flow

Two-phase flow in an ordinary porous medium has been considered in Chapter 5. We now consider the problem of approximating a highly nonlinear, dual-porosity model of the flow of two completely immiscible, incompressible fluid phases in a naturally fractured reservoir or aquifer. Further details can be found in the papers [Arb93a], [DA90], and [ADH91].

10.2.1 The Mesoscopic and Dual-Porosity Models

Although the effects of compressibility can be taken into account [Arb93a], we will assume incompressible media and fluids. In that case, the equations for the wetting fluid phase α and the nonwetting fluid phase β can be easily described when gravitational effects are taken into account.

Each fluid phase $\xi = \alpha$ or β has its own pressure p_ξ, (constant) density ρ_ξ, viscosity μ_ξ, and saturation S_ξ. We also define the phase *potentials* Φ_ξ as

$$\Phi_\xi = p_\xi - \rho_\xi \, g \, z(x),$$

where g is the gravitational constant and $z(x)$ is the depth of point x.

Again, the material properties are discontinuous over the domain Ω as we go from matrix to fracture. This includes the porosity ϕ_ε and permeability k_ε, but now also the functions giving the relative permeability $k_{r\xi,\varepsilon}(S_\alpha)$, $\xi = \alpha, \beta$, and the capillary pressure $p_{c,\varepsilon}(S_\alpha)$. We must also scale the absolute permeability by ε^2 as in Chapter 3 Eq. (4.3) and the gravitational terms by ε^{-1} in the matrix as we describe below.

The mesoscopic equations consist of two phase conservation equations, the definitions of the two phase potentials, a capillary pressure relation, and a volume balance equation. The two phase equations are a combination of the two-phase generalization of the Darcy law and conservation of mass (or volume). Because Darcy's law gives the phase velocities as

$$-\frac{k_\varepsilon k_{r\xi,\varepsilon}(S_\alpha^\varepsilon)}{\mu_\xi} \nabla \Phi_\xi^\varepsilon = -\frac{k_\varepsilon k_{r\xi,\varepsilon}(S_\alpha^\varepsilon)}{\mu_\xi} (\nabla p_\xi^\varepsilon - \rho_\xi g \nabla z_\varepsilon),$$

the full set of equations is

$$\begin{cases} \phi_\varepsilon \dfrac{\partial S_\xi^\varepsilon}{\partial t} - \nabla \cdot \left(\dfrac{k_\varepsilon k_{r\xi,\varepsilon}(S_\alpha^\varepsilon)}{\mu_\xi} \nabla \Phi_\xi^\varepsilon \right) = f_\xi, & \text{in } (0, T) \times \Omega, \; \xi = \alpha, \beta, \\[2ex] \Phi_\xi^\varepsilon = p_\xi^\varepsilon - \rho_\xi \, g \, z(x), & \text{in } (0, T) \times \mathfrak{F}^\varepsilon, \; \xi = \alpha, \beta, \\[2ex] \Phi_\xi^\varepsilon = p_\xi^\varepsilon - \rho_\xi \, g \, z(x_c^\varepsilon + \varepsilon^{-1}(x - x_c^\varepsilon)) + \Phi_{\xi,0}^\varepsilon, & \text{in } (0, T) \times \mathfrak{B}^\varepsilon, \; \xi = \alpha, \beta, \\[2ex] p_{c,\varepsilon}(S_\alpha^\varepsilon) = p_\beta^\varepsilon - p_\alpha^\varepsilon, & \text{in } (0, T) \times \Omega, \\[2ex] S_\alpha^\varepsilon + S_\beta^\varepsilon = 1, & \text{in } (0, T) \times \Omega, \end{cases}$$

$$(2.1)$$

plus appropriate initial and boundary conditions, where $x_c^\varepsilon(x)$ is the centroid of the block containing x and $\Phi_{\xi,0}^\varepsilon$ is a reference potential. We have scaled the depth in the definition of the potentials in the matrix so that gravitational effects do not lose strength as we homogenize [Arb93a]. We must also account for the fact that a potential has no absolute scale by explicitly defining the reference potentials as follows. Set $\Phi^\varepsilon_{\beta,0} = 0$, and define $\Phi^\varepsilon_{\alpha,0}(t, x)$ by requiring

$$
\begin{cases}
\displaystyle\int_{\mathfrak{B}^\varepsilon(x)} p^{-1}_{c,\mathfrak{B}}\left(\bar{\Phi}^\varepsilon_\beta - \bar{\Phi}^\varepsilon_\alpha + \Phi^\varepsilon_{\alpha,0}(t, x) + (\rho_\beta - \rho_\alpha)\, g\, z(x^\varepsilon_c + \varepsilon^{-1}(X - x^\varepsilon_c))\right) dX \\[2mm]
\quad = \displaystyle\int_{\mathfrak{B}^\varepsilon(x)} p^{-1}_{c,\mathfrak{B}}\left(\bar{\Phi}^\varepsilon_\beta - \bar{\Phi}^\varepsilon_\alpha + (\rho_\beta - \rho_\alpha)\, g\, z(X)\right) dX, \\[3mm]
\bar{\Phi}^\varepsilon_\xi = \dfrac{1}{|\partial\mathfrak{B}^\varepsilon(x)|}\displaystyle\int_{\partial\mathfrak{B}^\varepsilon(x)} \Phi^\varepsilon_\xi\, da(X),
\end{cases}
\tag{2.2}
$$

where $S^\varepsilon(x)$ is the matrix block containing x. This procedure maintains a proper relationship between the matrix and fracture potentials as we homogenize, by balancing mass through the capillary pressure relation. Note that, when $\varepsilon = 1$, we have the expected mesoscopic equations.

Formal homogenization [Arb93a] leads to a set of dual-porosity equations that govern the fracture $p_{\xi\mathfrak{F}}(t, x)$, $\Phi_{\xi\mathfrak{F}}(t, x)$, and $S_{\xi\mathfrak{F}}(t, x)$ and the matrix $p_{\xi\mathfrak{B}}(t, x, y)$, $\Phi_{\xi\mathfrak{B}}(t, x, y)$, and $S_{\xi\mathfrak{B}}(t, x, y)$. The fracture equations in $(0, T) \times \Omega$ are

$$
\begin{cases}
\phi^* \dfrac{\partial S_{\xi\mathfrak{F}}}{\partial t} + \dfrac{\phi_\mathfrak{B}}{|Y|}\displaystyle\int_\mathfrak{B} \dfrac{\partial S_{\xi\mathfrak{B}}}{\partial t}(t, x, y)\, dy \\[2mm]
\quad -\nabla \cdot \left(\dfrac{k^* k_{r\xi\mathfrak{F}}(S_{\alpha\mathfrak{F}})}{\mu_\xi}\nabla\Phi_{\xi\mathfrak{F}}\right) = f_\xi, & \xi = \alpha, \beta, \\[3mm]
\Phi^\varepsilon_{\xi\mathfrak{F}} = p_{\xi\mathfrak{F}} - \rho_\xi\, g\, z(x), & \xi = \alpha, \beta, \\[2mm]
p_{c\mathfrak{F}}(S_{\alpha\mathfrak{F}}) = p_{\beta\mathfrak{F}} - p_{\alpha\mathfrak{F}}, \\[2mm]
S_{\alpha\mathfrak{F}} + S_{\beta\mathfrak{F}} = 1,
\end{cases}
\tag{2.3}
$$

where ϕ^* and k^* are as in the single-phase case. The matrix equations are

$$
\begin{cases}
\phi_\mathfrak{B}\dfrac{\partial S_{\xi\mathfrak{B}}}{\partial t} - \nabla \cdot \left(\dfrac{k_\mathfrak{B} k_{r\xi\mathfrak{B}}(S_{\alpha\mathfrak{B}})}{\mu_\xi}\nabla\Phi_{\xi\mathfrak{B}}\right) \\[2mm]
\quad = f_\xi(t, x) & \text{in } (0, T) \times \Omega \times \mathfrak{B},\ \xi = \alpha, \beta, \\[2mm]
\Phi^\varepsilon_{\xi\mathfrak{B}} = p_{\xi\mathfrak{B}} - \rho_\xi\, g\, z(x + y - y_c) + \Phi_{\xi,0}, & \text{in } (0, T) \times \mathfrak{F}^\varepsilon,\ \xi = \alpha, \beta, \\[2mm]
p_{c\mathfrak{B}}(S_{\alpha\mathfrak{B}}) = p_{\beta\mathfrak{B}} - p_{\alpha\mathfrak{B}} & \text{in } (0, T) \times \Omega \times \mathfrak{B}, \\[2mm]
S_{\alpha\mathfrak{B}} + S_{\beta\mathfrak{B}} = 1 & \text{in } (0, T) \times \Omega \times \mathfrak{B}, \\[2mm]
\Phi_{\xi\mathfrak{B}}(t, x, y) = \Phi_{\xi\mathfrak{F}}(t, x) & \text{on } (0, T) \times \Omega \times \partial\mathfrak{B},
\end{cases}
\tag{2.4}
$$

where y_c is the centroid of Y. Note that the boundary condition on $\partial\mathfrak{B}$ requires continuity of potential, not pressure, as might be expected. The reference potential is defined by setting $\Phi_{\beta,0} = 0$ and requiring that $\Phi_{\alpha,0}(t, x)$ satisfies

$$
\int_\mathfrak{B} p^{-1}_{c,\mathfrak{B}}\left(\Phi_{\beta\mathfrak{F}} - \Phi_{\alpha\mathfrak{F}} + \Phi_{\alpha,0} + (\rho_\beta - \rho_\alpha)\, g\, z(x + y - y_c)\right) dy = |\mathfrak{B}| S_{\alpha\mathfrak{F}}. \tag{2.5}
$$

10.2.2 Numerical Solution

Although each of the systems (2.1)–(2.5) is posed with seven independent variables, two phase pressures, potentials, and saturations, and $\Phi_{\alpha,0}$, we can reduce this number to two. The reference potential is zero in the mesoscopic equations (with $\varepsilon = 1$), and it changes very slowly in the dual-porosity case. Thus we can use its old value when solving for the other variables and, then, update it at the end of the time step using (2.5). Using the volume balance relation, we can eliminate one of the saturations; moreover, the capillary pressure relation allows us to eliminate the other saturation (or either of the two pressures, but, for simplicity, let us assume that we eliminate both saturations), that is, the saturations are functions of the appropriate pressures:

$$S_\alpha = p_c^{-1}(p_\beta - p_\alpha),$$

and

$$S_\beta = 1 - S_\alpha.$$

Finally, the potentials are also functions of the pressures and can be eliminated. We are, thus, left only to discretize the nonlinear partial differential equations.

Again we use a backward Euler, time discretization on the complete system (2.3)–(2.4). For the matrix, we obtain equations similar to (1.5). For time level t^n, we have partial differential equations in y for each fixed x for the matrix pressures $p_{\alpha\mathfrak{B}}^n(x, y)$ and $p_{\beta\mathfrak{B}}^n(x, y)$ with $p_{\alpha\mathfrak{F}}^n(x)$ and $p_{\beta\mathfrak{F}}^n(x)$ as two parameters of the system. Our decoupling strategy cannot be applied directly to these equations, however, because they are nonlinear.

We further discretize (2.3)–(2.4) in space by applying Galerkin finite elements, finite differences, or some other appropriate method. This reduces the system to a fully discrete, finite-dimensional problem. Let the number of fracture unknowns be I, and denote them at time level t^n by

$$\vec{p}_{\xi\mathfrak{F}}^n = \{p_{\xi\mathfrak{F},i}^n, \ i = 1, 2, ..., I\}, \quad \xi = \alpha, \beta.$$

Usually, $p_{\xi\mathfrak{F},i}^n$ will be an approximation of $p_{\xi\mathfrak{F}}^n$ at grid point i. For the purposes of exposition, assume this to be the case, and also assume that the numerical method requires only a matrix block at each grid point, as in Fig. 10.1. Then, associated with each grid point $i = 1, 2, ..., I$, we have a series of matrix unknowns

$$\vec{p}_{\xi\mathfrak{B},i} = \{p_{\xi\mathfrak{B},ij}, \ j = 1, 2, ..., J_i\}, \quad \xi = \alpha, \beta,$$

in the ith matrix block. Let

$$\vec{p}_{\xi\mathfrak{B}} = \{p_{\xi\mathfrak{B},ij}, \ i = 1, 2, ..., I, \ j = 1, 2, ..., J_i\}, \quad \xi = \alpha, \beta.$$

The fully discrete, nonlinear equations, then, take the form,

$$\begin{cases} \mathcal{F}_i(\vec{p}_{\alpha\mathfrak{F}}^n, \vec{p}_{\beta\mathfrak{F}}^n, \vec{p}_{\alpha\mathfrak{B}}^n, \vec{p}_{\beta\mathfrak{B}}^n) = 0, & i = 1, 2, ..., I, \\ \mathcal{M}_{ij}(p_{\alpha\mathfrak{F},i}^n, p_{\beta\mathfrak{F},i}^n, \vec{p}_{\alpha\mathfrak{B},i}^n, \vec{p}_{\beta\mathfrak{B},i}^n) = 0, & i = 1, 2, ..., I, \ j = 1, 2, ..., J_i, \end{cases}$$

$$(2.6)$$

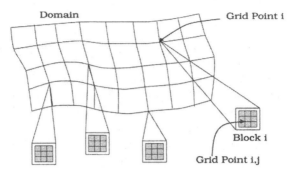

FIGURE 10.1. The discrete grid.

for some nonlinear functions $\mathcal{F}_i$ and $\mathcal{M}_{ij}$.

One way to solve nonlinear equations of this form is to use Newton's method to linearize it. Let

$$\vec{p}_{\xi\mathfrak{F}}^{\,n,m} \quad \text{and} \quad \vec{p}_{\xi\mathfrak{B}}^{\,n,m}, \quad \xi = \alpha, \beta,$$

denote the mth Newton iterate for the nth time level's solution, let D_π denote partial differentiation with respect to π, and let

$$\mathcal{F}_i^{n,m-1} = \mathcal{F}_i(\vec{p}_{\alpha\mathfrak{F}}^{\,n,m-1}, \vec{p}_{\beta\mathfrak{F}}^{\,n,m-1}, \vec{p}_{\alpha\mathfrak{B}}^{\,n,m-1}, \vec{p}_{\beta\mathfrak{B}}^{\,n,m-1}),$$

and

$$\mathcal{M}_{ij}^{n,m-1} = \mathcal{M}_{ij}(p_{\alpha\mathfrak{F},i}^{n,m-1}, p_{\beta\mathfrak{F},i}^{n,m-1}, \vec{p}_{\alpha\mathfrak{B},i}^{\,n,m-1}, \vec{p}_{\beta\mathfrak{B},i}^{\,n,m-1}),$$

and similarly for the derivatives. Then, we can describe the Newton procedure as the following iterative process:

1. Start with an initial guess for the solution

$$\vec{p}_{\xi\mathfrak{F}}^{\,n,0}, \quad \vec{p}_{\xi\mathfrak{B}}^{\,n,0}, \quad \xi = \alpha, \beta.$$

2. For each $m = 1, 2, \ldots$, until convergence is reached,

    (a) solve for

$$\delta\vec{p}_{\xi\mathfrak{F}}^{\,n,m}, \quad \delta\vec{p}_{\xi\mathfrak{B}}^{\,n,m}, \quad \xi = \alpha, \beta,$$

satisfying

$$
\begin{cases}
\mathcal{F}_i^{n,m-1} + \displaystyle\sum_{i'}\left\{ D_{p_{\alpha\mathfrak{F},i'}}\mathcal{F}_i^{n,m-1}\delta p_{\alpha\mathfrak{F},i'}^{n,m}\right.\\[2mm]
+ D_{p_{\beta\mathfrak{F},i'}}\mathcal{F}_i^{n,m-1}\delta p_{\beta\mathfrak{F},i'}^{n,m}\\[2mm]
+ \displaystyle\sum_{j'}\left[D_{p_{\alpha\mathfrak{B},i'j'}}\mathcal{F}_i^{n,m-1}\delta p_{\alpha\mathfrak{B},i'j'}^{n,m}\right.\\[2mm]
\left.\left.+ D_{p_{\beta\mathfrak{B},i'j'}}\mathcal{F}_i^{n,m-1}\delta p_{\beta\mathfrak{B},i'j'}^{n,m}\right]\right\}\\[2mm]
\qquad = 0, \quad i = 1,2,\ldots,I,\\[2mm]
\text{and}\\[2mm]
\mathcal{M}_{ij}^{n,m-1} + D_{p_{\alpha\mathfrak{F},i}}\mathcal{M}_{ij}^{n,m-1}\delta p_{\alpha\mathfrak{F},i}^{n,m}\\[2mm]
+ D_{p_{\beta\mathfrak{F},i}}\mathcal{M}_{ij}^{n,m-1}\delta p_{\beta\mathfrak{F},i}^{n,m}\\[2mm]
+ \displaystyle\sum_{j'}\left[D_{p_{\alpha\mathfrak{B},ij'}}\mathcal{M}_{ij}^{n,m-1}\delta p_{\alpha\mathfrak{B},ij'}^{n,m}\right.\\[2mm]
\left.+ D_{p_{\beta\mathfrak{B},ij'}}\mathcal{M}_{ij}^{n,m-1}\delta p_{\beta\mathfrak{B},ij'}^{n,m}\right]\\[2mm]
\qquad = 0, \quad i = 1,2,\ldots,I,\ \ j = 1,2,\ldots,J_i,
\end{cases}
\tag{2.7}
$$

(b) define

$$
\vec{p}_{\xi\mathfrak{F}}^{\,n,m} = \vec{p}_{\xi\mathfrak{F}}^{\,n,m-1} + \delta\vec{p}_{\xi\mathfrak{F}}^{\,n,m},\qquad
\vec{p}_{\xi\mathfrak{B}}^{\,n,m} = \vec{p}_{\xi\mathfrak{B}}^{\,n,m-1} + \delta\vec{p}_{\xi\mathfrak{B}}^{\,n,m},\qquad \xi = \alpha,\beta.
$$

Within the linearized Newton problem, the matrix solution in the ith block is an affine operator of $p_{\alpha\mathfrak{F},i}^{n,m}$ and $p_{\beta\mathfrak{F},i}^{n,m}$. Therefore, we can, again, decouple the matrix and fracture problems, as in the last section [DHA91]. We replace the matrix problem in (2.7) above by the following three problems for

$$
(\tilde{\delta}\vec{p}_{\alpha\mathfrak{B},i}^{\,n,m},\ \tilde{\delta}\vec{p}_{\beta\mathfrak{B},i}^{\,n,m}),\quad
(\hat{\delta}\vec{p}_{\alpha\mathfrak{B},i}^{\,n,m},\ \hat{\delta}\vec{p}_{\beta\mathfrak{B},i}^{\,n,m}),\quad \text{and}\quad
(\bar{\delta}\vec{p}_{\alpha\mathfrak{B},i}^{\,n,m},\ \bar{\delta}\vec{p}_{\beta\mathfrak{B},i}^{\,n,m}).
$$

For each $i = 1,2,\ldots,I$ and $j = 1,2,\ldots,J_i$,

$$
\begin{cases}
D_{p_{\alpha\mathfrak{F},i}}\mathcal{M}_{ij}^{n,m-1} + \displaystyle\sum_{j'}\left[D_{p_{\alpha\mathfrak{B},ij'}}\mathcal{M}_{ij}^{n,m-1}\tilde{\delta}p_{\alpha\mathfrak{B},ij'}^{n,m}\right.\\[2mm]
\left.+ D_{p_{\beta\mathfrak{B},ij'}}\mathcal{M}_{ij}^{n,m-1}\tilde{\delta}p_{\beta\mathfrak{B},ij'}^{n,m}\right] = 0,\\[3mm]
D_{p_{\beta\mathfrak{F},i}}\mathcal{M}_{ij}^{n,m-1} + \displaystyle\sum_{j'}\left[D_{p_{\alpha\mathfrak{B},ij'}}\mathcal{M}_{ij}^{n,m-1}\hat{\delta}p_{\alpha\mathfrak{B},ij'}^{n,m}\right.\\[2mm]
\left.+ D_{p_{\beta\mathfrak{B},ij'}}\mathcal{M}_{ij}^{n,m-1}\hat{\delta}p_{\beta\mathfrak{B},ij'}^{n,m}\right] = 0,\\[3mm]
\text{and}\\[2mm]
\mathcal{M}_{ij}^{n,m-1} + \displaystyle\sum_{j'}\left[D_{p_{\alpha\mathfrak{B},ij'}}\mathcal{M}_{ij}^{n,m-1}\bar{\delta}p_{\alpha\mathfrak{B},ij'}^{n,m}\right.\\[2mm]
\left.+ D_{p_{\beta\mathfrak{B},ij'}}\mathcal{M}_{ij}^{n,m-1}\bar{\delta}p_{\beta\mathfrak{B},ij'}^{n,m}\right] = 0.
\end{cases}
\tag{2.8}
$$

The result is that

$$\delta p_{\xi\mathfrak{B},ij}^{n,m} = \tilde{\delta} p_{\xi\mathfrak{B},ij}^{n,m} \delta p_{\alpha\mathfrak{F},i}^{n,m} + \hat{\delta} p_{\xi\mathfrak{B},ij}^{n,m} \delta p_{\beta\mathfrak{F},i}^{n,m} + \bar{\delta} p_{\xi\mathfrak{B},ij}^{n,m}. \tag{2.9}$$

Thus, we modify step 2(a) of the Newton algorithm by first solving (2.8). The fracture δ-pressures are then given by solving the fracture equations of (2.7), using, implicitly, the definition (2.9). Finally, we explicitly use the fracture δ-pressures and (2.9) to update the matrix δ-pressures.

10.3   Some Computational Results

In this section we present some computational results that illustrate the dual-porosity model and its ability to approximate the mesoscopic model. (See also [Arb93b].)

10.3.1   Single-Phase

We consider a simulation representing, for example, primary depletion of oil from a petroleum reservoir. To be able to solve the mesoscopic model, we restrict to two spatial dimensions and a relatively small system. However, we consider, more generally, the model of section 10.1.1 with gravitational effects, as described in detail in [Arb93a].

The flow is not given by a potential field. However, we can define the pseudopotential,

$$\Phi = \int_{p_0}^{p} \frac{dP}{\rho(P) g} - z(x) = \frac{1}{\rho_0 g c}\left[1 - e^{-c(p-p_0)}\right] - z(x) = \frac{1}{\rho_0 g c}\left(1 - \frac{\rho_0}{\rho}\right) - z(x), \tag{3.1}$$

using (1.1), so that

$$\rho = \frac{\rho_0}{1 - \rho_0 g c\left(\Phi + z(x)\right)}. \tag{3.2}$$

Then,

$$\rho(p) g \, \nabla\Phi = \nabla p - \rho(p) g \, \nabla z(x)$$

is a nonlinear relation giving the Darcy velocity after multiplication by k/μ. After multiplication by $\rho(p)$, we have the mass flux as $\frac{\rho(p)^2 g k}{\mu} \nabla\Phi$, which replaces the $\frac{k}{\mu c}\nabla\rho$ terms in (1.2)–(1.4). The fracture and matrix pseudopotentials are defined by

$$\Phi_{\mathfrak{F}}(t, x) = \frac{1}{\rho_0 g c}\left(1 - \frac{\rho_0}{\rho_{\mathfrak{F}}(t, x)}\right) - z(x) \tag{3.3}$$

and

$$\Phi_{\mathfrak{B}}(t, x, y) = \frac{1}{\rho_0 g c}\left(1 - \frac{\rho_0}{\rho_{\mathfrak{B}}(t, x, y)}\right) - z(x + y - y_c), \tag{3.4}$$

and the boundary condition in (1.4) is replaced by

$$\Phi_{\mathcal{B}}(t, x, y) = \Phi_{\mathfrak{F}}(t, x) \quad \text{on } (0, T) \times \Omega \times \partial\mathcal{B}, \tag{3.5}$$

giving continuity of pseudopotential, where the reference value is defined for $x \in \Omega$ by

$$\int_{\mathcal{B}} \frac{\rho_0}{1 - \rho_0 gc \left(\Phi_{\mathfrak{F}} - \Phi_0 + z(x + y - y_c)\right)} \, dy$$

$$= \int_{\mathcal{B}} \frac{\rho_0}{1 - \rho_0 gc \left(\Phi_{\mathfrak{F}} + z(x)\right)} \, dy = |\mathcal{B}| \rho_{\mathfrak{F}}. \tag{3.6}$$

We assume that the fluid is an oil with $\rho_0 = 0.8$ g/cm^3, $p_0 = 0$ psi, $c = 10^{-5}$ psi^{-1}, and $\mu = 2$ cp. Let the entire reservoir be 8 meters long and 8 meters deep. Let there be 16 matrix blocks filling the reservoir, each 2 meters by 2 meters, with $\phi_{\mathcal{B}} = 0.2$ and $k_{\mathcal{B}} = 1$ millidarcy. The fractures are 100 micrometers wide, with $\phi_{\mathfrak{F}} = 1$ and $k_{\mathfrak{F}} = 844$ darcy. We have used the usual formula for the permeability of the fracture as the width squared divided by 12, from considering Stoke's flow between parallel plates. For the dual-porosity model, the macroscopic fracture porosity ϕ^* is given by the geometry, and the permeability $k^* = 42.2$ millidarcy is approximately just $k_{\mathfrak{F}}$ times the fracture width divided by the matrix block length in two dimensions (see [Arb93b]).

A well is placed in the upper left corner of the reservoir. The initial pressure there is 4000 psi, and the fluid is in gravitational equilibrium. The well pressure is dropped linearly to 3000 psi in 10 days.

We show the results from four simulations, one on the mesoscopic scale and three on the dual-porosity, macroscopic scale. We used a Galerkin finite element spatial discretization with piecewise continuous, bilinear basis functions. There were 89 × 89 (7921) grid points for the mesoscopic solution. We used 3 × 3, 4 × 4 and, then, 8 × 8 grid points for the fracture calculation of the dual-porosity simulation, and 21 points for the calculation in each matrix block, yielding a total of 189, 336, and 1344 grid points, respectively. The linearization techniques introduced earlier in this chapter were used to solve the dual-porosity systems. It is difficult to compare the performance of different numerical techniques, because it depends on the specific way the code is implemented. However, it should be noted that the mesoscopic solution required about a half hour to solve accurately because it is so very ill-conditioned, but the much better conditioned dual-porosity simulations took only a few seconds.

In Fig. 10.2, we show the oil production rates as a function of time. All four simulations predicted roughly the same production history, both in the draw-down phase and in the reequilibration stage after 10 days.

In Fig. 10.3, we show the mesoscopic oil density contours (actually, the densities minus 0.824 and multiplied by 10,000). As can be seen, pressure is reduced most strongly in the fractures between the matrix blocks and also most strongly near the well in the upper left. Oil in the interior of the matrix blocks takes longer to

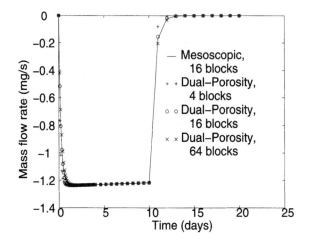

FIGURE 10.2. Oil production rates as a function of time.

be produced than oil in the fractures and near the surface of the blocks. Note also
the strong influence of gravity in these simulations.

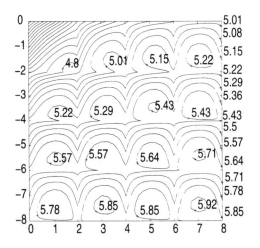

FIGURE 10.3. Mesoscopic density contours [in $(\times 10^{-4} + .824)$g/cm^3] as a function of
position (in m).

In Fig. 10.4, we show the dual-porosity matrix oil density contours when there
are 16 fracture grid points (so there are also the correct number of matrix blocks,
16). As can be seen, the dual-porosity model has captured the flow of the system
quite accurately, except near the well. Because the well itself is a spatially local
feature of the reservoir, it should not be expected to be approximated as well by the
dual-porosity model, however, there is *not* a pollution of the solution away from
the well.

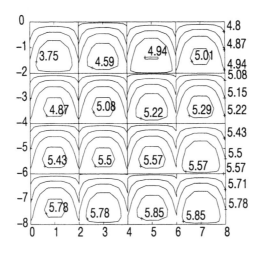

FIGURE 10.4. Dual-porosity matrix density contours [in $(\times 10^{-4}+.824)$g/cm^3] for the 4×4 fracture discretization.

In Figs. 10.5–10.6, we show the dual-porosity matrix oil density contours when there are 9 and 64 fracture grid points. Now, the number of matrix blocks in the dual-porosity model is at variance with the true number. Nevertheless, the average behavior of the system has been captured quite well.

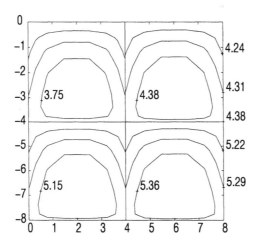

FIGURE 10.5. Dual-porosity matrix density contours [in $(\times 10^{-4}+.824)$g/cm^3] for the 3×3 fracture discretization.

10.3.2   Two-Phase

We now consider simulating a petroleum water flood. Water is injected into the reservoir at some rate through an injection well. The water displaces oil, and

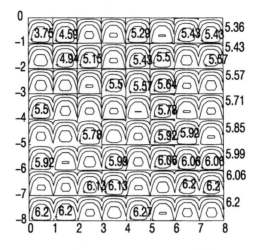

FIGURE 10.6. Dual-porosity matrix density contours [in $(\times 10^{-4} + .824)$g/cm^3] for the 8×8 fracture discretization.

oil and some of the water is recovered at a second, production well. We restrict attention to a single matrix block in two spatial dimensions so that the mesoscopic equations can be solved accurately. We consider three cases wherein the fracture width is either 200 or 100 micrometers and the wells are placed either horizontally or vertically.

Assume that the wetting phase α is water with $\rho_\alpha = 1.0$ g/cm^3 and $\mu_\alpha = 0.5$ cp, and the β phase is an oil with $\rho_\beta = 0.8$ g/cm^3 and $\mu_\beta = 2.0$ cp. The reservoir and its single block is a square of size ℓ by ℓ where $\ell = 2$ m. The fracture width δ is assumed to be either 200 μm or 100 μm (but note that, by our conventions, only half of the fractures actually reside in the reservoir). Assume that $\phi_\mathfrak{B} = 0.2$, $k_\mathfrak{B} = 4$ millidarcy, $\phi_\mathfrak{F} = 1.0$, ϕ^* is given by the geometry as either 0.0002 or 0.0001, $k_\mathfrak{F} = \delta^2/12$ is either 3, 377, 000 or 844, 000 millidarcy, and $k^* = \delta^3/12\ell$ is either 337 or 42 millidarcy. We take the residual saturations in the fractures as $S_{r\alpha\mathfrak{F}} = 0$ and $S_{r\beta\mathfrak{F}} = 0$, and in the matrix as $S_{r\alpha\mathfrak{B}} = 0.25$ and $S_{r\beta\mathfrak{B}} = 0.3$; thus, $0.25 \leq S_{\alpha,\mathfrak{B}} \leq 0.7$. The relative permeability and capillary pressure functions take the form

$$k_{r\alpha\mathfrak{F}}(s) = s, \qquad k_{r\alpha\mathfrak{B}}(s) = \left(\frac{s - 0.25}{0.75}\right)^2,$$

$$k_{r\beta\mathfrak{F}}(s) = 1 - s, \qquad k_{r\beta\mathfrak{B}}(s) = \left(\frac{0.7 - s}{0.7}\right)^2,$$

$$p_{c\mathfrak{F}}(s) = 4(1 - s)^{200} \text{ psi}, \quad p_{c\mathfrak{B}}(s) = 4\left(\frac{0.7 - s}{0.45}\right)^4 \text{ psi}.$$

The capillary pressure functions are plotted in Fig. 10.7.

The wells are configured to lie on opposite edges of the reservoir, either horizontally or vertically. Water is injected at a rate of 20 pore volumes per year from below if the wells are horizontal, and from the left if they are vertical.

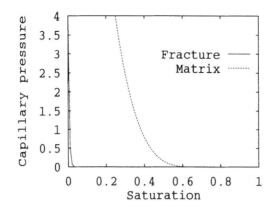

FIGURE 10.7. The capillary pressure functions (in psi).

As we will see in our simulations, there are several physical phenomena that influence the movement of the fluids. Oil is recovered from the matrix, primarily, through capillary imbibition of water. This displaces oil, which can, then, be recovered at the production well. As water is imbibed into the matrix, however, it lowers the relative permeability to oil; consequently, an insulating boundary layer is produced that inhibits further oil displacement. It is the trade-off between these two effects that, primarily, determines the behavior of the reservoir. Another significant phenomenon is the gravitational force acting on the fluids; oil tends to "float" above water. This helps to break down the insulating boundary layer. Finally, a less important, but, as we will see, a sometimes significant phenomenon is the presence of a pressure gradient across the block. This gives rise to "viscous" forces that tend to move the fluids to one side of the matrix block, again, breaking down the insulating boundary layer. This last effect is absent in the dual-porosity model, and we will need to introduce a "viscous" dual-porosity model as a simple modification (see also [Arb89a], [Arb89b], and [CS95]).

The oil recovery curves shown in Fig. 10.8 apply to Case 1, in which the fracture is relatively wide (200 micrometers) and the well is symmetric with respect to gravity (i.e., horizontal). Virtually no potential gradient can be supported in the fractures because they are so wide. Because the dual-porosity model assumes that the potentials do not vary in space over the boundary of the matrix block, we see very good agreement, in this case, between the mesoscopic and dual-porosity models in the amount of fluid produced from the reservoir. The viscous dual-porosity model is also in good agreement.

In Case 2 we restrict the fracture width to 100 micrometers (and keep horizontal wells). As shown in Fig. 10.9, the mesoscopic and dual-porosity models exhibit good agreement in oil recovery only for early time; at later times they predict significantly different recovery rates.

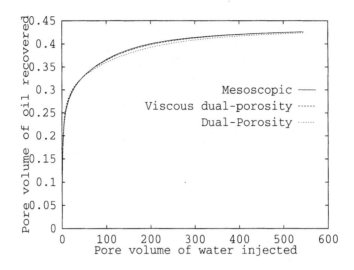

FIGURE 10.8. Oil produced as a function of water injected (200 μm fracture, horizontal wells).

In Fig. 10.10 we show the water saturation contours in the matrix for the simulation at 10 days. (The dual-porosity results are the average of all matrix blocks simulated across the reservoir.) The agreement looks quite good, but a careful inspection will reveal that water has been pushed upward from the injection well to

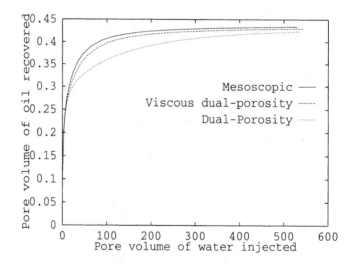

FIGURE 10.9. Oil produced as a function of water injected (100 μm fracture, horizontal wells).

the production well in the mesoscopic model. This is due to the slight, but significant pressure gradient across the block. Our viscous dual-porosity model accounts
properly for this effect, and we now define it.

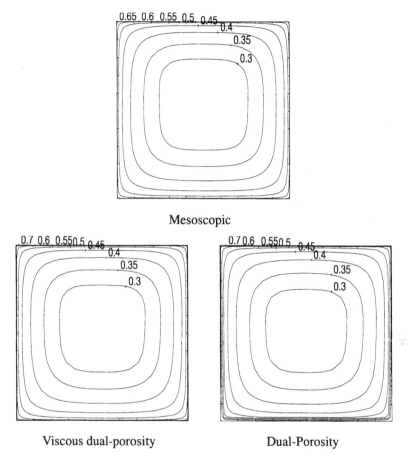

FIGURE 10.10. Water saturation at 10 days (100 μm fracture, horizontal wells).

The viscous dual-porosity model is given by replacing the boundary condition
on the matrix block in (2.4) by

$$\Phi_{\xi\mathfrak{B}}(t, x, y) = \Phi_{\xi\mathfrak{F}}(t, x) + \nabla_x \Phi_{\xi\mathfrak{F}}(t, x) \cdot (y - y_c), \quad \text{on } (0, T) \times \Omega \times \partial\mathfrak{B}. \quad (3.7)$$

Although the gradient creates a coupling of more than one fracture grid point to
each matrix block, the number is small, and the techniques introduced above can
be used to decouple the matrix computations from the fracture computation. It is
believed that the source term in the fracture equations should be modified as well.
However, this was not done in these simulations.

Finally, in Case 3, we consider a more challenging problem that exhibits all of
the main phenomena influencing fluid movement. Again, we take a 100 micrometer

fracture but drive the system with vertical wells. The wells are now asymmetric with respect to gravity. The recovery curves in Fig. 10.11 show that the dual-porosity model is only able to predict oil production accurately early when imbibition dominates. Later, the viscous model is needed to obtain accurate recovery curves.

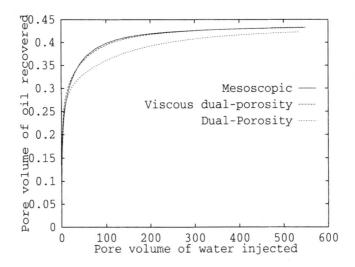

FIGURE 10.11. Oil produced as a function of water injected ($100\,\mu$m fracture, vertical wells).

In Figs. 10.12–10.16, we show the history of the simulation for the three models. Starting from gravitational equilibrium, we see seepage left to right into the fracture at 0.1 day, with water sliding preferentially under oil due to gravity. The dual-porosity model cannot see these effects, but the viscous model does an admirable job. The insulating layer of water around the boundary of the matrix block can be seen in all three models by 1 day. By 10 and 100 days, we start to see the gravitational and viscous effects showing up in the mesoscopic model; these are well approximated in the viscous model. The dual-porosity model must maintain a symmetric solution, and so it approximates only gravity well. By 1000 days, there is very little oil left in the matrix block. The mesoscopic model exhibits water coning, where water creeps up from below into the production well. The viscous model cannot predict this effect but is reasonable otherwise. The dual-porosity model itself cannot see the pronounced breakdown in the insulating boundary layer to the right of the block.

It is an open question how to derive the viscous dual-porosity model from homogenization. However, homogenization has given us the basic model which we can modify simply to improve the simulation of two-phase flow, so that later time predictions can be made accurately.

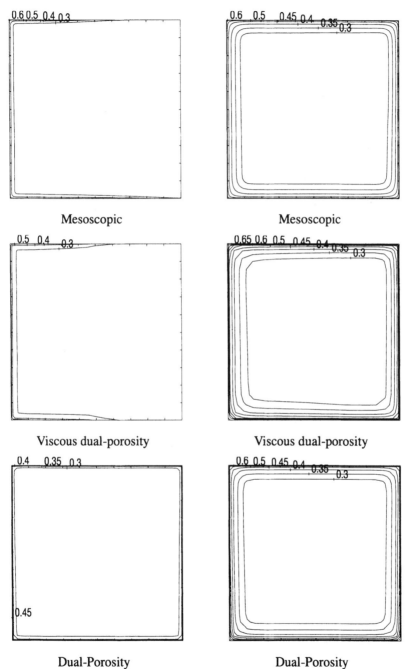

Mesoscopic

Viscous dual-porosity

Dual-Porosity

Mesoscopic

Viscous dual-porosity

Dual-Porosity

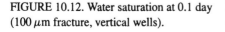

FIGURE 10.12. Water saturation at 0.1 day (100 μm fracture, vertical wells).

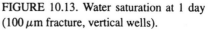

FIGURE 10.13. Water saturation at 1 day (100 μm fracture, vertical wells).

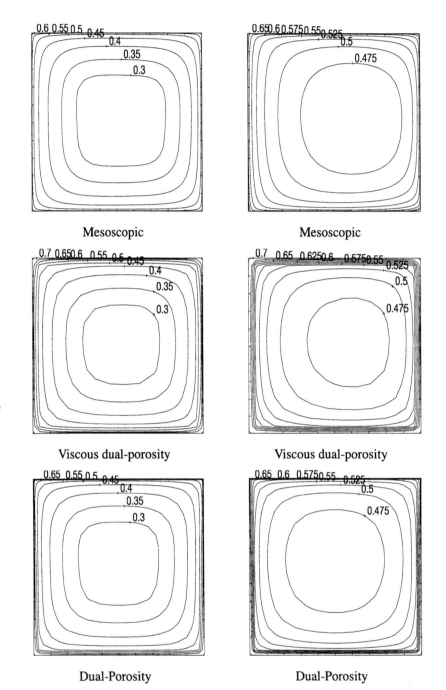

Mesoscopic

Mesoscopic

Viscous dual-porosity

Viscous dual-porosity

Dual-Porosity

Dual-Porosity

FIGURE 10.14. Water saturation at 10 days (100 μm fracture, vertical wells).

FIGURE 10.15. Water saturation at 100 days (100 μm fracture, vertical wells).

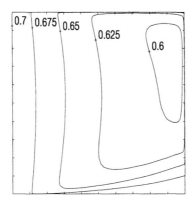

Mesoscopic

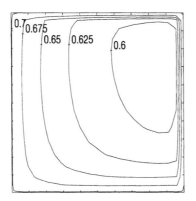

Viscous dual-porosity

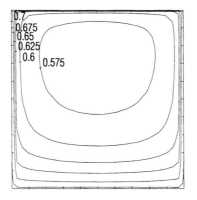

Dual-Porosity

FIGURE 10.16. Water saturation at 1000 days (100 μm fracture, vertical wells).

A

Mathematical Approaches and Methods

Grégoire Allaire

This appendix is devoted to a brief survey of the most popular methods and approaches in the mathematical theory of homogenization. This survey is definitely not complete and is merely a short introduction to that huge and fascinating field of homogenization. For a more advanced presentation of homogenization, the reader is referred to the classical books [BP89], [BLP78], [Mas93], [JKO94], and [SP80]. Many results below are given without proofs, but with many comments and references. As a deliberate parti pris, we have focused on periodic homogenization rather than on the general case of the theory. In particular, we do not say anything about the stochastic theory of homogenization. This means that complete proofs and detailed descriptions of the methods are only given for homogenization problems in periodic structures. It has the advantage of greatly simplifying the presentation and keeping the number of technically delicate points as few as possible. However, we emphasize that many of the homogenization methods below are not restricted to periodic problems and are valid without any geometrical or statistical assumptions. We hope this appendix is, therefore, easily accessible to nonmathematicians who can quickly understand the main ideas and approaches in homogenization and possibly consult the references for more details.

In the first section, various notions of convergence are introduced in relation to homogenization: the Γ-convergence of De Giorgi ([Gio75] and [Gio84]), the G-convergence of Spagnolo ([Spa68] and [Spa76]), and the H-convergence of Tartar ([Tar78], [MT77], and [MT95b]). The second section is dedicated to the so-called energy method of Tartar [Tar78] which is a very elegant and constructive approach to homogenization. The third section focuses on the two-scale convergence method, introduced in [All92a] and [Ngu89], which is dedicated to periodic homogenization. Finally the fourth section introduces the so-called reiterated homogenization.

A.1   Γ-, G-, and H-Convergence

A.1.1   Γ-Convergence

The Γ-convergence is an abstract notion of functional convergence which has been introduced by De Giorgi ([Gio75] and [Gio84]). It is not restricted to homogenization, and it has many applications in the calculus of variations, such as singular perturbation problems. A detailed presentation of Γ-convergence and several applications may be found in the books [But89] and [Mas93]. A close notion is that of *épiconvergence* developed by Attouch [Att84]. We first give the abstract definition of Γ-convergence and the fundamental theorem of Γ-convergence which motivates this definition.

Definition 1.1 *Let X be a metric space endowed with a distance d. Let ε be a sequence of positive indexes which goes to zero. Let F_ε be a sequence of functions defined on X with values in $\mathbb{R}$. The sequence F_ε is said to Γ-converge to a limit function F_0 if, for any point $x \in X$,*

1. *all sequences x_ε converging to x satisfy*

$$F_0(x) \leq \liminf_{\varepsilon \to 0} F_\varepsilon(x_\varepsilon), \text{ and}$$

2. *there exists at least one sequence x_ε converging to x, such that*

$$F_0(x) = \lim_{\varepsilon \to 0} F_\varepsilon(x_\varepsilon).$$

Remark The Γ-convergence and the Γ-limit depend on the choice of the distance d. It is easily checked that, by definition, a Γ-limit F_0 is a lower semicontinuous function on X, i.e., $F_0(x) \leq \liminf_{\varepsilon \to 0} F_0(x_\varepsilon)$ for any sequence x_ε converging to x. For example, if the sequence F_ε is constant equal to F, then, its Γ-limit F_0 is the lower semicontinuous envelope (or *relaxation*) of F.

Definition 1.2 *A sequence F_ε is said to be d-equicoercive on X if there exists a compact set K (independent of ε) such that*

$$\inf_{x \in X} F_\varepsilon(x) = \inf_{x \in K} F_\varepsilon(x).$$

The definition of Γ-convergence makes sense because of the following *fundamental theorem* which yields the convergence of the minimum values and of the minimizers for an equicoercive Γ-converging sequence.

Theorem 1.3 *Let F_ε be a d-equicoercive sequence on X which Γ-converges to a limit F_0. Then,*

1. *the minima of F_ε converge to that of F_0, i.e.,*

$$\min_{x \in X} F_0(x) = \lim_{\varepsilon \to 0} \left(\inf_{x \in X} F_\varepsilon(x) \right), \text{ and}$$

2. *the minimizers of F_ε converge to those of F_0, i.e., if x_ε converges to x and $\lim_{\varepsilon \to 0} F_\varepsilon(x_\varepsilon) = \lim_{\varepsilon \to 0} (\inf_{x \in X} F_\varepsilon(x))$, then, x is a minimizer of F_0.*

Theorem 1.4 *Assume that the metric space X (with the distance d) is separable (i.e., contains a dense countable subset). Let F_ε be a sequence of functions defined on X. Then, there exist a subsequence $F_{\varepsilon'}$ and a Γ-limit F_0 such that $F_{\varepsilon'}$ Γ-converges to F_0.*

A proof of the above theorems may be found in [Mas93]. Their main interest is to show that the notion of Γ-convergence is, roughly speaking, equivalent to the convergence of minimizers. Note, however, that they do not give any method, in practice, for computing the Γ-limit F_0.

Remark The relevance of Γ-convergence to homogenization is the following. Consider, for example, a linear diffusion process in a periodic domain Ω with period ε. Assume that the tensor of diffusion is $A(x/\varepsilon)$, where $A(y)$ is a symmetric, coercive, and bounded matrix which is Y-periodic. For a source term $f(x) \in L^2(\Omega)$, this problem of diffusion is a linear partial differential equation

$$\begin{cases} -\nabla \cdot \left(A \left(\tfrac{x}{\varepsilon} \right) \nabla u^\varepsilon \right) = f & \text{in } \Omega \\ u^\varepsilon = 0 & \text{on } \partial\Omega. \end{cases} \tag{1.1}$$

It is well-known that, when the matrix A is *symmetric*, the p.d.e. (1.1) is equivalent to the following variational problem: find $u^\varepsilon \in H_0^1(\Omega)$ which achieves the minimal value of

$$\min_{u \in H_0^1(\Omega)} \left(\frac{1}{2} \int_\Omega A \left(\frac{x}{\varepsilon} \right) \nabla u \cdot \nabla u \, dx - \int_\Omega f u \, dx \right). \tag{1.2}$$

Therefore, the Γ-convergence (with respect to the strong topology of $L^2(\Omega)$) of the functionals subject to minimization in (1.2) is equivalent to the homogenization of the p.d.e. (1.1). A definite advantage of the Γ-convergence is that it is not restricted to linear equations (or equivalently quadratic functionals), as we shall see below.

As an application of the Γ-convergence, we state a result of the homogenization of a nonlinear functional in a periodic domain (see [Bra85], [Mül87]). Let Ω be the periodic domain (a bounded open set in $\mathbb{R}^N$) with period εY, where ε is a small positive number and $Y = (0, 1)^N$ the rescaled unit cell. For some $p > 1$, let $W(y, \lambda)$ be a nonlinear energy with $|\lambda|^p$ growth at infinity. More precisely, W satisfies the following assumptions:

1. For any $\lambda \in \mathbb{R}^N$, the function $y \to W(y, \lambda)$ is measurable and Y-periodic.

2. Almost everywhere in $y \in Y$, the function $\lambda \to W(y, \lambda)$ is continuously differentiable.

3. There exist two constants $0 < c \leq C$, such that, almost everywhere in $y \in Y$ and for any $\lambda \in \mathbb{R}^N$,

$$0 \leq c|\lambda|^p \leq W(y, \lambda) \leq C \left(1 + |\lambda|^p \right),$$

and

$$\left| \frac{\partial W}{\partial \lambda}(y, \lambda) \right| \leq C \left(1 + |\lambda|^{p-1}\right).$$

Let $f(x) \in L^{p'}(\Omega)$ be a source term, with $1/p + 1/p' = 1$. We consider a family of functionals depending on $v(x) \in W_0^{1,p}(\Omega)$

$$I_\varepsilon(v) = \int_\Omega \left(W \left(\frac{x}{\varepsilon}, \nabla v(x) \right) - f(x)v(x) \right) dx. \tag{1.3}$$

Theorem 1.5 *Let I denote the homogenized functional defined by*

$$I(v) = \int_\Omega \left(W_{hom} \left(\nabla v(x) \right) - f(x)v(x) \right) dx, \tag{1.4}$$

where the homogenized energy W_{hom} is given by

$$W_{hom}(\lambda) = \inf_{k \in \mathbb{N}} \inf_{w(y) \in W_\#^{1,p}(kY)} \frac{1}{k^N} \int_{kY} W(y, \lambda + \nabla w(y)) \, dy. \tag{1.5}$$

(In (1.5), kY denotes the cube $(0, k)^N$, i.e., the union of k unit cells Y in each direction.)

Then, the sequence I_ε Γ-converges to I. This implies the following fact on the minimizers : for each value of $\varepsilon > 0$, let $u^\varepsilon(x) \in W_0^{1,p}(\Omega)$ be an almost minimizer of the functional I_ε in the sense that

$$I_\varepsilon(u^\varepsilon) \leq \inf_{v \in W_0^{1,p}(\Omega)} I_\varepsilon(v) + \varepsilon.$$

Then, up to a subsequence, u^ε converges weakly in $W_0^{1,p}(\Omega)$ to a limit u which is precisely a minimizer of the homogenized functional I, i.e.,

$$u^\varepsilon \rightharpoonup u \text{ weakly in } W_0^{1,p}(\Omega),$$

$$I_\varepsilon(u^\varepsilon) \to I(u), \quad \inf_{v \in W_0^{1,p}(\Omega)} I_\varepsilon(v) \to \min_{v \in W_0^{1,p}(\Omega)} I(v),$$

and

$$I(u) = \min_{v \in W_0^{1,p}(\Omega)} I(v).$$

In the above theorem, we do not assume that there exists a minimizer of I_ε for each value of ε (this motivates the definition of the almost minimizer). On the contrary, there always exists a minimizer of the homogenized functional I, possibly nonunique (which explains why the convergence of u^ε holds true only up to a subsequence).

Remark The striking aspect of Theorem 1.5 is that definition (1.5) of the homogenized energy involves a minimization over a supplementary integer variable k, the number of unit cells on which a local problem has to be solved. Recall that, for a linear problem (corresponding here to a quadratic energy), the computation of the homogenized coefficients (equivalently, the homogenized quadratic energy) requires solving the so-called cell problems in a single cell Y (see Chapter 3 or section A.2 below). In the nonlinear case, the situation is much more complex: it is not known, a priori, on how many cells the local solutions which give the homogenized energy have to be computed. In practice, one needs to calculate the local solutions $w_k^\lambda(y)$ for all possible sizes k and, then, to minimize over k the deduced energies $\int_{kY} W\left(y, \lambda + \nabla w_k^\lambda(y)\right) dy$. A convincing example, where the homogenized energy has to be computed on more than a single unit cell, is given by Müller in [Mül87]. This strange nonlinear effect is typical of Γ-convergence and cannot be obtained by the other methods described below.

However, there is a special case of nonlinear energies for which Theorem 1.5 can be simplified in the sense that the minimum in (1.5) over k is attained for $k = 1$: that of convex energies in the variable λ. As stated in the next theorem, for such convex energies, the local problems is posed in a single unit cell (this includes, in particular, the case of quadratic energies corresponding to linear problems).

Theorem 1.6 *Assume, in addition, that the energy function $\lambda \to W(y, \lambda)$ is convex (in this case, there exists a minimizer u^ε for the functional I_ε). Then, Theorem 1.5 applies, and, furthermore, the homogenized energy W_{hom} has a simpler definition*

$$W_{hom}(\lambda) = \inf_{w(y) \in W_\#^{1,p}(Y)} \int_Y W(y, \lambda + \nabla w(y)) \, dy. \tag{1.6}$$

A proof of Theorem 1.6, using Γ-convergence, is due to Marcellini [Mar78], but it can also be obtained with the energy method [Tar78] or the two-scale convergence method [All92a]. The key point in Theorem 1.6 is that the homogenized coefficients for a convex energy are computed on a single unit cell, as in the linear case.

A.1.2   G-Convergence

The G-convergence is a notion of convergence associated with sequences of symmetric, second-order, elliptic operators. It was introduced in the late sixties by Spagnolo [Spa68]. The G means Green because this type of convergence corresponds roughly to the convergence of the associated Green functions. The main result of the G-convergence is a compactness theorem in the homogenization theory which states that, for any bounded and uniformly coercive sequence of coefficients of a symmetric, second-order, elliptic equation, there exist a subsequence and a G-limit (i.e., homogenized coefficients) such that, for any source term, the corresponding subsequence of solutions converges to the solution of the homogenized equation. In physical terms, it means that the physical properties of a heterogeneous medium (such as its permeability, conductivity, or elastic moduli) can be well approximated by the properties of a homogeneous or homogenized medium if the size of the heterogeneities is small compared to the overall size of the medium.

For simplicity, we introduce the notion of G-convergence for a specific example of operators, namely, a scalar diffusion process with a Dirichlet boundary condition, but all the results hold for a large class of second-order, elliptic operators and boundary conditions. Let Ω be a bounded open set in $\mathbb{R}^N$, and let α, β be two positive constants, such that $0 < \alpha \leq \beta$. We introduce the set $\mathcal{M}_s(\alpha, \beta, \Omega)$ of all possible symmetric diffusion tensors defined on Ω with uniform coercivity constant α and $L^\infty(\Omega)$-bound β. More precisely, denoting by $M_s^{N \times N}$ the set of $N \times N$ symmetric matrices,

$$\mathcal{M}_s(\alpha, \beta, \Omega) = \left\{ A(x) \in L^\infty(\Omega; M_s^{N \times N}) \right.$$
$$\text{such that } \alpha |\xi|^2 \leq A(x)\xi \cdot \xi \leq \beta |\xi|^2 \text{ for any } \xi \in \mathbb{R}^N \left. \right\}.$$

We consider a sequence $A_\varepsilon(x)$ of diffusion tensors in $\mathcal{M}_s(\alpha, \beta, \Omega)$, indexed by a sequence of positive numbers ε going to 0. Here, ε is not associated with any specific length scale or statistical property of the diffusion process. In other words, no special assumptions (like periodicity or stationarity) are placed on the sequence A_ε.

For a given source term $f(x) \in L^2(\Omega)$, the following diffusion equation admits a unique solution u^ε in the space $H_0^1(\Omega)$ by application of Lax–Milgram lemma:

$$\begin{cases} -\nabla \cdot (A_\varepsilon \nabla u^\varepsilon) = f & \text{in } \Omega, \\ u^\varepsilon = 0 & \text{on } \partial\Omega. \end{cases} \tag{1.7}$$

The G-convergence of operators associated with the sequence A_ε is defined below as the convergence of the corresponding solutions u^ε.

Definition 1.7 *The sequence of tensors $A_\varepsilon(x)$ is said to G-converge to a limit $A^*(x)$, as ε goes to 0, if, for any right-hand side $f \in L^2(\Omega)$ in (1.7), the sequence of solutions u^ε converges weakly in $H_0^1(\Omega)$ to a limit u which is the unique solution of the homogenized equation associated with A^*:*

$$\begin{cases} -\nabla \cdot (A^* \nabla u) = f & \text{in } \Omega, \\ u = 0 & \text{on } \partial\Omega. \end{cases} \tag{1.8}$$

This definition makes sense because of the compactness of the set $\mathcal{M}_s(\alpha, \beta, \Omega)$ with respect to the G-convergence, as stated in the next theorem.

Theorem 1.8 *For any sequence A_ε in $\mathcal{M}_s(\alpha, \beta, \Omega)$, there exist a subsequence (still denoted by ε) and a homogenized limit A^*, belonging to $\mathcal{M}_s(\alpha, \beta, \Omega)$, such that A_ε G-converges to A^*.*

The original proof of Theorem 1.8 (due to Spagnolo [Spa68]) was based on the convergence of the Green functions associated with (1.7). Another proof uses the Γ-convergence of De Giorgi. A simpler proof was found by Tartar in the framework of the H-convergence which is a generalization of the G-convergence to the case of nonsymmetric operators. The interested reader is referred to the next subsection on H-convergence for a discussion of such a proof.

Remark If a sequence A_ε converges strongly in $L^\infty(\Omega)^{N^2}$ to a limit A, then its G-limit A^* coincides with A. In general the G-limit A^* of a sequence A_ε has nothing to do with its weak-* $L^\infty(\Omega)$-limit. For example, a straightforward computation in one dimension shows that the G-limit of a sequence A_ε is given as the inverse of the weak-* $L^\infty(\Omega)$-limit of A_ε^{-1} (the so-called harmonic limit). This last result holds true only in one dimension, and no explicit formula is available in higher dimensions.

The G-convergence enjoys a few useful properties as enumerated in the following proposition.

Proposition 1.9   *1. If a sequence A_ε G-converges, its G-limit is unique.*

2. *Let A_ε and B_ε be two sequences which G-converge to A^* and B^*, respectively. Let $\omega \subset \Omega$ be a subset strictly included in Ω, such that $A_\varepsilon = B_\varepsilon$ in ω. Then, $A^* = B^*$ in ω (this property is called the locality of G-convergence).*

3. *The G-limit of a sequence A_ε is independent of the source term f and of the boundary condition on $\partial\Omega$.*

4. *Let A_ε be a sequence which G-converges to A^*. Then, the associated density of energy $A_\varepsilon \nabla u^\varepsilon \cdot \nabla u^\varepsilon$ also converges to the homogenized density of energy $A^* \nabla u \cdot \nabla u$ in the sense of distributions in Ω.*

A.1.3   H-Convergence

The H-convergence is a generalization of the G-convergence to the case of non-symmetric problems. More than that, it provides a new constructive proof (the so-called energy method) of the main compactness theorem, which is both simpler and more general than the previous proofs. The H-convergence (H stands for "homogenization") was introduced by Murat and Tartar [MT77], [MT95b], [MT95a], [Tar78] in the mid-seventies.

As for the G-convergence, we simply introduce the notion of H-convergence for a scalar diffusion process with a Dirichlet boundary condition, although all the results hold for any second-order, elliptic operators and boundary conditions. Let Ω be a bounded open set in $\mathbb{R}^N$, and let α, β be two positive constants, such that $0 < \alpha \le \beta$. We denote by $\mathcal{M}(\alpha, \beta, \Omega)$ the set of all (*possibly nonsymmetric*) diffusion tensors defined on Ω with uniform coercivity constant α and $L^\infty(\Omega)$-bound β. More precisely, denoting by $M^{N \times N}$ the set of $N \times N$ matrices,

$$\mathcal{M}(\alpha, \beta, \Omega) = \left\{ A(x) \in L^\infty(\Omega; M^{N \times N}) \right.$$
$$\text{such that } \alpha|\xi|^2 \le A(x)\xi \cdot \xi \le \beta|\xi|^2 \text{ for any } \xi \in \mathbb{R}^N \left. \right\}.$$

We consider a sequence $A_\varepsilon(x)$ of diffusion tensors in $\mathcal{M}(\alpha, \beta, \Omega)$, indexed by a sequence of positive numbers ε going to 0. Once again, ε is not associated with any specific length scale or statistical property of the diffusion process. We emphasize that the tensors A_ε are not necessarily symmetric. This corresponds to a possible drift in the diffusion process.

For a given source term $f(x) \in L^2(\Omega)$ there exists a unique solution u^ε in $H_0^1(\Omega)$ of the following diffusion equation by the Lax–Milgram lemma:

$$\begin{cases} -\nabla \cdot (A_\varepsilon \nabla u^\varepsilon) = f & \text{in } \Omega, \\ u^\varepsilon = 0 & \text{on } \partial\Omega. \end{cases} \tag{1.9}$$

The H-convergence of the sequence A_ε differs from the previous G-convergence in the sense that it requires more than the mere convergence of the sequence of solutions u^ε.

Definition 1.10 *The sequence of tensors $A_\varepsilon(x)$ is said to H-converge to a limit $A^*(x)$, as ε goes to 0, if, for any right-hand side $f \in L^2(\Omega)$ in (1.9), the sequence of solutions u^ε converges weakly in $H_0^1(\Omega)$ to a limit u, and the sequence of fluxes $A_\varepsilon \nabla u^\varepsilon$ converges weakly in $L^2(\Omega)^N$ to $A^* \nabla u$, where u is the unique solution of the homogenized equation associated with A^*:*

$$\begin{cases} -\nabla \cdot (A^* \nabla u) = f & \text{in } \Omega, \\ u = 0 & \text{on } \partial\Omega. \end{cases} \tag{1.10}$$

This definition makes sense because of the following compactness result.

Theorem 1.11 *For any sequence A_ε in $\mathcal{M}(\alpha, \beta, \Omega)$, there exist a subsequence (still denoted by ε) and a homogenized limit A^*, belonging to $\mathcal{M}(\alpha, \frac{\beta^2}{\alpha}, \Omega)$, such that A_ε H-converges to A^*.*

Remark Notice that the set $\mathcal{M}(\alpha, \beta, \Omega)$ is not stable with respect to the H-convergence (as is the case for the G-convergence), because the $L^\infty(\Omega)$-bound of the H-limit can be increased by a factor of $\beta/\alpha \geq 1$. This is a specific effect of the nonsymmetry of a sequence A_ε. In physical terms, it means that microscopic convective phenomena can yield macroscopic diffusive effects.

The proof of Theorem 1.11 is constructive and based on the so-called energy method described in the next section. Beyond its theoretical interest for proving the above compactness result, the energy method of Tartar is of paramount importance in practical applications because it gives a convenient recipe for homogenizing any second-order, elliptic system. A detailed proof of Theorem 1.11 may be found in [MT77] and [MT95b].

Like G-convergence, H-convergence satisfies the same properties as stated in Proposition 1.9, namely, uniqueness of the H-limit, locality, independence of the H-limit with respect to the boundary condition, and convergence of the energy density.

To conclude this section, we give a simple example which demonstrates the necessity of requiring the convergence of the fluxes $A_\varepsilon \nabla u^\varepsilon$ on top of that of the solutions u^ε to have a coherent definition of H-convergence. Let B be a constant skew-symmetric matrix, i.e., such that its entries satisfy

$$b_{ij} = -b_{ji} \quad \text{for all } 1 \leq i, j \leq N.$$

Then, for any real-valued function $u(x)$,

$$\nabla \cdot (B\nabla u(x)) = \sum_{1 \le i,j \le N} b_{ij} \frac{\partial^2 u}{\partial x_i \partial x_j} = 0.$$

Therefore, if u is a solution of the homogenized equation (1.10), it is also a solution of the following equation

$$\begin{cases} -\nabla \cdot ((A^* + B)\nabla u) = f & \text{in } \Omega, \\ u = 0 & \text{on } \partial\Omega. \end{cases}$$

Assume for a moment that the definition of H-convergence is the same as that of G-convergence (i.e., only the convergence of solutions is required). Then, if A^* is a H-limit of a sequence A_ε, so is $A^* + B$ for any constant, skew-symmetric matrix B, which contradicts the uniqueness of the H-limit (a highly desirable feature of any type of convergence). Therefore, in the nonsymmetric case, the definition of H-convergence must include an additional condition compared to that of G-convergence. This is precisely the role of the convergence of fluxes $A_\varepsilon \nabla u^\varepsilon$.

A.2   The Energy Method

A.2.1   Setting of a Model Problem

A very elegant and efficient method for homogenizing partial differential equations has been devised by Tartar [Tar78], [MT77], and [MT95b], which has later been called the *energy method* although it has nothing to do with any kind of energy. It is sometimes more appropriately called the *oscillating test function method*, but it is most commonly referred to as the energy method, and we shall stick to this name.

The energy method is a very general method in homogenization. It does not require any geometric assumptions about the behavior of the p.d.e. coefficients: neither periodicity nor statistical properties like stationarity or ergodicity. Actually, it encompasses all other approaches in the framework of H-convergence. As was already mentioned in the previous section, the energy method is a constructive proof for the compactness theorem of H-convergence. However, to expose the energy method in its full generality may hide the key ideas of the method in a lot of technicalities. Therefore, for clarity, we prefer to present the energy method on a model problem of periodic homogenization. Nevertheless, we reemphasize that the energy method works also for nonperiodic homogenization, as the reader can be convinced by referring to [MT77], [MT95b], [CM82], and [CM95].

We consider a model problem of diffusion in a periodic medium, a usual example in all textbooks on homogenization, but, of course, the energy method covers many other problems with slight changes.

To fix ideas, the periodic domain is called Ω (a bounded open set in $\mathbb{R}^N$), its period ε (a positive number which is assumed to be very small), and the rescaled unit cell Y (i.e., $Y = (0, 1)^N$). The tensor of diffusion in Ω is given by an $N \times N$ matrix $A(x, \frac{x}{\varepsilon})$, not necessarily symmetric (if some drift is allowed in the diffusion process), where $A(x, y)$ is a smooth function of the slow variable $x \in \Omega$ and a Y-periodic function of the fast variable $y \in Y$ (e.g., $A(x, y) \in C(\Omega; L_{\#}^{\infty}(Y))^{N^2}$). Furthermore, this matrix A satisfies the usual coerciveness and boundedness assumptions (which simply mean that diffusion is a dissipative process): there exist two positive constants α and β, satisfying $0 < \alpha \le \beta$, such that, for any constant vector $\xi \in \mathbb{R}^N$ and at any point $(x, y) \in \Omega \times Y$,

$$\alpha |\xi|^2 \le \sum_{i,j=1}^{N} A_{i,j}(x, y)\xi_i \xi_j \le \beta |\xi|^2,$$

where $A_{i,j}$ denotes the entries of the matrix A.

Denoting the source term (a function in $L^2(\Omega)$) by $f(x)$ and enforcing a Dirichlet boundary condition (for simplicity), our model problem of diffusion reads

$$\begin{cases} -\nabla \cdot \left(A(x, \frac{x}{\varepsilon})\nabla u^{\varepsilon}\right) = f & \text{in } \Omega, \\ u^{\varepsilon} = 0 & \text{on } \partial\Omega. \end{cases} \tag{2.1}$$

It is well-known that, by application of the Lax–Milgram lemma, equation (2.1) admits a unique solution u^{ε} in the space $H_0^1(\Omega)$ (more smoothness can be obtained for the solution u^{ε}, but it is not necessary in the sequel). Moreover, u^{ε} satisfies the following a priori estimate:

$$\|u^{\varepsilon}\|_{H_0^1(\Omega)} \le C\|f\|_{L^2(\Omega)}, \tag{2.2}$$

where C is a positive constant which does not depend on ε (this estimate is obtained by multiplying equation (2.1) by u^{ε}, integrating by parts, and using the Poincaré inequality). It implies that the sequence u^{ε}, indexed by a sequence of periods ε which goes to 0, is bounded in the Sobolev space $H_0^1(\Omega)$.

The goal of the rest of this section is to prove the following homogenization result for the model problem (2.1) by the energy method:

Theorem 2.1 *The sequence $u^{\varepsilon}(x)$ of solutions of (2.1) converges weakly in $H_0^1(\Omega)$ to a limit $u(x)$ which is the unique solution of the homogenized problem,*

$$\begin{cases} -\nabla \cdot (A^*(x)\nabla u(x)) = f(x) & \text{in } \Omega, \\ u = 0 & \text{on } \partial\Omega. \end{cases} \tag{2.3}$$

In (2.3), the homogenized diffusion tensor, A^*, is defined by its entries

$$A_{ij}^*(x) = \int_Y A(x, y)\left(\vec{e}_i + \nabla_y w_i(x, y)\right) \cdot \left(\vec{e}_j + \nabla_y w_j(x, y)\right) dy, \tag{2.4}$$

where $w_i(x, y)$ are defined, at each point $x \in \Omega$, as the unique solutions in $H_{\#}^1(Y)/\mathbb{R}$ of the so-called cell problems

$$\begin{cases} -\nabla_y \cdot \left(A(x, y)\left(\vec{e}_i + \nabla_y w_i(x, y)\right)\right) = 0 & \text{in } Y, \\ y \to w_i(x, y) & Y\text{-periodic,} \end{cases} \tag{2.5}$$

with $(\vec{e}_i)_{1 \le i \le N}$ the canonical basis of $\mathbb{R}^N$.

Remark Of course, the homogenized system defined by (2.3) is the same as that obtained by formal two-scale asymptotic expansions (see formula (3.6) in Chapter 1). Recall that the asymptotic expansion for u^ε is

$$u^\varepsilon(x) = u(x) + \varepsilon u_1(x, \frac{x}{\varepsilon}) + \cdots,$$

where u is nothing but the homogenized solution of (2.3) and u_1 is related to the solutions of the local problems (2.5) through

$$u_1(x, y) = \sum_{i=1}^{N} \frac{\partial u}{\partial x_i}(x)w_i(x, y). \tag{2.6}$$

To shed some light on the principles of the energy method, let us begin with a naive attempt to prove Theorem 2.1 by passing to the limit in the variational formulation. The original problem (2.1) admits the following variational formulation:

$$\int_\Omega A(x, \frac{x}{\varepsilon})\nabla u^\varepsilon(x) \cdot \nabla\varphi(x)dx = \int_\Omega f(x)\varphi(x)dx, \tag{2.7}$$

for any test function $\varphi \in H_0^1(\Omega)$. By estimate (2.2), we can extract a subsequence, still denoted by ε, such that u^ε converges weakly in $H_0^1(\Omega)$ to a limit u. Unfortunately, the left-hand side of (2.7) involves the product of two weakly converging sequences in $L^2(\Omega)$, $A(x, \frac{x}{\varepsilon})$ and $\nabla u^\varepsilon(x)$, and it is not true that it converges to the product of the weak limits. Therefore, we cannot pass to the limit in (2.7) without further argument.

The main idea of the energy method is to replace the fixed test function φ in (2.7) by a weakly converging sequence φ_ε (the so-called *oscillating test function*), chosen so that the left-hand side of (2.7) miraculously passes to the limit. This phenomenon is an example of the *compensated compactness* theory, developed by Murat and Tartar ([Mur78] and [Tar79]), which, under additional conditions, permits passing to the limit in some products of weak convergences.

A.2.2 Proof of the Results

This subsection is devoted to the proof of Theorem 2.1 by the energy method. The key idea is the choice of an oscillating test function $\varphi_\varepsilon(x)$. Let $\varphi(x) \in \mathcal{D}(\Omega)$ be a smooth function with compact support in Ω. Copying the two first terms of the asymptotic expansion of u^ε, the oscillating test function φ_ε is defined by

$$\varphi_\varepsilon(x) = \varphi(x) + \varepsilon\sum_{i=1}^{N} \frac{\partial\varphi}{\partial x_i}(x)w_i^*(x, \frac{x}{\varepsilon}), \tag{2.8}$$

where $w_i^*(x, y)$ are not the solutions of the cell problems, defined in (2.5), but that of the *dual cell problems*

$$\begin{cases} -\nabla_y \cdot \left(A^t(x, y)\left(\vec{e}_i + \nabla_y w_i^*(x, y)\right)\right) = 0 & \text{in } Y, \\ y \rightarrow w_i^*(x, y) & Y\text{-periodic.} \end{cases} \tag{2.9}$$

The difference between (2.5) and (2.9) is that the matrix $A(x, y)$ has been replaced by its *transpose* $A^t(x, y)$. For simplicity, we assume that $A(x, y)$ is continuously differentiable in x with values in $L^\infty(Y)^{N^2}$. Then, because $w_i^*(x, y)$ has the same regularity in x as A and because φ is smooth, φ_ε belongs to $H_0^1(\Omega)$. If $A(x, y)$ is simply continuous in x (and not differentiable), the functions w_i^* need to be regularized in x, and, afterwards, this regularization has to be justified (it is purely a technical matter that we skip hereafter). Furthermore, by the periodicity in y of w_i^*, it is easily seen that $\varepsilon w_i^*(x, \frac{x}{\varepsilon})$ is a bounded sequence in $H^1(\Omega)$, which converges weakly to 0 (see Lemma 2.2 below, if necessary).

The next step is to insert this oscillating test function φ_ε in the variational formulation (2.7)

$$\int_\Omega A(x, \frac{x}{\varepsilon})\nabla u^\varepsilon(x) \cdot \nabla \varphi_\varepsilon(x)dx = \int_\Omega f(x)\varphi_\varepsilon(x)dx. \tag{2.10}$$

To take advantage of our knowledge of Eq. (2.9), we develop and integrate by parts in (2.10). Noting that

$$
\nabla\varphi_\varepsilon \; = \; \sum_{i=1}^N \frac{\partial\varphi}{\partial x_i}(x)\left(\vec{e}_i + \nabla_y w_i^*(x, \frac{x}{\varepsilon})\right)
$$
$$
+ \varepsilon \sum_{i=1}^N \left(\frac{\partial\nabla\varphi}{\partial x_i}(x)w_i^*(x, \frac{x}{\varepsilon}) + \frac{\partial\varphi}{\partial x_i}(x)\nabla_x w_i^*(x, \frac{x}{\varepsilon})\right),
$$

yields

$$\int_\Omega A(x, \frac{x}{\varepsilon})\nabla u^\varepsilon(x) \cdot \nabla\varphi_\varepsilon(x)dx \tag{2.11}$$
$$
= \int_\Omega A(x, \frac{x}{\varepsilon})\nabla u^\varepsilon(x) \cdot \sum_{i=1}^N \frac{\partial\varphi}{\partial x_i}(x)\left(\vec{e}_i + \nabla_y w_i^*(x, \frac{x}{\varepsilon})\right)dx
$$
$$
+\varepsilon \int_\Omega A(x, \frac{x}{\varepsilon})\nabla u^\varepsilon(x) \cdot \sum_{i=1}^N \left(\frac{\partial\nabla\varphi}{\partial x_i}(x)w_i^*(x, \frac{x}{\varepsilon}) + \frac{\partial\varphi}{\partial x_i}(x)\nabla_x w_i^*(x, \frac{x}{\varepsilon})\right).
$$

The last term in (2.11) is easily seen to be bounded by a constant time ε and, thus, cancels out in the limit. In the first term of (2.11), an integration by parts gives

$$\int_\Omega A(x, \frac{x}{\varepsilon})\nabla u^\varepsilon(x) \cdot \sum_{i=1}^N \frac{\partial\varphi}{\partial x_i}(x)\left(\vec{e}_i + \nabla_y w_i^*(x, \frac{x}{\varepsilon})\right)dx = \tag{2.12}$$
$$
- \int_\Omega u^\varepsilon(x)\nabla \cdot \left(A^t(x, \frac{x}{\varepsilon})\sum_{i=1}^N \frac{\partial\varphi}{\partial x_i}(x)\left(\vec{e}_i + \nabla_y w_i^*(x, \frac{x}{\varepsilon})\right)\right)dx.
$$

Let us compute the divergence in the right-hand side of (2.12), which is actually a function of x and $y = x/\varepsilon$:

$$d_\varepsilon(x) = \nabla \cdot \left(A^t(x, \frac{x}{\varepsilon})\sum_{i=1}^N \frac{\partial\varphi}{\partial x_i}(x)\left(\vec{e}_i + \nabla_y w_i^*(x, \frac{x}{\varepsilon})\right)\right) \tag{2.13}$$

$$= \nabla_x \cdot \left(A^I(x, y) \sum_{i=1}^{N} \frac{\partial \varphi}{\partial x_i}(x) \left(\vec{e}_i + \nabla_y w_i^*(x, y) \right) \right)$$

$$+ \frac{1}{\varepsilon} \sum_{i=1}^{N} \frac{\partial \varphi}{\partial x_i}(x) \nabla_y \cdot \left(A^I(x, y) \left(\vec{e}_i + \nabla_y w_i^*(x, y) \right) \right).$$

The last term of order ε^{-1} in the right-hand side of (2.13) is simply zero by definition (2.9) of w_i^*. Therefore, $d_\varepsilon(x)$ is bounded in $L^2(\Omega)$, and, because it is a periodically oscillating function, it converges weakly to its average by virtue of Lemma 2.2.

The main point of this simplification is that we are now able to pass to the limit in the right-hand side of (2.12). Recall that u^ε is bounded in $H_0^1(\Omega)$. By applying the Rellich theorem, there exists a subsequence (still indexed by ε, for simplicity) and a limit $u \in H_0^1(\Omega)$ such that u^ε converges *strongly* to u in $L^2(\Omega)$. The right-hand side of (2.12) is the product of a weak convergence (d_ε) and a strong convergence (u^ε), and, thus, its limit is the product of the two limits. In other words,

$$\lim_{\varepsilon \to 0} \int_\Omega A(x, \frac{x}{\varepsilon}) \nabla u^\varepsilon(x) \cdot \nabla \varphi_\varepsilon(x) dx =$$

$$- \int_\Omega u(x) \nabla_x \cdot \left(\int_Y A^I(x, y) \sum_{i=1}^{N} \frac{\partial \varphi}{\partial x_i}(x) \left(\vec{e}_i + \nabla_y w_i^*(x, y) \right) dy \right) dx. \quad (2.14)$$

By definition (2.4) of A^*, it is easily seen that the right-hand side of (2.14) is nothing other than

$$- \int_\Omega u(x) \nabla \cdot \left(A^{*I}(x) \nabla \varphi(x) \right) dx.$$

Finally, a last integration by parts yields the limit variational formulation of (2.10),

$$\int_\Omega A^*(x) \nabla u(x) \cdot \nabla \varphi(x) dx = \int_\Omega f(x) \varphi(x) dx. \quad (2.15)$$

By density of smooth functions in $H_0^1(\Omega)$, (2.15) is valid for any test function $\varphi \in H_0^1(\Omega)$. Because A^* satisfies the same coercivity condition as A, the Lax–Milgram lemma shows that (2.15) admits a unique solution in $H_0^1(\Omega)$. This last result proves that any subsequence of u^ε converges to the same limit u. Therefore, the entire sequence u^ε, and not only a subsequence, converges to the homogenized solution u. This concludes the proof of Theorem 2.1.   Q.E.D.

In the course of the proof of Theorem 2.1, the following lemma on periodically oscillating functions was used several times. Its proof is elementary, at least for smooth functions, by using a covering of the domain Ω in small cubes of size ε and the notion of Riemann integration (approximation of integrals by discrete sums). A detailed proof may be found, e.g., in [BM84] or [All92a].

Lemma 2.2 *Let $w(x, y)$ be a continuous function in x, square integrable and Y-periodic in y, i.e., $w(x, y) \in L_{\#}^2(Y; C(\overline{\Omega}))$. Then, the sequence $w(x, \frac{x}{\varepsilon})$ converges weakly in $L^2(\Omega)$ to $\int_Y w(x, y) dy$.*

Remark As a final comment, let us reemphasize that the energy method is not restricted to the periodic case and works without any assumption about the behavior of the sequence of the diffusion tensor. The energy method is also valid for some nonlinear problems involving convex minimization (see Subsection A.1.1 and references therein), and monotone operators (corresponding to nonsymmetric problems).

A.3   Two-Scale Convergence

A.3.1   A Brief Presentation

Contrary to the previous homogenization methods, the two-scale convergence method is devoted only to periodic homogenization problems. It is, therefore, a less general method than the Γ-, G-, and H-convergence, but, because it is dedicated to periodic homogenization, it is also more efficient and simple in this context. Two-scale convergence was recently introduced by Nguetseng [Ngu89] and Allaire [All92a] and is exposed in a self-contained fashion below. It has been further generalized to the stochastic setting of homogenization in [BMW94], thus, considerably extending its scope. The next subsection is concerned with the main theoretical results which are at the root of this method, whereas the last subsection contains a detailed application of the method on a simple model problem.

Before going into the details of the two-scale convergence method, let us explain its main idea and the reasons for its success. In periodic homogenization problems, it is well-known that the homogenized problem can be heuristically obtained by using the two-scale asymptotic expansions as described in many textbooks (see, e.g., [BP89], [BLP78], [SP80], and Chapter 1). Denoting the size of the periodic heterogeneities (a small number which goes to zero in this asymptotic process) by ε and the sequence (indexed by ε) of solutions of the considered partial differential equation with periodically oscillating coefficients by u^ε, a two-scale asymptotic expansion is an ansatz of the form,

$$u^\varepsilon(x) = u_0(x, \frac{x}{\varepsilon}) + \varepsilon u_1(x, \frac{x}{\varepsilon}) + \varepsilon^2 u_2(x, \frac{x}{\varepsilon}) + \cdots, \tag{3.1}$$

where each function $u_i(x, y)$ in this series depends on two variables, x the macroscopic (or slow) variable and y the microscopic (or fast) variable, and is Y-periodic in y (Y is the unit period). Inserting the ansatz (3.1) in the equation satisfied by u^ε and identifying powers of ε leads to a cascade of equations for each term $u_i(x, y)$. In general, averaging with respect to y yields the homogenized equation for u_0 (see Chapters 1 and 3 for details). Unfortunately, mathematically, this method of two-scale asymptotic expansions is only formal because, a priori, there is no reason for the ansatz (3.1) to hold true. Thus, another step is required to rigorously justify the homogenization result obtained heuristically with this two-scale asymptotic expansion (see, for example, the previous section on the energy method). Despite its frequent success in homogenizing many different types of equations, this method is not entirely satisfactory because it involves two steps, formal derivation and rigorous justification of the homogenized problem, which have little in common and are partly redundant.

Consequently, there is room for a more efficient method which will combine these two steps in a single, simpler one. This is exactly the purpose of the *two-scale convergence* method which is based on a new type of convergence (see Definition 3.1). Roughly speaking, two-scale convergence is a *rigorous* justification of the first term of the ansatz (3.1) for any bounded sequence u^ε, in the sense that it asserts the existence of a *two-scale* limit $u_0(x, y)$, such that u^ε, tested again any periodically oscillating test function, converges to $u_0(x, y)$:

$$\int_\Omega u^\varepsilon(x)\varphi(x, \frac{x}{\varepsilon})dx \to \int_\Omega \int_Y u_0(x, y)\varphi(x, y)dxdy. \tag{3.2}$$

Two-scale convergence is an improvement over the usual weak convergence because equation (3.2) measures the periodic oscillations of the sequence u^ε. The two-scale convergence method is based on this result: multiplying the equation satisfied by u^ε with an oscillating test function $\varphi(x, \frac{x}{\varepsilon})$ and passing to the *two-scale* limit automatically yields the homogenized problem.

A.3.2   Statement of the Principal Results

Let us begin this section with a few notations. Ω is an open set of $\mathbb{R}^N$ (not necessarily bounded), and $Y = (0, 1)^N$ is the unit cube. We denote, by $C_\#^\infty(Y)$, the space of infinitely differentiable functions in $\mathbb{R}^N$ which are periodic of period Y and, by $C_\#(Y)$, the Banach space of continuous and Y-periodic functions. Eventually, $\mathcal{D}(\Omega; C_\#^\infty(Y))$ denotes the space of infinitely smooth and compactly supported functions in Ω with values in the space $C_\#^\infty(Y)$.

Definition 3.1 *A sequence of functions u^ε in $L^2(\Omega)$ is said to two-scale converge to a limit $u_0(x, y)$ belonging to $L^2(\Omega \times Y)$ if, for any function $\varphi(x, y)$ in $\mathcal{D}(\Omega; C_\#^\infty(Y))$, it satisfies*

$$\lim_{\varepsilon \to 0} \int_\Omega u^\varepsilon(x)\varphi(x, \frac{x}{\varepsilon})dx = \int_\Omega \int_Y u_0(x, y)\varphi(x, y)dxdy.$$

This notion of "two-scale convergence" makes sense because of the next compactness theorem.

Theorem 3.2 *From each bounded sequence u^ε in $L^2(\Omega)$, one can extract a subsequence, and there exists a limit $u_0(x, y) \in L^2(\Omega \times Y)$ such that this subsequence two-scale converges to u_0.*

Before sketching the proof of Theorem 3.2, we give a few examples of two-scale convergences.

1. Any sequence u^ε which converges strongly in $L^2(\Omega)$ to a limit $u(x)$, two-scale converges to the same limit $u(x)$.

2. For any smooth function $u_0(x, y)$, Y-periodic in y, the associated sequence $u^\varepsilon(x) = u_0(x, \frac{x}{\varepsilon})$ two-scale converges to $u_0(x, y)$.

3. For the same smooth and Y-periodic function $u_0(x, y)$, the sequence defined by $v_\varepsilon(x) = u_0(x, \frac{x}{\varepsilon^2})$ has the same two-scale limit and weak-L^2 limit, namely, $\int_Y u_0(x, y)dy$ (this is a consequence of the difference of orders in the speed of oscillations for v_ε and the test functions $\varphi(x, \frac{x}{\varepsilon})$). Clearly, the two-scale limit captures only the oscillations which are in resonance with those of the test functions $\varphi(x, \frac{x}{\varepsilon})$.

4. Any sequence u^ε which admits an asymptotic expansion of the type $u^\varepsilon(x) = u_0(x, \frac{x}{\varepsilon}) + \varepsilon u_1(x, \frac{x}{\varepsilon}) + \varepsilon^2 u_2(x, \frac{x}{\varepsilon}) + \cdots$, where the functions $u_i(x, y)$ are smooth and Y-periodic in y, two-scale converges to the first term of the expansion, namely, $u_0(x, y)$.

Briefly, we now give the main ideas of the proof of Theorem 3.2 (it was first proven by Nguetseng [Ngu89], but we follow the proof in [All92a] to which the interested reader is referred for more details). For simplicity in the exposition, we assume, during the proof, that Ω is a bounded set. The following elementary lemma is first required (the proof of which is left to the reader).

Lemma 3.3 *Let* $B = C(\bar{\Omega}; C_\#(Y))$ *be the space of continuous functions* $\varphi(x, y)$ *on* $\bar{\Omega} \times Y$ *which are* Y-*periodic in* y. *Then,* B *is a separable Banach space (i.e., it contains a dense countable family), is dense in* $L^2(\Omega \times Y)$, *and any of its elements* $\varphi(x, y)$ *satisfies*

$$\int_\Omega |\varphi(x, \frac{x}{\varepsilon})|^2 dx \leq C\|\varphi\|_B^2$$

and

$$\lim_{\varepsilon \to 0} \int_\Omega |\varphi(x, \frac{x}{\varepsilon})|^2 dx = \int_\Omega \int_Y |\varphi(x, y)|^2 dxdy.$$

Proof of Theorem 3.2. By the Schwarz inequality,

$$\left| \int_\Omega u^\varepsilon(x)\varphi(x, \frac{x}{\varepsilon})dx \right| \leq C \left| \int_\Omega \varphi(x, \frac{x}{\varepsilon})dx \right|^{\frac{1}{2}} \leq C\|\varphi\|_B. \tag{3.3}$$

This implies that the left-hand side of (3.3) is a continuous linear form on B which can be identified as a duality product $\langle \mu_\varepsilon, \varphi \rangle_{B',B}$ for some bounded sequence of measures μ_ε. Because B is separable, one can extract a subsequence and there exists a limit μ_0, such μ_ε converges to μ_0 in the weak * topology of B' (the dual of B). On the other hand, Lemma 3.3 allows us to pass to the limit in the middle term of (3.3). Combining these two results yields

$$|\langle \mu_0, \varphi \rangle_{B',B}| \leq C \left| \int_\Omega \int_Y |\varphi(x, y)|^2 dxdy \right|^{\frac{1}{2}}. \tag{3.4}$$

Eq. (3.4) shows that μ_0 is actually a continuous form on $L^2(\Omega \times Y)$, by density of B in this space. Thus, there exists $u_0(x, y) \in L^2(\Omega \times Y)$, such that

$$\langle \mu_0, \varphi \rangle_{B', B} = \int_\Omega \int_Y u_0(x, y)\varphi(x, y)dxdy,$$

which concludes the proof of Theorem 3.2.   Q.E.D.

The next theorem shows that more information is contained in a two-scale limit than in a weak-L^2 limit; some of the oscillations of a sequence are contained in its two-scale limit. When all of them are captured by the two-scale limit (condition (3.6) below), one can even obtain a strong convergence (a corrector result in the vocabulary of homogenization).

Theorem 3.4 *Let u^ε be a sequence of functions in $L^2(\Omega)$ which two-scale converges to a limit $u_0(x, y) \in L^2(\Omega \times Y)$.*

1. Then, u^ε converges weakly in $L^2(\Omega)$ to $u(x) = \int_Y u_0(x, y)dy$, and

$$\lim_{\varepsilon \to 0} \|u^\varepsilon\|^2_{L^2(\Omega)} \geq \|u_0\|^2_{L^2(\Omega \times Y)} \geq \|u\|^2_{L^2(\Omega)}. \tag{3.5}$$

2. Assume, further, that $u_0(x, y)$ is smooth and that

$$\lim_{\varepsilon \to 0} \|u^\varepsilon\|^2_{L^2(\Omega)} = \|u_0\|^2_{L^2(\Omega \times Y)}. \tag{3.6}$$

Then,

$$\|u^\varepsilon(x) - u_0(x, \frac{x}{\varepsilon})\|^2_{L^2(\Omega)} \to 0. \tag{3.7}$$

Proof of Theorem 3.4. By taking test functions depending only on x in Definition 3.1, the weak convergence in $L^2(\Omega)$ of the sequence u^ε is established. Then, developing the inequality,

$$\int_\Omega |u^\varepsilon(x) - \varphi(x, \frac{x}{\varepsilon})|^2 dx \geq 0,$$

easily yields formula (3.5). Furthermore, under assumption (3.6), it is easily obtained that

$$\lim_{\varepsilon \to 0} \int_\Omega |u^\varepsilon(x) - \varphi(x, \frac{x}{\varepsilon})|^2 dx = \int_\Omega \int_Y |u_0(x, y) - \varphi(x, y)|^2 dxdy.$$

If u_0 is smooth enough to be a test function φ, it yields (3.7).   Q.E.D.

Remark The smoothness assumption on u_0 in the second part of Theorem 3.4 is needed only to ensure the measurability of $u_0(x, \frac{x}{\varepsilon})$ (which otherwise is not guaranteed for a function of $L^2(\Omega \times Y)$). One can further check that any function in $L^2(\Omega \times Y)$ is attained as a two-scale limit (see Lemma 1.13 in [All92a]), which implies that two-scale limits have no extra regularity.

So far we have considered only bounded sequences in $L^2(\Omega)$. The next theorem investigates the case of a bounded sequence in $H^1(\Omega)$.

Theorem 3.5 *Let u^ε be a bounded sequence in $H^1(\Omega)$. Then, up to a subsequence, u^ε two-scale converges to a limit $u(x) \in H^1(\Omega)$, and ∇u^ε two-scale converges to $\nabla_x u(x) + \nabla_y u_1(x, y)$, where the function $u_1(x, y)$ belongs to $L^2(\Omega; H^1_\#(Y)/\mathbb{R})$.*

Proof Because u^ε (resp. ∇u^ε) is bounded in $L^2(\Omega)$ (resp. $L^2(\Omega)^N$), up to a subsequence, it two-scale converges to a limit $u_0(x, y) \in L^2(\Omega \times Y)$ (respectively $\xi_0(x, y) \in L^2(\Omega \times Y)^N$). Thus, for any $\psi(x, y) \in \mathcal{D}\left(\Omega; C^\infty_\#(Y)^N\right)$,

$$\lim_{\varepsilon \to 0} \int_\Omega \nabla u^\varepsilon(x) \cdot \psi(x, \frac{x}{\varepsilon}) dx = \int_\Omega \int_Y \xi_0(x, y) \cdot \psi(x, y) dx dy. \qquad (3.8)$$

Integrating the left-hand side of (3.8) by parts gives

$$\varepsilon \int_\Omega \nabla u^\varepsilon(x) \cdot \psi(x, \frac{x}{\varepsilon}) dx = -\int_\Omega u^\varepsilon(x) \left(\nabla_y \cdot \psi(x, \frac{x}{\varepsilon}) + \varepsilon \nabla_x \cdot \psi(x, \frac{x}{\varepsilon})\right) dx. \qquad (3.9)$$

Passing to the limit yields

$$0 = -\int_\Omega \int_Y u_0(x, y) \nabla_y \cdot \psi(x, y) dx dy. \qquad (3.10)$$

This implies that $u_0(x, y)$ does not depend on y. Thus, there exists $u(x) \in L^2(\Omega)$, such that $u_0 = u$. Next, in (3.8), we choose a function ψ, such that $\nabla_y \cdot \psi(x, y) = 0$. Integrating by parts, we obtain

$$\lim_{\varepsilon \to 0} \int_\Omega u^\varepsilon(x) \nabla_x \cdot \psi(x, \frac{x}{\varepsilon}) dx \begin{aligned} &= -\int_\Omega \int_Y \xi_0(x, y) \cdot \psi(x, y) dx dy \\ &= \int_\Omega \int_Y u(x) \nabla_x \cdot \psi(x, y) dx dy. \end{aligned} \qquad (3.11)$$

If ψ does not depend on y, (3.11) proves that $u(x)$ belongs to $H^1(\Omega)$. Furthermore, we deduce from (3.11) that

$$\int_\Omega \int_Y (\xi_0(x, y) - \nabla u(x)) \cdot \psi(x, y) dx dy = 0 \qquad (3.12)$$

for any function $\psi(x, y) \in \mathcal{D}\left(\Omega; C^\infty_\#(Y)^N\right)$ with $\nabla_y \cdot \psi(x, y) = 0$. Recall that the function orthogonal to divergence-free functions are exactly the gradients (this well-known result can be very easily proven in the present context by Fourier analysis in Y). Thus, there exists a unique function $u_1(x, y)$ in $L^2(\Omega; H^1_\#(Y)/\mathbb{R})$, such that

$$\xi_0(x, y) = \nabla u(x) + \nabla_y u_1(x, y). \qquad (3.13)$$

Q.E.D.

There are many generalizations of Theorem 3.5 which gives the precise form of the two-scale limit of a sequence of functions for which some extra estimates on part of their derivatives are available. To obtain as much information as possible on the two-scale limit is a key point in applying the two-scale convergence method, as described in the next subsection. For completeness, we give two examples below of such generalizations of Theorem 3.5, the proofs of which may be found in [All92a].

Theorem 3.6   *1. Let u^ε be a bounded sequence in $L^2(\Omega)$, such that $\varepsilon\nabla u^\varepsilon$ is also bounded in $L^2(\Omega)^N$. Then, there exists a two-scale limit $u_0(x, y) \in L^2(\Omega; H^1_\#(Y)/\mathbb{R})$ such that, up to a subsequence, u^ε two-scale converges to $u_0(x, y)$, and $\varepsilon\nabla u^\varepsilon$ to $\nabla_y u_0(x, y)$.*

2. Let u^ε be a bounded sequence of vector-valued functions in $L^2(\Omega)^N$, such that its divergence $\nabla \cdot u^\varepsilon$ is also bounded in $L^2(\Omega)$. Then, there exists a two-scale limit $u_0(x, y) \in L^2(\Omega \times Y)^N$ which is divergence-free with respect to y, i.e., $\nabla_y \cdot u_0 = 0$, has a divergence with respect to x, $\nabla_x \cdot u_0$, in $L^2(\Omega \times Y)$, and such that, up to a subsequence, u^ε two-scale converges to $u_0(x, y)$ and $\nabla \cdot u^\varepsilon$ to $\nabla_x \cdot u_0(x, y)$.

We conclude this subsection by discussing the possibility of passing to the two-scale limit in nonlinear expressions. It is well-known that, in general, nonlinear functionals are not continuous with respect to the weak topologies of $L^p(\Omega)$ spaces ($1 \leq p \leq +\infty$). Unfortunately, the same is true for the two-scale convergence which is also a weak-type convergence. As for the usual weak $L^p(\Omega)$ topology, we can merely establish a lower semicontinuity result for *convex* functionals of the same type rather than inequality (3.5) on the norm of the two-scale limit.

Theorem 3.7 *Let u^ε be a sequence of functions in $L^2(\Omega)$ which two-scale converges to a limit $u_0(x, y) \in L^2(\Omega \times Y)$. Let $F(x, y, \lambda)$ be a functional depending continuously on $x \in \Omega$ and measurably on $y \in Y$, which is Y-periodic in y, and convex and continuously differentiable in $\lambda \in \mathbb{R}$. Assume that F has quadratic growth in λ, i.e., there exists two constants $0 < c \leq C$ such that, for any $(x, y) \in \Omega \times Y$,*

$$c|\lambda|^2 \leq F(x, y, \lambda) \leq C\left(1 + |\lambda|^2\right).$$

Then, the sequence $F(x, x/\varepsilon, u^\varepsilon)$ is lower semicontinuous in the following sense

$$\lim_{\varepsilon \to 0} \int_\Omega F(x, \frac{x}{\varepsilon}, u^\varepsilon(x))dx \geq \int_\Omega \int_Y F(x, y, u_0(x, y))dxdy. \tag{3.14}$$

The proof of Theorem 3.7 is not difficult by noting that a convex function is the supremum of all linear functions which remain below it and passing to the two-scale limit in this class of linear functions (see section 3 in [All92a] for details).

Remark Theorem 3.7 has been stated for a functional with quadratic growth. Of course, a similar result holds for other growth conditions by simply generalizing Definition 3.1 and Theorem 3.2 on two-scale convergence to sequences bounded in $L^p(\Omega)$ spaces ($1 < p \leq +\infty$) (see also section 4.3).

A.3.3   Application to a Model Problem

This subsection shows how the notion of two-scale convergence can be used for homogenizing partial differential equations with periodically oscillating coefficients. Our purpose is to give a tutorial on the two-scale convergence method, demonstrating its power and simplicity rather than being complete and investigating all possible applications. Therefore, the usual model problem of diffusion in a periodic medium is reconsidered, as in section A.2. Of course, the principles of the two-scale convergence method are valid in many other cases with only slight changes, including nonlinear (monotone or convex) problems.

We recall that Ω is the periodic domain (a bounded open set in $\mathbb{R}^N$), ε its period, and $Y = (0, 1)^N$ the rescaled unit cell. The tensor of diffusion in Ω is an $N \times N$ matrix $A(x, \frac{x}{\varepsilon})$, not necessarily symmetric, where $A(x, y)$ belongs to $C(\bar{\Omega}; L_\#^\infty(Y))^{N^2}$ and, everywhere in $\Omega \times Y$, satisfies

$$\alpha|\xi|^2 \le \sum_{i,j=1}^N A_{i,j}(x, y)\xi_i\xi_j \le \beta|\xi|^2,$$

for any vector $\xi \in \mathbb{R}^N$, where α and β are two constants, such that $0 < \alpha \le \beta$.

We consider the following model problem of diffusion:

$$\begin{cases} -\nabla \cdot \left(A(x, \frac{x}{\varepsilon})\nabla u^\varepsilon\right) = f & \text{in } \Omega, \\ u^\varepsilon = 0 & \text{on } \partial\Omega. \end{cases} \tag{3.15}$$

If the source term $f(x)$ belongs to $L^2(\Omega)$, equation (3.15) admits a unique solution u^ε in $H_0^1(\Omega)$ by applying the Lax–Milgram lemma. Moreover, u^ε satisfies the following a priori estimate:

$$\|u^\varepsilon\|_{H_0^1(\Omega)} \le C\|f\|_{L^2(\Omega)}, \tag{3.16}$$

where C is a positive constant which does not depend on ε.

We now describe the so-called "two-scale convergence method" for homogenizing problem (3.15). In a **first step**, we deduce the precise form of the two-scale limit of the sequence u^ε from the a priori estimate (3.16). By application of Theorem 3.5, there exist two functions, $u(x) \in H_0^1(\Omega)$ and $u_1(x, y) \in L^2(\Omega; H_\#^1(Y)/\mathbb{R})$, such that, up to a subsequence, u^ε two-scale converges to $u(x)$, and ∇u^ε two-scale converges to $\nabla_x u(x) + \nabla_y u_1(x, y)$. In view of these limits, u^ε is expected to behave as $u(x) + \varepsilon u_1(x, \frac{x}{\varepsilon})$.

Then, in a **second step**, we multiply equation (3.15) by a test function similar to the limit of u^ε, namely, $\varphi(x) + \varepsilon\varphi_1(x, \frac{x}{\varepsilon})$, where $\varphi(x) \in \mathcal{D}(\Omega)$ and $\varphi_1(x, y) \in \mathcal{D}(\Omega; C_\#^\infty(Y))$. This yields

$$\int_\Omega A(x, \frac{x}{\varepsilon})\nabla u^\varepsilon \cdot \left(\nabla\varphi(x) + \nabla_y\varphi_1(x, \frac{x}{\varepsilon}) + \varepsilon\nabla_x\varphi_1(x, \frac{x}{\varepsilon})\right) dx$$
$$= \int_\Omega f(x)\left(\varphi(x) + \varepsilon\varphi_1(x, \frac{x}{\varepsilon})\right) dx. \tag{3.17}$$

Regarding $A^t(x, \frac{x}{\varepsilon})\left(\nabla\varphi(x) + \nabla_y\varphi_1(x, \frac{x}{\varepsilon})\right)$ as a test function for the two-scale convergence (see Definition 3.1), we pass to the two-scale limit in (3.17) for the sequence ∇u^ε. Although this test function is not necessarily very smooth, as required by Definition 3.1, it belongs at least to $C\left(\bar{\Omega}; L_\#^2(Y)\right)$ which, as can be shown, is sufficient for the two-scale convergence Theorem 3.2 to hold (see [All92a] for details). Thus, the two-scale limit of equation (3.17) is given by

$$\int_\Omega \int_Y A(x, y) \left(\nabla u(x) + \nabla_y u_1(x, y)\right) \cdot \left(\nabla \varphi(x) + \nabla_y \varphi_1(x, y)\right) dx dy$$
$$= \int_\Omega f(x) \varphi(x) dx. \quad (3.18)$$

In a **third step**, we read off a variational formulation for (u, u_1) in (3.18). Note that (3.18) holds true for any (φ, φ_1) in the Hilbert space $H_0^1(\Omega) \times L^2 \left(\Omega; H_\#^1(Y)/\mathbb{R}\right)$ by density of smooth functions in this space. Endowing it with the norm $\sqrt{(\|\nabla u(x)\|_{L^2(\Omega)}^2 + \|\nabla_y u_1(x, y)\|_{L^2(\Omega \times Y)}^2)}$, the assumptions of the Lax–Milgram lemma are easily checked for the variational formulation (3.18). The main point is the coercivity of the bilinear form defined by the left-hand side of (3.18). The coercivity of A yields

$$\int_\Omega \int_Y A(x, y) \left(\nabla \varphi(x) + \nabla_y \varphi_1(x, y)\right) \cdot \left(\nabla \varphi(x) + \nabla_y \varphi_1(x, y)\right) dx dy$$
$$\geq \alpha \int_\Omega \int_Y |\nabla \varphi(x) + \nabla_y \varphi_1(x, y)|^2 dx dy$$
$$= \alpha \int_\Omega |\nabla \varphi(x)|^2 dx + \alpha \int_\Omega \int_Y |\nabla_y \varphi_1(x, y)|^2 dx dy.$$

By applying the Lax–Milgram lemma, we conclude that there exists a unique solution (u, u_1) of the variational formulation (3.18) in $H_0^1(\Omega) \times L^2 \left(\Omega; H_\#^1(Y)/\mathbb{R}\right)$. Consequently, the entire sequences u^ε and ∇u^ε converge to $u(x)$ and $\nabla u(x) + \nabla_y u_1(x, y)$. An easy integration by parts shows that (3.18) is a variational formulation associated with the following system of equations, the so-called "two-scale homogenized problem":

$$\begin{cases} -\nabla_y \cdot \left(A(x, y) \left(\nabla u(x) + \nabla_y u_1(x, y)\right)\right) = 0 & \text{in } \Omega \times Y, \\ -\nabla_x \cdot \left(\int_Y A(x, y) \left(\nabla u(x) + \nabla_y u_1(x, y)\right) dy\right) = f(x) & \text{in } \Omega, \\ y \to u_1(x, y) & Y\text{-periodic,} \\ u = 0 & \text{on } \partial\Omega. \end{cases} \quad (3.19)$$

At this point, the homogenization process could be considered achieved because the entire sequence of solutions u^ε converges to the solution of a well-posed limit problem, namely, the two-scale homogenized problem (3.19). However, it is usually preferable, from a physical or numerical point of view, to eliminate the microscopic variable y (one does not want to solve the small scale structure). In other words, we want to extract and decouple the usual homogenized and local (or cell) equations from the two-scale homogenized problem.

Thus, in a **fourth (and optional) step**, the y variable and the u_1 unknown are eliminated from (3.19). It is an easy algebraic exercise to prove that u_1 can be computed in terms of the gradient of u through the relationship

$$u_1(x, y) = \sum_{i=1}^{N} \frac{\partial u}{\partial x_i}(x) w_i(x, y), \tag{3.20}$$

where $w_i(x, y)$ are defined, at each point $x \in \Omega$, as the unique solutions in $H^1_\#(Y)/\mathbb{R}$ of the cell problems (see formulas (3.6) in Chapter 1 and (2.5) in this chapter):

$$\begin{cases} -\nabla_y \cdot \left(A(x, y) \left(\vec{e}_i + \nabla_y w_i(x, y) \right) \right) = 0 & \text{in } Y, \\ y \rightarrow w_i(x, y) & Y\text{-periodic,} \end{cases} \tag{3.21}$$

with $(\vec{e}_i)_{1 \le i \le N}$ the canonical basis of $\mathbb{R}^N$. Then, plugging formula (3.20) into (3.19) yields the usual homogenized problem for u:

$$\begin{cases} -\nabla_x \cdot (A^*(x) \nabla u(x)) = f(x) & \text{in } \Omega, \\ u = 0 & \text{on } \partial\Omega, \end{cases} \tag{3.22}$$

where the homogenized diffusion tensor is given by its entries

$$A^*_{ij}(x) = \int_Y A(x, y) \left(\vec{e}_i + \nabla_y w_i(x, y) \right) \cdot \left(\vec{e}_j + \nabla_y w_j(x, y) \right) dy. \tag{3.23}$$

Of course, all the above formulas coincide with those usually obtained by using asymptotic exansions (see section 1.3.1).

Due to the simple form of our model problem, the two equations of (3.19) can be decoupled in a microscopic and a macroscopic equation, (3.21) and (3.22) respectively, but we emphasize that it is not always possible. Sometimes, it leads to very complicated forms of the homogenized equation, including integro-differential operators (see for example the double-porosity model in Chapter 3). Thus, the homogenized equation does not always belong to a class for which an existence and uniqueness theory is easily available, contrary to the two-scale homogenized system, which, in most cases, is of the same type as the original problem, but with double the number of variables (x and y) and unknowns (u and u_1). The supplementary microscopic variable and unknown play the role of "hidden" variables in the vocabulary of mechanics. Although their presence doubles the size of the limit problem, it greatly simplifies its structure (which could be useful for numerical purposes, too), whereas eliminating them introduces "strange" effects (such as memory or nonlocal effects) in the usual homogenized problem.

It is often very useful to obtain so-called "corrector" results which permit obtaining strong (or pointwise) convergences, instead of just weak ones, by adding some extra information stemming from the local equations. Typically, in the above example, we simply proved that the sequence u^ε converges weakly to the homogenized solution u in $H^1_0(\Omega)$. Introducing the local solution u_1, this weak convergence can be improved as follows:

$$\left(u^\varepsilon(x) - u(x) - \varepsilon u_1\left(x, \frac{x}{\varepsilon}\right) \right) \rightarrow 0 \text{ in } H^1_0(\Omega) \text{ strongly.} \tag{3.24}$$

This type of result is easily obtained with the two-scale convergence method. This rigorously justifies the two first terms in the usual asymptotic expansion (see chapter 1) of the sequence u^ε. Note that, by standard regularity results for the solutions $w_i(x, y)$ of the cell problem (3.21), the term $u_1(x, \frac{x}{\varepsilon})$ does actually belong to $L^2(\Omega)$ and can be seen as a test function for the two-scale convergence. Furthermore, if the matrix A is smooth in x, say in $W^{1,\infty}(\Omega)$, then, u and w_i are also smooth in x which implies that $u_1(x, \frac{x}{\varepsilon})$ belongs to $H^1(\Omega)$. Under this mild assumption, we can write

$$\int_\Omega A(x, \frac{x}{\varepsilon}) \left(\nabla u^\varepsilon(x) - \nabla u(x) - \nabla_y u_1(x, \frac{x}{\varepsilon}) \right)$$
$$\cdot \left(\nabla u^\varepsilon(x) - \nabla u(x) - \nabla_y u_1(x, \frac{x}{\varepsilon}) \right) dx$$
$$= \int_\Omega f(x) u^\varepsilon(x) dx$$
$$+ \int_\Omega A(x, \frac{x}{\varepsilon}) \left(\nabla u(x) + \nabla_y u_1(x, \frac{x}{\varepsilon}) \right) \cdot \left(\nabla u(x) + \nabla_y u_1(x, \frac{x}{\varepsilon}) \right) dx$$
$$- \int_\Omega A(x, \frac{x}{\varepsilon}) \nabla u^\varepsilon(x) \cdot \left(\nabla u(x) + \nabla_y u_1(x, \frac{x}{\varepsilon}) \right) dx$$
$$- \int_\Omega A(x, \frac{x}{\varepsilon}) \left(\nabla u(x) + \nabla_y u_1(x, \frac{x}{\varepsilon}) \right) \cdot \nabla u^\varepsilon(x) dx. \tag{3.25}$$

Using the coercivity condition for A and passing to the two-scale limit yields

$$\alpha \lim_{\varepsilon \to 0} \| \nabla u^\varepsilon(x) - \nabla u(x) - \nabla_y u_1(x, \frac{x}{\varepsilon}) \|^2_{L^2(\Omega)} \le \int_\Omega f(x) u(x) dx$$
$$- \int_\Omega \int_Y A(x, y) \left(\nabla u(x) + \nabla_y u_1(x, y) \right) \cdot \left(\nabla u(x) + \nabla_y u_1(x, y) \right) dx dy.$$

In view of (3.19), the right-hand side of the above equation is equal to zero, which gives the desired result.

A.4 Iterated Homogenization

In the previous section, we considered homogenization problems in periodic media where only two different length scales were considered, namely, the macroscopic (of the order of the domain size) and the microscopic (of the order of the heterogeneities period), which have a ratio denoted by ε. Of course, in the real world, porous media are far from being periodic and usually exhibit many different length scales of heterogeneities. The very crude modeling of section A.3 can be further improved by considering not only a single scale of heterogeneities but several periodic scales of heterogeneities (up to a countable infinite number of scales). This type of homogenization problem is called *reiterated homogenization* (following a terminology of [BLP78]) because, under a mild assumption on the separation of scales, it amounts to successively homogenizing the smallest scale while keeping the larger ones fixed.

Here, we shall simply state the main result of this process of reiterated homogenization on a model problem to explain the main ideas without dwelling too much on technicalities. Our model problem is a diffusion equation in a multiply periodic domain Ω (a bounded open set in $\mathbb{R}^N$). We asume that there are n scales of heterogeneities $\varepsilon_1, \varepsilon_2, \cdots, \varepsilon_n$ which depend on a single positive parameter ε which tends to zero. The key assumption is that all scales go to zero as ε does, while remaining ordered, ε_1 being the largest and ε_n the smallest, i.e.,

$$\lim_{\varepsilon \to 0} \varepsilon_i(\varepsilon) = 0, \text{ for } 1 \leq i \leq n, \tag{4.1}$$

and

$$\lim_{\varepsilon \to 0} \frac{\varepsilon_i(\varepsilon)}{\varepsilon_{i-1}(\varepsilon)} = 0, \text{ for } 2 \leq i \leq n. \tag{4.2}$$

For simplicity, the rescaled unit cell Y_i at each scale is assumed to be the same, equal to the unit cube $Y = (0, 1)^N$. The tensor of diffusion in Ω is given by an $N \times N$ matrix $A(x, \frac{x}{\varepsilon_1}, \cdots, \frac{x}{\varepsilon_n})$, not necessarily symmetric (if some drift is allowed in the diffusion process), where $A(x, y_1, \cdots, y_n)$ is a continuous function of all variables $x \in \Omega$ and $y_i \in Y_i$, which is Y_i-periodic in y_i (less smoothness is sometime acceptable, see [AB96]). Furthermore, this matrix A satisfies the usual coerciveness and boundedness assumptions: there exist two positive constants α and β, satisfying $0 < \alpha \leq \beta$, such that, for any constant vector $\xi \in \mathbb{R}^N$ and at any point $(x, y_1, \cdots, y_n)$,

$$\alpha|\xi|^2 \leq \sum_{i,j=1}^{N} A_{i,j}(x, y_1, \cdots, y_n)\xi_i\xi_j \leq \beta|\xi|^2,$$

where $A_{i,j}$ denotes the entries of the matrix A.

Denoting the source term by $f(x) \in L^2(\Omega)$ and enforcing a Dirichlet boundary condition, our model problem of diffusion in a multiply periodic medium reads

$$\begin{cases} -\nabla \cdot \left(A(x, \frac{x}{\varepsilon_1}, \cdots, \frac{x}{\varepsilon_n})\nabla u^\varepsilon \right) = f & \text{in } \Omega, \\ u^\varepsilon = 0 & \text{on } \partial\Omega. \end{cases} \tag{4.3}$$

By applying the Lax–Milgram lemma, equation (4.3) admits a unique solution u^ε in $H_0^1(\Omega)$. Moreover, u^ε satisfies the following a priori estimate:

$$\|u^\varepsilon\|_{H_0^1(\Omega)} \leq C\|f\|_{L^2(\Omega)}, \tag{4.4}$$

where C is a positive constant which does not depend on ε. It implies that the sequence u^ε is bounded in the Sobolev space $H_0^1(\Omega)$.

To compute the homogenized diffusion tensor we need the following notations. Let

$$A_n(y_0, y_1, \cdots, y_n)$$

be the original tensor $A(x, y_1, \cdots, y_n)$ (for convenience, the macroscopic variable x is denoted by y_0). For $0 \leq i \leq n - 1$, a tensor $A_i(y_0, y_1, \cdots, y_i)$ is defined as the homogenized tensor of $A_{i+1}(y_0, y_1, \cdots, y_i, \frac{x}{\varepsilon})$ where all the larger scales $y_0, y_1, \cdots, y_i$ (including the macroscopic one y_0) are kept fixed. We also denote the last homogenized tensor $A_0(y_0)$ by $A^*(x)$, for which there is no more microscale. In other words, the rule for computing the final homogenized tensor $A^*(x)$ is to separately and sequentially homogenize the different scales from the smallest to the largest. More precisely, at each scale $1 \leq i \leq n$, we introduce the solutions $w_p^i(y_0, y_1, \cdots, y_i)$, with $1 \leq p \leq N$, defined, at each point $(y_0, y_1, \cdots, y_{i-1})$, as the unique solutions in $H_\#^1(Y_i)/\mathbb{R}$ of the local problems:

$$
\begin{cases}
-\nabla_{y_i} \cdot \left(A_i(y_0, y_1, \cdots, y_i) \left(e_p + \nabla_{y_i} w_p^i(y_0, y_1, \cdots, y_i) \right) \right) = 0 & \text{in } Y_i, \\
y_i \to w_p^i(y_0, y_1, \cdots, y_i) & Y_i\text{-periodic,}
\end{cases}
$$
(4.5)

with $(e_p)_{1 \leq p \leq N}$ the canonical basis of $\mathbb{R}^N$. Then, the sequence $A_i(y_0, y_1, \cdots, y_i)$ is defined by its entries,

$$
\begin{aligned}
A_i^{pq}&(y_0, y_1, \cdots, y_i) \\
&= \int_{Y_{i+1}} A_{i+1}(y_0, y_1, \cdots, y_i, y_{i+1}) \\
&\quad \left(e_p + \nabla_{y_{i+1}} w_p^{i+1} \right) \cdot \left(e_q + \nabla_{y_{i+1}} w_q^{i+1} \right) dy_{i+1}.
\end{aligned}
$$
(4.6)

Formulas (4.5) and (4.6) are usually used for computing the homogenized coefficients of a single-scale periodic medium. Finally, the main result of this reiterated homogenization process is the following theorem.

Theorem 4.1 *The sequence $u^\varepsilon(x)$ of solutions of equation(4.3) converges weakly in $H_0^1(\Omega)$ to $u(x)$, the unique solution of the homogenized problem,*

$$
\begin{cases}
-\nabla_x \cdot (A^*(x)\nabla u(x)) = f(x) & \text{in } \Omega, \\
u = 0 & \text{on } \partial\Omega,
\end{cases}
$$
(4.7)

where the homogenized diffusion tensor is given by the last term A_0 of the sequence defined by (4.6).

Theorem 4.1 was first proven in [BLP78] when the scales are successive powers of ε, i.e., $\varepsilon_i = \varepsilon^i$ (this assumption favors the use of multiple-scale asymptotic expansions). A general proof of Theorem 4.1 (including the case of an infinite number of scales) is given in [AB96], where a notion of multiple-scale convergence is introduced (generalizing the two-scale convergence described in section A.3).

Reiterated homogenization has been used in [Ave87] for rigorously justifying the so-called *differential effective scheme* for computing effective coefficients in a heterogeneous medium with an infinite number of length scales. The differential effective scheme is a well-known method for estimating mechanical properties of composite materials (see, e.g., [Nor85]). Loosely speaking, it amounts to computing homogenized coefficients as the solution of an ordinary differential equation where the "time" variable is the increment in the volume fraction of an inclusion phase. This differential effective scheme could also be applied to evaluate diffusion constants in porous media, but, to our knowledge, it has never been done so far.

B

Mathematical Symbols and Definitions

B.1 List of Symbols

Time and space variables:
Macrovariables:

t	= time
N	= space dimension
$\Omega \subset \mathbb{R}^N$	= domain in space, a bounded open set
x	= point in Ω
$\partial\Omega$	= boundary of Ω
$\vec{\nu}$	= outer normal vector on the boundary $\partial\Omega$

Geometric microvariables:

Y	= periodicity cell of unit volume
y	= point in Y
$\mathfrak{P}$	= standard pore volume occupied by fluid
$\mathfrak{S}$	= standard solid particle, pellet, grain
Γ	= standard pore-particle interface
$\vec{\nu}$	= normal vector on the interface Γ directed out of $\mathfrak{S}$
$\mathcal{O}$	= standard obstacle

Geometric mesovariables:

Y	= periodicity cell of unit volume
y	= point in Y
$\mathfrak{F}$	= standard fracture, fissure
$\mathfrak{B}$	= standard block, aggregate formed by the solid matrix
Γ	= standard interface between fractures and blocks
$\vec{\nu}$	= normal vector on the interface Γ directed out of $\mathfrak{B}$

Geometric ε-dependent variables for micromodels:

ε	= small length scale parameter
$\mathfrak{P}^{\varepsilon}$	= pores filled with fluid (liquid and/or gas)
$\mathfrak{S}^{\varepsilon}$	= solid matrix
Γ^{ε}	= interface between pores and solid matrix
$\vec{\nu}$	= normal vector on the interface Γ^{ε} directed out of $\mathfrak{S}^{\varepsilon}$
$\mathcal{O}^{\varepsilon}$	= the space occupied by obstacles

Geometric ε-dependent variables for mesomodels:

ε = small length scale parameter
$\mathfrak{F}^\varepsilon$ = fractures, fissures
$\mathfrak{B}^\varepsilon$ = blocks, aggregates formed by the solid matrix
Γ^ε = interface between fractures and blocks
$\vec{v}$ = normal vector on the interface Γ^ε directed out of $\mathfrak{B}^\varepsilon$

Micro- or mesoscopic functions:

p^ε = pressure
u^ε = generic global (fast) function
U^ε = generic local (slow) function
$\vec{v}^\varepsilon$ = fluid velocity

Solutions of the cell problems:

w_j = generic cell solution
π_j = cell pressure
$\vec{w}_j$ = cell velocity

Macroscopic functions:

p = pressure
u = generic global (fast) function
U = generic local (slow) function
$\vec{v}$ = fluid velocity

General mathematical symbols and notations:

∇ = gradient operator
$\nabla\cdot$ = divergence operator
$\Delta = \nabla \cdot \nabla$ = Laplace delta operator
$d\Gamma$ = surface measure on Γ
$\vec{e}_i$ = ith unit vector
$\mathbb{N}$ = set of natural numbers
$\mathbb{R}$ = set of real numbers
$\mathbb{Z}$ = set of integer numbers
o, O = Landau symbols

The Landau symbols have the following meaning: $f(\varepsilon) = o(g(\varepsilon))$ for $\varepsilon \to 0$ means

$$\lim_{\varepsilon \to 0} \frac{f(\varepsilon)}{|g(\varepsilon)|} = 0,$$

and $f(\varepsilon) = O(g(\varepsilon))$ for $\varepsilon \to 0$ means that there is an ε-independent constant C and an $\varepsilon_0 > 0$, such that

$$\left|\frac{f(\varepsilon)}{g(\varepsilon)}\right| \le C \text{ for all } 0 < \varepsilon < \varepsilon_0.$$

Consequently, $f(\varepsilon) = o(1)$ means that $f(\varepsilon)$ tends to zero, and $f(\varepsilon) = O(1)$ means that $f(\varepsilon)$ remains bounded for small ε.

Stochastic notations:

$$\mathcal{M} \quad = \text{set of random events}$$

Physical quantitites and constitutive functions:

C	= contrast of property number, see section 8.2
D	= rate of strain tensor, see section 4.2
μ	= dynamic fluid viscosity, see sections 3.1, 5.1.1, and 8.2
Nu	= Nusselt number, see section 7.4
ω	= pulsation, see section 8.1
Pe	= Péclet number, see section 7.5
Ra	= Rayleigh number, see section 7.3
Re	= Reynolds number, see section 4.2
Sh	= Strouhal number, see section 8.2

B.2   Function Spaces

Here we give only short descriptions of the function spaces used in this book. The Lebesgue spaces are studied in detail in any textbook on functional analysis (see, e.g., [DL90]). A general introduction to Sobolev spaces can be found in [Ada75].

B.2.1   Macroscopic Function Spaces

Banach space of continuous functions:

$$C(\overline{\Omega}) = \{u : u : (\overline{\Omega} \to \mathbb{R}, \ u \text{ is continuous}\}$$

with the norm

$$\|u\|_{C(\Omega)} = \sup_{x \in \Omega} |u(x)|.$$

Banach space of essentially bounded functions:

$$L^{\infty}(\Omega) = \{u : u : \Omega \to \mathbb{R}, \ u \text{ is essentially bounded}\}$$

with the norm

$$\|u\|_{L^{\infty}(\Omega)} = \text{ess sup}_{x \in \Omega} |u(x)|.$$

Space of test functions:

$$\mathcal{D}(\Omega) = \{u : u : \Omega \to \mathbb{R},$$
$$\text{supp } u \text{ is compact, } u \text{ has derivatives of all orders}\}$$

where supp u denotes the support

$$\text{supp } u = \text{closure } (\{x : x \in \Omega, u(x) \neq 0\}).$$

The Lebesgue space (a Banach-space) of p-integrable functions:

$$L^p(\Omega) = \{u : u : \Omega \to \mathbb{R}, \int_\Omega |u(x)|^p \, dx < \infty\}$$

with the norm

$$\|u\|_{L^p(\Omega)} = (\int_\Omega |u(x)|^p \, dx)^{\frac{1}{p}}.$$

The Lebesgue space (a Hilbert-space) of square-integrable functions:

$$L^2(\Omega) = \{u : u : \Omega \to \mathbb{R}, \int_\Omega |u(x)|^2 \, dx < \infty\}$$

with the norm

$$\|u\|_{L^2(\Omega)} = (\int_\Omega |u(x)|^2 \, dx)^{\frac{1}{2}}.$$

The Lebesgue space (a Banach-space) of vector-valued, p-integrable functions:

$$(L^p(\Omega))^N = \{u : u : \Omega \to \mathbb{R}^N, \int_\Omega |u(x)|^p \, dx < \infty\}$$

with the norm

$$\|u\|_{(L^p(\Omega))^N} = (\int_\Omega |u(x)|^p \, dx)^{\frac{1}{p}},$$

where now $|z|$ denotes the p-norm

$$|(z_1, ..., z_N)| = (\sum_i^N |z_i|^p)^{\frac{1}{p}}$$

in $\mathbb{R}^N$. The Lebesgue space (a Hilbert-space) of vector-valued, square-integrable functions:

$$(L^2(\Omega))^N = \{u : u : \Omega \to \mathbb{R}^N, \int_\Omega |u(x)|^2 \, dx < \infty\}$$

with the norm

$$\|u\|_{(L^2(\Omega))^N} = (\int_\Omega |u(x)|^2 \, dx)^{\frac{1}{2}}$$

where now $|z|$ denotes the Euclidian norm

$$|(z_1, ..., z_N)| = (\sum_i^N |z_i|^2)^{\frac{1}{2}}$$

in $\mathbb{R}^N$.

The Sobolev space (a Banach-space) of functions having p-integrable derivatives:

$$W^{1,p}(\Omega) = \{u : u \in L^p(\Omega), \nabla u \in (L^p(\Omega))^N\}$$

with the norm

$$\|u\|_{W^{1,p}(\Omega)} = (\|u\|^p_{L^p(\Omega)} + \|\nabla u\|^p_{(L^p(\Omega))^N})^{\frac{1}{p}}.$$

The Sobolev space (a Hilbert space) of functions having square-integrable derivatives:

$$H^1(\Omega) = W^{1,2}(\Omega)$$

with the norm

$$\|u\|_{H^1(\Omega)} = (\|u\|^2_{L^2(\Omega)} + \|\nabla u\|^2_{(L^2(\Omega))^N})^{\frac{1}{2}}.$$

The Sobolev space of functions having zero trace:

$$W^{1,p}_0(\Omega) = \{u : u \in W^{1,p}(\Omega), \text{ trace } u = 0\}$$

where *trace* denotes the linear operator which maps functions $u \in H^1(\Omega)$ onto their boundary values trace $u = u|_{\partial\Omega}$. In other words, the functions $u \in W^{1,p}_0(\Omega)$ vanish on the boundary $\partial\Omega$.

$$H^1_0(\Omega) = W^{1,2}_0(\Omega).$$

The dual space of $H^1_0(\Omega)$, i.e., the space of all continuous linear functionals on $H^1_0(\Omega)$, is denoted by $H^{-1}(\Omega)$, where the duality is denoted by

$$\langle \cdot, \cdot \rangle_{H^{-1}(\Omega), H^1_0(\Omega)}.$$

B.2.2 Micro- and Mesoscopic Function Spaces

Banach space of continuous functions:

$$C_\#(Y) = \{u : u : Y \to \mathbb{R}, u \text{ is continuous and } Y\text{--periodic}\}$$

with the norm

$$\|u\|_{C_\#(Y)} = \sup_{y \in Y} |u(y)|.$$

This function space often is identified with the space of continuous and Y-periodic functions defined on all $\mathbb{R}^n$; it may also be constructed as a space of restrictions of these functions to the cell Y. The same remark applies to the spaces that follow.
Banach space of essentially bounded functions:

$$L^\infty_\#(Y) = \{u : u : Y \to \mathbb{R}, u \text{ is essentially bounded and } Y\text{--periodic}\}$$

with the norm

$$\|u\|_{L^\infty_\#(Y)} = \text{ess sup}_{y \in Y} |u(y)|.$$

Sobolev spaces of Y-periodic functions:

$$W_\#^{1,p}(Y) = \{u : u \in W^{1,p}(Y), u \text{ is } Y\text{--periodic}\}$$

with the norm

$$\|u\|_{W_\#^{1,p}(Y)} = (\|u\|_{L^p(Y)}^p + (\|\nabla u\|_{(L^p(Y))^N}^p)^{\frac{1}{p}}.$$

$$W_\#^{1,p}(\mathfrak{P}) = \{u : u \in W^{1,p}(\mathfrak{P}), u \text{ is } Y\text{--periodic}\}$$

with the norm

$$\|u\|_{W_\#^{1,p}(\mathfrak{P})} = (\|u\|_{L^p(\mathfrak{P})}^p + (\|\nabla u\|_{(L^p(\mathfrak{P}))^N}^p)^{\frac{1}{p}}.$$

$$W_\Gamma^{1,p}(\mathfrak{P}) = \{u : u \in W^{1,p}(\mathfrak{P}), u \text{ vanishes on } \Gamma, u \text{ is } Y\text{--periodic}\},$$

$$H_\#^1(Y) = W_\#^{1,2}(Y),$$

and

$$H_\#^1(\mathfrak{P}) = W_\#^{1,2}(\mathfrak{P}).$$

B.2.3 Two-Scale Function Spaces

In the following, we give short definitions of spaces of functions that have values in Banach- or Hilbert-spaces. Details about such spaces can be found in [DL90] and in the appendix of [Bre73].

Lebesgue spaces of functions with a micro- (or meso-) and a macrovariable:

$$L^2(\Omega \times Y) = L^2(\Omega; L^2(Y))$$
$$= \{u : u : \Omega \to L^2(Y), \int_\Omega \|u(x, .)\|_{L^2(Y)}^2 \, dx < \infty\}$$

with the norm

$$\|u\|_{L^2(\Omega;L^2(Y))} = (\int_\Omega |u(x, .)|_{L^2(Y)}^2 \, dx)^{\frac{1}{2}}.$$

$$L^2(\Omega \times \mathfrak{P}) = L^2(\Omega; L^2(\mathfrak{P}))$$
$$= \{u : u : \Omega \to L^2(\mathfrak{P}), \int_\Omega \|u(x, .)\|_{L^2(\mathfrak{P})}^2 \, dx < \infty\}$$

with the norm

$$\|u\|_{L^2(\Omega;L^2(\mathfrak{P}))} = (\int_\Omega |u(x, .)|_{L^2(\mathfrak{P})}^2 \, dx)^{\frac{1}{2}}.$$

Space of test functions with a micro- (or meso-) and a macrovariable:

$$\mathcal{D}(\Omega; C_\#^\infty(Y)) = \{u : u : \Omega \to C_\#^\infty(Y),$$
$$\text{supp } u \text{ is compact}, u \text{ has derivatives of all orders}\}$$

Sobolev space of functions with a micro- (or meso-) and a macrovariable:

$$L^2(\Omega; H_\#^1(Y)) = \{u : u : \Omega \to H_\#^1(Y), \int_\Omega \|u(x, .)\|_{H^1(Y)}^2 \, dx < \infty\}$$

with the norm

$$\|u\|_{L^2(\Omega;H_\#^1(Y))} = (\int_\Omega |u(x, .)|_{H^1(Y)}^2 \, dx)^{\frac{1}{2}}.$$

B.2.4 Time-Dependent Function Spaces

Almost all the spaces that were mentioned in the preceding sections have generalizations to the time-dependent case. We give only a few examples. We assume that a number $T > 0$ is given.

Space of functions that are continuous with respect to time and square-integrable with respect to space:

$$C([0, T]; L^2(\Omega))$$
$$= \{u : u : [0, T] \to L^2(\Omega), u \text{ is continuous on the time interval } [0, T]\}$$

with norm

$$u_{C([0,T];L^2(\Omega))} = \sup_{t \in [0,T]} \|u(t, .)\|_{L^2(\Omega)}.$$

Space of functions that are square-integrable with respect to time and have square-integrable derivatives with respect to space:

$$L^2(0, T; H^1(\Omega)) = \{u : u : [0, T] \to H^1(\Omega), \int_0^T \|u(t, .)\|_{H^1(\Omega)}^2 \, dt < \infty\}$$

with norm

$$u_{L^2(0,T;H^1(\Omega))} = \left(\int_0^T \|u(t, .)\|_{H^1(\Omega)}^2 \, dt\right)^{\frac{1}{2}}.$$

C

References

[AB88]   B. Amaziane and A. Bourgeat. Effective behavior of two-phase flow in hetero-geneous reservoir, numerical simulation in oil recovery. In M. F. Wheeler, ed-itor, *Numerical Simulation in Oil Recovery*, volume 11 of *The IMA Volumes in Mathematics and Its Applications*, pages 1–22, Berlin, 1988. Springer-Verlag.

[AB92]   J.-L. Auriault and C. Boutin. Deformable porous media with double porosity. Quasi-statics. I. Coupling effects. *Transp. Porous Media*, 7.1:63–82, 1992.

[AB93]   J.-L. Auriault and C. Boutin. Deformable porous media with double porosity. Quasi-statics. II. Memory effects. *Transp. Porous Media*, 10.2:153–169, 1993.

[AB94]   J.-L. Auriault and C. Boutin. Deformable porous media with double porosity. III: Acoustics. *Transp. Porous Media*, 14:143–162, 1994.

[AB96]   G. Allaire and M. Briane. Multiscale convergence and reiterated homogeniza-tion. *Proc. Roy. Soc. Edinburgh*, 126A:297–342, 1996. To appear.

[ABEA90] B. Amaziane, A. Bourgeat, and H. El Amri. Un résultat d'existence pour un modèle d'écoulements diphasiques dans un milieu poreux à plusieurs types de roches. *Publication du Laboratoire d'Analyse Numérique, Université de Pau*, 90–117, 1990.

[ABK91]   B. Amaziane, A. Bourgeat, and J. Koebbe. Numerical simulation and homog-enization of two-phase flow in heterogeneous porou media. *Transp. Porous Media*, 6.5-6:519–547, 1991.

[ACDP92] E. Acerbi, V. Chiado Piat, G. Dal Maso, and D. Percivale. An extension the-orem from connected sets and homogenization in general periodic domains. *Nonlinear Anal.*, TMA 18:481–496, 1992.

[AD85]   H. W. Alt and E. Di Benedetto. Nonsteady flow of water and oil through inhomogeneous porous media. *Ann. Scuola Norm. Sup. Pisa*, 12(4):335–392, 1985.

[Ada75]   R. A. Adams. *Sobolev Spaces*. Academic Press, New York, 1975.

[ADH90]   T. Arbogast, J. Douglas, Jr., and U. Hornung. Derivation of the double porosity model of single phase flow via homogenization theory. *SIAM J. Math. Anal.*, 21:823–836, 1990.

[ADH91]   T. Arbogast, J. Douglas, Jr., and U. Hornung. Modeling of naturally frac-tured petroleum reservoirs by formal homogenization techniques. In Dautray

R., editor, *Frontiers in Pure and Applied Mathematics*, pages 1–19. Elsevier, Amsterdam, 1991.

[ADH95]   G. Allaire, A. Damlamian, and U. Hornung. Two-scale convergence on periodic surfaces and applications. In A. Bourgeat, C. Carasso, S. Luckhaus, and A. Mikelić, editors, *Mathematical Modelling of Flow through Porous Media*, pages 15–25, Singapore, 1995. World Scientific.

[AG91]   M. Aizenman and G. Grimmett. Strict monotonicity for critical points in percolation and ferromagnetic models. *J. Statist. Phys.*, 63:817–835, 1991.

[Aga93]   I. Aganovic. Homogenization of free boundary oscillations of an inviscid fluid in a porous reservoir. *Math. Mod. Numer. Anal.*, 27.1:65–76, 1993.

[AHL71]   V. Ambegaokar, B. I. Halperin, and J. S. Langer. Hopping conductivity in disordered systems. *Phys. Rev. B*, 4:2612–2620, 1971.

[AHZ90]   Y. Amirat, K. Hamdache, and A. Ziani. Homogenization of a model of miscible displacements in porous media. *Asymptotic Anal.*, 3.1:77–89, 1990.

[AKM90]   S.N. Antontsev, A.V. Kazhikhov, and V.N. Monakhov. *Boundary Value Problems in Mechanics of Nonhomogeneous Fluids*. Elsevier Science, Amsterdam, 1990.

[ALB89]   J.-L. Auriault, O. Lebaigue, and G. Bonnet. Dynamics of two immiscible fluids flowing through deformable porous media. *Transp. Porous Media*, 4:105–128, 1989.

[All89]   G. Allaire. Homogenization of the Stokes flow in a connected porous medium. *Asymptotic Anal.*, 2:203–222, 1989.

[All91a] G. Allaire. Continuity of the Darcy's law in the low-volume fraction limit. *Ann. Scuola Norm. Sup. Pisa*, 18:475–499, 1991.

[All91b] G. Allaire. Homogenization of the Navier–Stokes equation with a slip boundary condition. *Comm. Pure Appl. Math.*, 44:605–641, 1991.

[All91c] G. Allaire. Homogenization of the Navier–Stokes equations in open sets with tiny holes I: Abstract framework, a volume distribution of holes. *Arch. Rat. Mech. Anal.*, 113:209–259, 1991.

[All91d] G. Allaire. Homogenization of the Navier–Stokes equations in open sets with tiny holes II: Non-critical sizes of the holes for a volume distribution and a surface distribution of holes. *Arch. Rat. Mech. Anal.*, 113:261–298, 1991.

[All92a] G. Allaire. Homogenization and two-scale convergence. *SIAM J. Math. Anal.*, 23.6:1482–1518, 1992.

[All92b] G. Allaire. Homogenization of the unsteady Stokes equation in porous media. In C. Bandle et al., editor, *Progress in partial differential equations: calculus of variations, applications*, volume 267 of *Pitman Research Notes in Mathematics Series*, pages 109–123, New York, 1992. Longman Higher Education.

[AMA⁺90] J. Adler, Y. Meir, A. Aharony, A. B. Harris, and L. Klein. Low-concentration series in general dimension. *J. Statist. Phys.*, 58:511–538, 1990.

[Anz26]   A. Anzelius. Über Erwärmung vermittels durchströmender Medien. *Zeit. Ang. Math. Mech.*, 6:291–294, 1926.

[AR93a]   J.-L. Auriault and P. Royer. Double conductivity media: A comparison between phenomenological and homogenization approaches. *Int. J. Heat Mass Transfer*, 36.10:2613–2621, 1993.

[AR93b]   J.-L. Auriault and P. Royer. Ecoulement d'un fluide très compressible dans un milieu poreux à double porosité. *CRAS, T. 317, Série II*, 4:431–436, 1993.

[Arb88]   T. Arbogast. The double porosity model for single phase flow in naturally fractured reservoirs. In M. F. Wheeler, editor, *Numerical Simulation in Oil*

Recovery, volume 11 of *The IMA Volumes in Mathematics and Its Applications*, pages 23–45, Berlin, 1988. Springer-Verlag.

[Arb89a]   T. Arbogast. Analysis of the simulation of single phase flow through a naturally fractured reservoir. *SIAM J. Numer. Anal.*, 26:12–29, 1989.

[Arb89b]   T. Arbogast. On the simulation of incompressible, miscible displacement in a naturally fractured petroleum reservoir. *RAIRO Modél. Math. Anal. Numér.*, 23:5–51, 1989.

[Arb91]   T. Arbogast. Gravitational forces in dual-porosity models of single phase flow. In *Proc. Thirteenth IMACS World Congress on Computation and Applied Mathematics*, pages 607–608, Dublin, Ireland, 1991. Trinity College.

[Arb93a]   T. Arbogast. Gravitational forces in dual-porosity systems. I. Model derivation by homogenization. *Transp. Porous Media*, 13:179–203, 1993.

[Arb93b]   T. Arbogast. Gravitational forces in dual-porosity systems. II. Computational validation of the homogenized model. *Transp. Porous Media*, 13:205–220, 1993.

[ASBH90]   J.-L. Auriault, T. Strzelecki, J. Bauer, and S. He. Porous deformable media saturated by a very compressible fluid: quasi-statics. *Eur. J. Mech. A/Solids*, 9.4:373–392, 1990.

[ASP77]   J.-L. Auriault and E. Sanchez-Palencia. Etude du comportement macroscopique d'un mileu poreux saturé déformable. *J. Mécanique*, 16:575–603, 1977.

[ASP86]   J.-L. Auriault and E. Sanchez-Palencia. Remarques sur la loi de Darcy pour les écoulements biphasiques en milieu poreux. *J.M.T.A., Numéro spécial "Modélisation asymptotique d'écoulements de fluides"*, pages 141–156, 1986.

[AT91]   M. Avellaneda and S. Torquato. Rigorous link between fluid permeability, electrical conductivity, and relaxation times for transport in porous media. *Phys. Fluids, A3*, pages 2529–2540, 1991.

[Att84]   H. Attouch. *Variational Convergence for Functions and Operators*. Pitman, Boston, 1984.

[Aur80]   J.-L. Auriault. Dynamic behaviour of a porous medium saturated by a Newtonian fluid. *Int. J. Eng. Sci.*, 16.6:775–785, 1980.

[Aur83]   J.-L. Auriault. Effective macroscopic description for heat conduction in periodic composites. *Int. J. Heat Mass Transfer*, 26.6:861–869, 1983.

[Aur91a]   J.-L. Auriault. Dynamic behaviour of porous media. In J. Bear and M. Y. Corapcioglu, editors, *Transport Processes in Porous Media*, pages 471–519. Kluwer, 1991.

[Aur91b]   J.-L. Auriault. Heterogeneous medium. Is an equivalent macroscopic description possible? *Int. J. Eng. Sci.*, 29.7:785–795, 1991.

[Aur92]   J.-L. Auriault. Acoustics of three phase porous media. In Todorovic Quintard, M., editor, *Heat and Mass Transfer in Porous Media*, pages 221–234, New York, 1992. Elsevier.

[Ave87]   M. Avellaneda. Iterated homogenization, differential effective medium theory and applications. *Comm. Pure Appl. Math.*, 40:527–554, 1987.

[BAH87]   R. B. Bird, R. C. Armstrong, and O. Hassager. *Dynamics of Polymeric Liquids, Vol. 1 Fluid Mechanics*. John Wiley & Sons, New York, 1987.

[Bar85]   J. A. Barker. Block-geometry functions characterizing transport in densely fissured media. *J. Hydrology*, 77:263–279, 1985.

[BB93]   B. Berkowitz and I. Balberg. Percolation theory and its application to groundwater hydrology. *Water Resour. Res.*, 29:775–794, 1993.

[BB95]   A. Badea and A. Bourgeat. Homogenization of two phase flow through randomly heterogeneous porous media. In A. Bourgeat, C. Carasso, S. Luckhaus,
and A. Mikelić, editors, *Mathematical Modelling of Flow through Porous Media*, pages 44–58, Singapore, 1995. World Scientific.

[BG90]   O. Bruno and K. Golden. Interchangeability and bounds on the effective conductivity of the square lattice. *J. Statist. Phys.*, 61:365, 1990.

[BG94]   L. Berlyand and K. Golden. Exact result for the effective conductivity of a
continuum percolation model. *Phys. Rev. B*, 50:2114–2117, 1994.

[BH57]   S. R. Broadbent and J. M. Hammersly. Percolation processes I. Crystals and
mazes. *Proc. Cambridge Philos. Soc.*, 53:629–641, 1957.

[BH95a]   A. Bourgeat and A. Hidani. Effective model of two phase flow in a porous
medium made of different rock types. *Applicable Anal.*, 58:1–29, 1995.

[BH95b]   A. Bourgeat and A. Hidani. A result of existence for a model of two-phase
flow in a porous medium made of different rock types. *Applicable Anal.*,
56:381–399, 1995.

[Bio56]   M. A. Biot. Theory of propagation of elastic waves in a fluid-saturated porous
solid. I. Low frequence range II. Higher frequency range. *J. Acoustic Soc.
Amer.*, 28:168–178, 179–191, 1956.

[BJ67]   G. S. Beavers and D. D. Joseph. Boundary conditions at a naturally permeable
wall. *J. Fluid Mech.*, 30:197–207, 1967.

[BJ87]   J. R. Banavar and D. L. Johnson. Characteristic pore sizes and transport in
porous media. *Phys. Rev. B*, 35:7283, 1987.

[BK95]   A. Yu. Beliaev and S. M. Kozlov. Darcy equation for random porous media.
Comm. Pure Appl. Math., 49.1:1–34, 1995.

[BKM95]   A. Bourgeat, S. Kozlov, and A. Mikelić. Effective equations of two-phase flow
in random media. *Calc. Variations PDEs*, 3:385–406, 1995.

[BLM95]   A. Bourgeat, S. Luckhaus, and A. Mikelić. Convergence of the homogenization process for a double porosity model of immiscible two phase flow. *SIAM
J. Appl. Math.*, 1995. To appear.

[BLP78]   A. Bensoussan, J. L. Lions, and G. Papanicolaou. *Asymptotic Analysis for
Periodic Structures*. North Holland–Elsevier Science Publishers, Amsterdam,
1978.

[BM84]   J. Ball and F. Murat. $W^{1,p}$-quasiconvexity and variational problems for multiple integrals. *J. Funct. Anal.*, 58:225–253, 1984.

[BM85]   J. G. Berryman and G. W. Milton. Normalization constraint for variational
bounds on fluid permeability. *J. Chem. Phys.*, 83:754–760, 1985.

[BM93]   A. Bourgeat and A. Mikelić. A note on homogenization of Bingham flow
through a porous medium. *J. Math. Pures Appl.*, 72:405–414, 1993.

[BM94a]   J. Baranger and A. Mikelić. Solutions stationnaires pour un écoulement quasi-
Newtonien avec l'échauffement visqueux. *C. R. Acad. Sci. Paris, Série I*,
319:637–642, 1994.

[BM94b]   A. Bourgeat and A. Mikelić. Homogenization of the two-phase immiscible
flow in one dimensional porous medium. *Asymptotic Anal.*, 9:359–380, 1994.

[BM95a]   J. Baranger and A. Mikelić. Stationary solutions to a quasi–Newtonian flow
with viscous heating. *Math. Models Methods Appl. Sci.*, 5:725–738, 1995.

[BM95b]   E. Blavier and A. Mikelić. On the stationary quasi-Newtonian flow through
porous medium. *Math. Methods Appl. Sci.*, 18:927–948, 1995.

[BM95c]   A. Bourgeat and A. Mikelić. Homogenization of the non-Newtonian flow through porous medium. *Nonlinear Analysis, Theory, Methods and Applications*, 1995. To appear.

[BMP95]   A. Bourgeat and E. Marušić-Paloka. Loi d'écoulement non linéaire entre deux plaques ondulées. *C. R. Acad. Sci. Paris, Série I*, t. 321 Série I:1115–1120, 1995.

[BMPM95] A. Bourgeat, E. Marušić-Paloka, and A. Mikelić. Effective behavior for a fluid flow in porous medium containing a thin fissure. *Asymptotic Anal.*, 11:241–262, 1995.

[BMW94]   A. Bourgeat, A. Mikelić, and S. Wright. Stochastic two-scale convergence in the mean and applications. *J. reine angew. Math.*, 456:19–51, 1994.

[BOJG86]   D. Berman, B. G. Orr, H. M. Jaeger, and A. M. Goldman. Conductances of filled two-dimensional networks. *Phys. Rev. B*, 33:4301, 1986.

[Bou84a]   A. Bourgeat. Homogenization method applied to the behavior of a naturally fissured reservoir. In K. J. Gross, editor, *Mathematical Method in Energy Research*, pages 181–193, Philadelphia, 1984. SIAM Publ.

[Bou84b]   A. Bourgeat. Homogenized behavior of two phase flows in naturally fractured reservoir with uniform fracture distribution. *Comput. Meth. Appl. Mech. Eng.*, 47:205–217, 1984.

[Bou85]   A. Bourgeat. Nonlinear homogenization of two-phase flow equations. In J. H. Lightbourne and S. M. Rankin, editors, *Physical Mathematics and Nonlinear Partial Differential Equations*, pages 207–212, New York, 1985. Dekker.

[Bou86]   A. Bourgeat. Homogenization of two-phase flow equations. *Proc. Symp. Pure Mathem.*, 45:157–163, 1986.

[Bou87]   A. Bourgeat. An effective dead oil model for two-phase flow in heterogeneous media. In R. Eaton, K. S. Udell, and M. Kaviany, editors, *Multiphase Transport in Porous Media*, volume 60, HTD 91, pages 1–4. ASME-FED, 1987.

[Bow76]   R. M. Bowen. Theory of mixtures. In E. C. Eringen, editor, *Continuum Physics*, pages 1–127, New York, 1976. Academic Press.

[BP89]   N. Bakhvalov and G. Panasenko. *Homogenization: Averaging Processes in Periodic Media*. Kluwer, Dordrecht, 1989.

[BQW88]   A. Bourgeat, M. Quintard, and S. Whitaker. Eléments de comparaison entre la méthode d'homogénéisation et la méthode de prise de moyenne avec fermeture. *C. R. Acad. Sci. Paris Mécanique analytique*, 30II:463–466, 1988.

[Bra85]   A. Braides. Homogenization of some almost periodic coercive functional. *Rend. Acad. Naz. Sci. XL*, 103:313–322, 1985.

[Bre73]   H. Brezis. *Operateurs Maximaux Monotones et semi-groupes de contractions dans les espaces de Hilbert*. North Holland–Elsevier Science Publishers, Amsterdam, 1973.

[Bri86]   A. Brillard. Asymptotic analysis of incompressible and viscous fluid flow through porous media. Brinkman's law via epi-convergence methods. *Ann. Fac. Sci. Toulouse*, 8:225–252, 1986.

[BS89]   M.-P. Bosse and R. E. Showalter. Homogenization of the layered medium equation. *Applicable Anal.*, 32:183–202, 1989.

[BSA78]   N. S. Boulton and T. D. Streltsova-Adams. Unsteady flow to a pumped well in an unconfined fissured aquifer. *Advances in Hydroscience*, pages 357–423, 1978.

[BSL60]   R. B. Bird, W. E. Stewart, and E. N. Lightfoot. *Transport Phenomena*. John Wiley & Sons, New York, 1960.

[But89]   G. Buttazzo. *Semicontinuity, relaxation and integral representation in the calculus of variations*. Pitman, London, 1989.

[BZK60]   G. I. Barenblatt, I. P. Zheltov, and I. N. Kochina. Basic concepts in the theory of seepage of homogeneous liquids in fissured rocks (strata). *J. Appl. Math. Mech.*, 24:1286–1303, 1960.

[Car37]   P. C. Carman. *Trans. Int. Chem. Eng.*, 15:150–166, 1937.

[Cat61]   L. Cattabriga. Su un problema al contorno relativo al sistema di equazioni di Stokes. *Rend. Sem. Mat. Padova*, 31:308–340, 1961.

[CC86]   J. T. Chayes and L. Chayes. Bulk transport properties and exponent inequalities for random resistor and flow networks. *Comm. Math. Phys.*, 105:133–152, 1986.

[CCS83]   P. Collela, P. Concus, and J. Sethian. Some numerical methods for discontinuous flows in porous media. Frontiers in Applied Mathematics, pages 161–186. SIAM Publ., 1983.

[CD88]   D. Cioranescu and P. Donato. Homogénéisation du problème de Neumann non homogène dans des ouverts perforés. *Asymptotic Anal.*, 1:115–138, 1988.

[CGR87]   E. Charlaix, E. Guyon, and S. Roux. Permeability of a random array of fractures of widely varying apertures. *Transp. Porous Media*, 2:31–43, 1987.

[Che78]   P. Cheng. Heat transfer in geothermal systems. *Adv. Heat Transfer*, 14:1–105, 1978.

[CHP87]   E. Charlaix, J. P. Hulin, and T. J. Plona. Experimental study of tracer dispersion in sintered glass porous materials of variable compation. *Phys. Fluids*, 30:1690–1698, 1987.

[Cio92]   D. Cioranescu. Quelques exemples de fluides Newtoniens généralisés. In J. F. Rodrigues and A. Sequeira, editors, *Mathematical topics in fluid mechanics*, volume 274 of *Pitman Research Notes in Mathematics Series*, pages 1–31, Harlow, 1992. Longman.

[CJ]   E. Canon and W. Jäger. Homogenization for nonlinear adsorption-diffusion processes in porous media. To appear.

[CJ86]   G. Chavent and J. Jaffré. *Mathematical models and finite elements for reservoir simulation*. Elsevier Science, Amsterdam, 1986.

[CM65]   R. H. Christopher and S. Middleman. Power-law through a packed tube. *I & EC Fundamentals*, 4:422–426, 1965.

[CM82]   D. Cioranescu and F. Murat. Un terme étrange venu d'ailleurs (i), (ii). In Brézis and Lions, editors, *Nonlinear partial differential equations and their applications. Research notes in mathematics*, volume 60, 70 of *Collège de France Seminar*, pages 98–138, 154–178, London, 1982. Pitman.

[CM95]   D. Cioranescu and F. Murat. Un terme étrange venu d'ailleurs (i), (ii). In R. V. Kohn, editor, *Topics in the mathematical modeling of composite materials*, Progress in Nonlinear Differential Equations and their Applications, Boston, 1995. Birkhäuser.

[Con82]   A. Coniglio. Cluster structure near the percolation threshold. *J. Phys. A*, 15:3829–3844, 1982.

[Con85]   C. Conca. Numerical results of the homogenization of Stokes and Navier–Stokes equations modeling a class of problems from fluid mechanics. *Comput. Meth. Appl. Mech. Eng.*, 53:223–258, 1985.

[Con87a]   C. Conca. Etude d'un fluide traversant une paroi perforée I. Comportement limite près de la paroi. *J. Math. Pures Appl.*, 66:1–43, 1987.

[Con87b]   C. Conca. Etude d'un fluide traversant une paroi perforée II. Comportement limite loin de la paroi. *J. Math. Pures Appl.*, 66:45–69, 1987.

[Con88]   C. Conca. The Stokes sieve problem. *Comm. Appl. Numer. Meth.*, 4.1:113–121, 1988.

[CP79]   P. Concus and W. Proskurowski. Numerical solution of a nonlinear hyperbolic equation by the random choice method. *J. Comp. Phys.*, 30:161–186, 1979.

[CR84]   R. E. Caflisch and J. Rubinstein. Lectures on the mathematical theory of multiphase flow. Technical report, Courant Institute of Mathematical Sciences, New York, 1984.

[CS64]   J. D. Coats and B. D. Smith. Dead-end pore volume and dispersion in porous media. *Trans. Soc. Petr. Eng.*, 231:73–84, 1964.

[CS76]   R. W. Carrol and R. E. Showalter. *Singular and Degenerate Cauchy Problems.* Academic Press, New York, 1976.

[CS82]   G. Chavent and G. Salzano. A finite element method for a 1–d water flooding problem with gravity. *J. Comp. Phys.*, 45:307–344, 1982.

[CS93]   J. D. Cook and R. E. Showalter. Distributed systems of pde in Hilbert space. *Differential & Integral Equations*, 6:981–994, 1993.

[CS94]   G. W. Clark and R. E. Showalter. Fluid flow in a layered medium. *Quarterly Appl. Math.*, 52:777–795, 1994.

[CS95]   J. D. Cook and B. D. Showalter. Microstructure diffusion models with secondary flux. *J. Math. Anal. Appl.*, 189:731–756, 1995.

[CVV59]   R.L. Chouke, P. Van Meurs, and C. Van der Poel. The instability of slow immiscible, viscous liquid-liquid displacements in permeable media. *Trans. AIME*, 216:188–194, 1959.

[DA90]   J. Douglas, Jr. and T. Arbogast. Dual-porosity models for flow in naturally fractured reservoirs. In J. Cushman, editor, *Dynamics of Fluids in Hierarchical Porous Media*, pages 177–221, New York, 1990. Academic Press.

[Dag89]   G. Dagan. *Flow and Transport in Porous Formations.* Springer-Verlag, New York, 1989.

[Dar56]   H. Darcy. *Les Fontaines Publiques de la Ville de Dijon.* Dalmont, Paris, 1856.

[Dea63]   H. A. Deans. A mathematical model for dispersion in the direction of flow in porous media. *Trans. Soc. Petr. Eng.*, 228:49–52, 1963.

[DHA91]   J. Douglas, Jr., J. L. Hensley, and T. Arbogast. A dual-porosity model for waterflooding in naturally fractured reservoirs. *Comput. Meth. Appl. Mech. Eng.*, 873:157–174, 1991.

[DJW53]   P. F. Diesler Jr. and R. H. Wilhelm. Diffusion in beds of porous solids: measurement by frequency response techniques. *Ind. Eng. Chem.*, 45:1219–1227, 1953.

[DK91]   C. J. van Duijn and P. Knabner. Solute transport in porous media with equilibrium and non-equilibrium multiple site adsorption: Travelling waves. *J. reine angew. Math.*, 415:1–49, 1991.

[DK92]   C. J. van Duijn and P. Knabner. Travelling waves in the transport of reactive solutes through porous media: Adsorption and binary ion exchange - Part 1. *Transp. Porous Media*, 8.2:167–194, 1992.

[DL90]   R. Dautray and J.-L. Lions. *Mathematical Analysis and Numerical Methods for Science and Technology Vol. 1–6.* Springer-Verlag, New York, 1990.

[Dou89]   P. le Doussal. Permeability versus conductivity for porous media with wide distribution of pore sizes. *Phys. Rev. B*, 39:4816, 1989.

[DPLAS87] J. Douglas, Jr., P. J. Paes Leme, T. Arbogast, and T. Schmitt. Simulation of flow in naturally fractured reservoirs. In *Ninth SPE Symposium on Reservoir Simulation Society of Petroleum Engineers, Dallas, Texas, 1987*, pages 271–279. Society of Petroleum Engineers, 1987. Paper SPE 16019.

[DPS] J. Douglas, Jr., M. Peszyńska, and R. E. Showalter. Single phase flow in partially fissured media. *Transp. Porous Media*. To appear.

[Dul92] F. A. L. Dullien. *Porous Media, Fluid Transport and Pore Structure*. Academic Press, New York, 1992.

[Dyk71] A. M. Dykhne. Conductivity of a two-dimensional two-phase system. *Sov. Phys. JETP*, 32:63, 1971.

[Ein06] A. Einstein. *Ann. Phys.*, 19:289, 1906.

[Ene91] H. I. Ene. Effects of anisotropy on the free convection from a vertical plate embedded in a porous medium. *Transp. Porous Media*, 6:183–194, 1991.

[Ene96] H. Ene. Homogenization method applied to dispersion of heat in periodic porous media. *Adv. Wat. Res.*, 1996. To appear.

[EP87] H. I. Ene and D. Polişevski. *Thermal Flow in Porous Media*. D. Reidel, Dordrecht, 1987.

[EP90] H. I. Ene and D. Polişevski. Steady convection in a porous layer with translational flow. *Acta Mechanica*, 84:13–18, 1990.

[ESP75] H. I. Ene and E. Sanchez-Palencia. Equations et phénomènes de surface pour l'écoulement dans un modèle de milieu poreux. *J. Mécanique*, 14:73–108, 1975.

[ESP81] H. I. Ene and E. Sanchez-Palencia. Some thermal problems in flow through a periodic model of porous media. *Int. J. Eng. Sci.*, 19.1:117–127, 1981.

[ESP82] H. I. Ene and E. Sanchez-Palencia. On thermal equation for flow in porous media. *Int. J. Eng. Sci.*, 20:623–630, 1982.

[Ewi83] R. E. Ewing. *The Mathematics of Reservoir Simulation*. Frontiers in Applied Mathematics. SIAM Publ., Philadelphia, 1983.

[Fat56] I. Fatt. The network model of porous media. *Trans. Am. Inst. Min. Metall. Pet. Eng.*, 207:144–177, 1956.

[FHS87] S. Feng, B. I. Halperin, and P. N. Sen. Transport properties of continuum systems near the percolation threshold. *Phys. Rev. B*, 35:197–214, 1987.

[FK92] A. Friedman and P. Knabner. A transport model with micro- and macro-structure. *J. Differential Equations*, 98.2:328–354, 1992.

[FT87] A. Friedman and A. Tzavaras. A quasilinear parabolic system arising in modelling of catalytic reactors. *J. Differential Equations*, 70:167–196, 1987.

[Gag80] G. Gagneux. Déplacements de fluides non miscibles incompressibles dans un cylindre poreux. *J. Mécanique*, 19:295–325, 1980.

[GD86] M. T. van Genuchten and F. N. Dalton. Models for simulating salt movement in aggregated field soils. *Geoderma*, 38:165–183, 1986.

[GG93a] M. T. van Genuchten and H. H. Gerke. Dual-porosity models for simulating solute transport in structured media. In *Colloquium "Porous or Fractured Unsaturated Media: Transport and Behaviour" October, 1992, Monte Verità, Ascona, Switzerland*, pages 182–205, 1993.

[GG93b] H. H. Gerke and M. T. van Genuchten. A dual-porosity model for simulating the preferential movement of water and solutes in structured porous media. *Water Resour. Res.*, 29:305–319, 1993.

[GG93c]   H. H. Gerke and M. T. van Genuchten. Evaluation of a first-order water transfer term for variably saturated dual-porosity flow models. *Water Resour. Res.*, 29:1225–1238, 1993.

[Gio75]   E. de Giorgi. Sulla convergenza di alcune successioni di integrali del tipo dell'area. *Rendi Conti di Mat.*, 8:277–294, 1975.

[Gio84]   E. de Giorgi. *G*-operators and Γ-convergence. In *Proc. Int. Congr. Math. (Warszawa, August 1983)*, pages 1175–1191. PWN Polish Scientific Publishers and North–Holland, 1984.

[GK]   K. Golden and S. M. Kozlov. Critical path analysis of transport in highly disordered random media. Submitted.

[GK84]   G. Grimmett and H. Kesten. First-passage percolation, network flows and electrical resistances. *Z. Wahr.*, 66:335–366, 1984.

[GL90]   D. B. Gingold and C. J. Lobb. Percolative conduction in three dimensions. *Phys. Rev. B*, 42:8220–8224, 1990.

[GM62]   S. R. de Groot and P. Mazur. *Non Equilibrium Thermodynamics*. North Holland–Elsevier Science Publishers, Amsterdam, 1962.

[Gol89]   K. Golden. Convexity in random resistor networks. In *Random Media and Composites*, pages 149–170, Philadelphia, 1989. SIAM Publ.

[Gol90]   K. Golden. Convexity and exponent inequalities for conduction near percolation. *Phys. Rev. Lett.*, 65:2923–2926, 1990.

[Gol91]   K. Golden. Bulk conductivity of the square lattice for complex volume fraction. *Int. Series of Num. Math.*, 102:71, 1991.

[Gol92]   K. Golden. Exponent inequalities for the bulk conductivity of a hierarchical model. *Comm. Math. Phys.*, 43:467–499, 1992.

[Gol94a]   K. Golden. The interaction of microwaves with sea ice. In *Proc. IMA Workshop on Waves in Complex and Other Random Media*, 1994. To appear.

[Gol94b]   K. Golden. Scaling law for conduction in partially connected systems. *Physica A*, 207:213–218, 1994.

[Gol95a]   K. Golden. Bounds on the complex permittivity of sea ice. *J. Geophys. Res. (Oceans)*, 100:13,699–13,711, 1995.

[Gol95b]   K. Golden. Statistical mechanics of conducting phase transitions. *J. Math. Phys.*, 1995. In press.

[Gor58]   L. P. Gor'kov. Stationary convection in a plane liquid layer near the critical heat transfer point. *Soviet Phys. JETP*, 6:311–323, 1958.

[GP83]   K. Golden and G. Papanicolaou. Bounds for effective parameters of heterogeneous media by analytic continuation. *Comm. Math. Phys.*, 90:473–491, 1983.

[Gri72]   R. B. Griffiths. Rigorous results and theorems. In C. Domb and M. S. Green, editors, *Phase Transitions and Critical Phenomena*, volume 1, pages 7–109, New York, 1972. Academic Press.

[Gri89]   G. Grimmett. *Percolation*. Springer-Verlag, New York, 1989.

[GW76]   M. T. van Genuchten and P. J. Wierenga. Mass transfer studies in sorbing porous media I. Analytical solutions. *Soil Sci. Soc. Amer. J.*, 40:473–480, 1976.

[Har83]   A. B. Harris. Field-theoretic approach to biconnectedness in percolating systems. *Phys. Rev. B*, 28:2614–2629, 1983.

[Has59]   H. Hasimoto. On the periodic fundamental solutions of the Stokes equations and their application to viscous flow past a cubic array of spheres. *J. Fluid Mech.*, 5(317–327), 1959.

268   C. References

[Hid93]   A. Hidani. Modélisation des écoulements diphasiques en milieu poreux à plusieurs types de roches. Thèse Doctorale en Analyse Numérique, Equation aux Dérivées Partielles et Calcul Scientifique, Université de Saint-Etienne, 1993.

[HJ87]   U. Hornung and W. Jäger. A model for chemical reactions in porous media. In W. Jäger J. Warnatz, editor, *Complex Chemical Reaction Systems. Mathematical Modeling and Simulation*, volume 47 of *Chemical Physics*, pages 318–334, Berlin, 1987. Springer-Verlag.

[HJ91]   U. Hornung and W. Jäger. Diffusion, convection, adsorption, and reaction of chemicals in porous media. *J. Differential Equations*, 92:199–225, 1991.

[HJ93a]   U. Hornung and W. Jäger. Homogenization of reactive transport through porous media. In C. Perelló, C. Simó, and Solà-Morales J., editors, *Int. Conf. on Differential Equations, Barcelona 1991*, pages 136–152. World Scientific, 1993.

[HJ93b]   U. Hornung and W. Jäger. Homogenization of reactive transport through porous media. In C. Perelló, C. Simó, and J. Solà-Morales, editors, *International Conference on Differential Equations, Barcelona 1991*, pages 136–152, Singapore, 1993. World Scientific.

[HJM94]   U. Hornung, W. Jäger, and A. Mikelić. Reactive transport through an array of cells with semipermeable membranes. *RAIRO Modél. Math. Anal. Numér.*, 28:59–94, 1994.

[HKL84]   A. B. Harris, S. Kim, and T. C. Lubensky. ε-expansion for the conductivity of a random resistor network. *Phys. Rev. Lett.*, 53:743–746, 1984.

[Hom87]   G. M. Homsy. Viscous fingering in porous media. *Ann. Rev. Fluid Mech.*, 19:271–311, 1987.

[Hor91a]   U. Hornung. Homogenization of miscible displacement in unsaturated aggregated soils. In G. dal Maso and F. Dell'Antonio, editors, *Composite Media and Homogenization Theory*, Progress in Nonlinear Differential Equations and Their Applications, pages 143–153, Boston, 1991. Birkhäuser.

[Hor91b]   U. Hornung. Miscible displacement in porous media influenced by mobile and immobile water. *Rocky Mountain J. Math.*, 21:645–669 corr. 1153–1158, 1991.

[Hor92a]   U. Hornung. Applications of the homogenization method to flow and transport in porous media. In Shutie Xiao, editor, *Summer School on Flow and Transport in Porous Media*, pages 167–222, Singapore, 1992. World Scientific.

[Hor92b]   U. Hornung. Modellierung von Stofftransport in aggregierten und geklüfteten Böden. *Wasserwirtschaft*, 92.1, 1992.

[Hor95]   U. Hornung. Models for flow and transport through porous media derived by homogenization. In M. F. Wheeler, editor, *Environmental Studies: Mathematical, Computational, and Statistical Analysis*, IMA Volumes on Mathematics, New York, 1995. Springer-Verlag.

[HS90]   U. Hornung and R. E. Showalter. Diffusion models for fractured media. *J. Math. Anal. Applics.*, 147:69–80, 1990.

[HS95a]   B. Hollingworth and R. E. Showalter. Semilinear degenerate parabolic systems and distributed capacitance models. *Discrete and Continuous Dynamical Systems*, 1:59–76, 1995.

[HS95b]   U. Hornung and R. Showalter. Elliptic-parabolic equations with hysteresis boundary conditions. *SIAM J. Math. Anal.*, 26:775–790, 1995.

[JHSD84] G. R. Jerauld, J. C. Hatfield, L. E. Scriven, and H. T. Davis. Percolation and conduction on Voronoi and triangular networks: A case study in topological disorder. *J. Phys. C*, 17:1519–1529, 1984.

[JKO94] V. V. Jikov, S. M. Kozlov, and O. A. Oleinik. *Homogenization of Differential Operators and Integral Functions*. Springer-Verlag, Berlin, 1994.

[JKS86] D. L. Johnson, J. Koplik, and L. M. Schwartz. New pore-size parameter characterizing transport in porous media. *Phys. Rev. Lett.*, 57:2564–2567, 1986.

[JM93] W. Jäger and A. Mikelić. Homogenization of the Laplace equation in a partially perforated domain. Technical report, Equipe d'Analyse Numerique Lyon - St. Etienne, 1993.

[JM94] W. Jäger and A. Mikelić. On the flow conditions at the boundary between a porous medium and an impervious solid. In M. Chipot, J. Saint Jean Paulin, and I. Shafrir, editors, *Progress in Partial Differential Equations*, volume 314 of *Pitman Research Notes in Mathematics*, pages 145–161, London, 1994. Longman Scientific and Technical.

[JM95] W. Jäger and A. Mikelić. On the boundary conditions at the contact interface between a porous medium and a free fluid. *Annali della Scuola Normale Superiore di Pisa, Classe di Scienze*, 1995.

[JNP82] D. Joseph, D. Nield, and G. Papanicolaou. Nonlinear equation governing flow in a saturated porous medium. *Water Resour. Res.*, 18:1049–1052, 1982.

[Kal86] F. Kalaydjian. Description macroscopique des écoulements diphasiques en milieu poreux en tenant compte de l'évolution spatio temporelle de l'interface fluide / fluide. Technical report, Institut Francais du Petrole, 1986.

[Kaz69] H. Kazemi. Pressure transient analysis of naturally fractured reservoirs with uniform fracture distributions. *Soc. Petrol. Eng. J.*, 9:451–462, 1969.

[Kel80] J. B. Keller. Darcy's law for flow in porous media and the two-space method. In R. L. Sternberg, editor, *Nonlinear Partial Differential Equations in Engineering and Applied Sciences*, pages 429–443, New York, 1980. Dekker.

[Kel87] J. B. Keller. Effective conductivity of periodic composites composed of two very unequal conductors. *J. Math. Phys.*, 28:2516, 1987.

[Ker83] A. R. Kerstein. Equivalence of the void percolation problem for overlapping spheres and a network problem. *J. Phys. A*, 16:3071–3075, 1983.

[Kes82] H. Kesten. *Percolation Theory for Mathematicians*. Birkhäuser, Boston, 1982.

[Kir71] S. Kirkpatrick. Classical transport in disordered media: Scaling and effective medium theories. *Phys. Rev. Lett.*, 27:1722, 1971.

[Kir83] S. Kirkpatrick. Models of disordered materials. In *Ill-Condensed Matter*, New York, 1983. North Holland–Elsevier Science Publishers.

[KL84] D. Kröner and S. Luckhaus. Flow of oil and water in a porous medium. *J. Differential Equations*, 55:276–288, 1984.

[KM86] R. V. Kohn and G. W. Milton. On bounding the effective conductivity of anisitropic composites. In J. L. Ericksen, D. Kinderlehrer, R. Kohn, and J.-L. Lions, editors, *Homogenization and Effective Moduli of Materials and Media*, pages 97–125, Berlin, 1986. Springer-Verlag.

[KMJPZ76] H. Kazemi, L. S. Merrill Jr., K. L. Porterfield, and P. R. Zeman. Numerical simulation of water-oil flow in naturally fractured reservoir. *Soc. Petrol. Eng. J.*, 267:317–326, 1976.

[Kna89] P. Knabner. Mathematische Modelle für den Transport gelöster Stoffe in sorbierenden porösen Medien. Technical Report 121, Univ. Augsburg, 1989.

[Kop82]   J. Koplik. Creeping flow in two-dimensional networks. *J. Fluid Mech.*, 119:219–247, 1982.

[Koz27]   J. S. Kozeny. *Ber. Wiener Akad. Abt IIa*, 136:271, 1927.

[Koz78]   S. M. Kozlov. Averaging random structures (in Russian). *Dokl. Akad. Nauk SSSR*, 241:1016, 1978.

[Koz80]   S. M. Kozlov. Averaging of random operators. *Math. USSR Sbornik*, 37:167–180, 1980.

[Koz89]   S. M. Kozlov. Geometric aspects of homogenization. *Russ. Math. Surv*, 44:91, 1989.

[Kru70]   S. N. Kruzkov. First order quasilinear equations with several independent variables. *Math. USSR, Sbornik*, 10:217–243, 1970.

[KS77]   S. N. Kruzkov and S. M. Sukorjanskii. Boundary value for systems of equations of two phase porous flow type, statement of the problems, questions of solvability, justification of approximate methods. *Math. USSR, Sbornik*, 33:62–80, 1977.

[KT86]   A. J. Katz and A. H. Thompson. Quantitative prediction of permeability in porous rock. *Phys. Rev. B*, 34:8179, 1986.

[KT87]   A. J. Katz and A. H. Thompson. Prediction of rock electrical conductivity from mercury injection experiments. *J. Geophys. Res. B*, 92:599, 1987.

[Kün83]   R. Künnemann. *Diffusionslimes und effektive Leitfähigkeit in einem zufälligen Gitter*. Ph. D. dissertation, Heidelberg, 1983.

[LA90]   R. Lipton and A. Avellaneda. A Darcy law for slow viscous flow past a stationary array of bubbles. *Proc. Roy. Soc. Edinburgh*, 114A:71–79, 1990.

[LC86]   K. A. Lurie and A. V. Cherkaev. Exact estimates of the conductivity of a binary mixture of isotropic materials. *Proc. Roy. Soc. Edinburgh*, 104A:21–38, 1986.

[Lev38]   M. C. Leverett. Steady flow of gas-oil-water mixtures through unconsolidated sands. *Trans. AIME*, 132:149, 1938.

[Lev40]   M. C. Leverett. Capillary behavior in porous solids. *J. Petr. Technol.*, pages 1–17, 1940.

[Lev77]   T. Levy. Acoustic phenomena in elastic porous media. *Mech. Res. Comm.*, 4.4:253–257, 1977.

[Lev83]   T. Levy. Fluid flow through an array of fixed particles. *Int. J. Eng. Sci.*, 21:11–23, 1983.

[LG]   V. I. Lytle and K. M. Golden. Microwave backscatter measurements from first year pack ice in the eastern Weddell Sea. *Antarctic J. U.S.* To appear.

[Lio69]   J.-L. Lions. *Quelques méthodes de résolution des problèmes aux limites non linéaires*. Dunod Gauthier-Villars, Paris, 1969.

[Lio81]   J.-L. Lions. *Some Methods in the Mathematical Analysis of Systems and Their Control*. Science Press and Gordon and Breach, Beijing and New York, 1981.

[LM68]   J.-L. Lions and E. Magenes. *Problèmes aux Limites non Homogènes et Applications*, volume 1. Dunod, Paris, 1968.

[LN73]   F. T. Lindstrom and M. N. L. Narasimhan. Mathematical theory of a kinetic model for dispersion of previously distributed chemicals in a sorbing porous medium. *SIAM J. Appl. Math.*, 24:496–510, 1973.

[Low34]   A. N. Lowan. On the problem of the heat recuperator. *Phil. Mag.*, 17:914–933, 1934.

[LSP75]   T. Levy and E. Sanchez-Palencia. On boundary conditions for fluid flow in porous media. *Int. J. Eng. Sci.*, 13:923–940, 1975.

[LSP81]   J.-L. Lions and E. Sanchez-Palencia. Ecoulement d'un fluide viscoplastique de Bingham dans un milieu poreux. *J. Math. Pures Appl.*, 60:341–360, 1981.

[LY52]   T. D. Lee and C. N. Yang. Statistical theory of equations of state and phase transitions. II. Lattice gas and Ising model. *Phys. Rev.*, 87:410, 1952.

[MA87]   A. Mikelić and I. Aganović. Homogenization in a porous medium under a nonhomogeneous boundary condition. *Boll. Un. Mat. Ital. (A)*, 1:171–180, 1987.

[MA88]   A. Mikelić and I. Aganović. Homogenization of stationary flow of miscible fluids in a domain with a grained boundary. *SIAM J. Math. Anal.*, 19:287–294, 1988.

[Mar78]   P. Marcellini. Periodic solutions and homogenization of non linear variational problems. *Ann. Mat. Pura Appl. (IV)*, CXVII:139–152, 1978.

[Mar81]   C. M. Marle. *Multiphase Flow in Porous Media*. Technique, Paris, 1981.

[Mar82]   C. M. Marle. On macroscopic equations governing multiphase flow with diffusion and chemical reactions in porous media. *Int. J. Eng. Sci.*, 20:643–662, 1982.

[Mas93]   G. dal Maso. *An Introduction to Γ-Convergence*. Birkhäuser, Boston, 1993.

[Mat67]   G. Matheron. *Eléments pour une Théorie des Milieux Poreux*. Masson et Cie, Paris, 1967.

[Max81]   C. Maxwell. *A Treatise on Electricity and Magnetismus*. Clarendon Press, Oxford, 1881.

[MBN90]   D. S. McLachlan, M. Blaszkiewicz, and R. E. Newnham. Electrical resistivity of composites. *J. Am. Ceram. Soc.*, 73:2187, 1990.

[Mik89]   A. Mikelić. A convergence theorem for homogenization of two-phase miscible flow through fractured reservoirs with uniform fracture distribution. *Applicable Anal.*, 33:203–214, 1989.

[Mik91]   A. Mikelić. Homogenization of nonstationary Navier–Stokes equations in a domain with a grained boundary. *Ann. Mat. Pura Appl. (IV)*, 158:167–179, 1991.

[Mik95]   A. Mikelić. Effets inertiels pour un écoulement stationnaire visqueux incompressible dans un milieu poreux. *C. R. Acad. Sci. Paris, Série I*, Série I, 320:1289–1294, 1995.

[Mil78]   R. K. Miller. An integrodifferential equation for rigid heat conductors with memory. *J. Math. Anal. Appl.*, 66:313–332, 1978.

[Mil91]   G. M. Milton. The field equation recursion method. In G. dal Maso and G. F. Dell'Antonio, editors, *Composite Media and Homogenization Theory, ICTP, Trieste 1990*, Progress in Nonlinear Differential Equations and Their Applications, pages 223–245. Birkhäuser, Boston, 1991.

[MK64]   V. A. Marchenko and E. J. Khruslov. Boundary value problems with a fine-grained boundary (Russian). *Mat. Sbornik*, 65 (107):458–472, 1964.

[MK74]   V. A. Marchenko and E. J. Khruslov. Boundary problems in domains with finely granulated boundaries (Russian). *Naukova Dumka, Kiev*, 1974.

[MM71]   V. P. Mjasnikov and P. P. Mosolov. A proof of Korn inequality. *Sov. Math. Dokl.*, 12:1618–1622, 1971.

[Mol91]   S. A. Molchanov. Ideas in the theory of random media. *Acta Appl. Math.*, 22:139, 1991.

[MP92]   A. Mikelić and M. Primicerio. Homogenization of heat conduction in materials with periodic inclusions of a perfect conductor. In C. Bandle, J. Bemelmans, M. Chipot, M. Gruter, and J. Saint Jean Paulin, editors, *Progress in PDE's,*

Calculus of Variations, and Applications, volume 267 of *Pitman Res. Notes in Math.*, pages 244–257, London, 1992.

[MP94]   A. Mikelić and Primicerio. Homogenization of the heat equation for a domain with a network of pipes with a well-mixed fluid. *Ann. Mat. Pura Appl. (IV)*, CLXVI:227–251, 1994.

[MT77]   F. Murat and L. Tartar. *H*-convergence. In R. V. Kohn, editor, *Séminaire d'Analyse Fonctionnelle et Numérique de l'Université d'Alger (1977)*, 1977.

[MT95a]   F. Murat and L. Tartar. Calcul des variations et homogénéisation. In R. V. Kohn, editor, *Cours de l'Ecole d'Eté d'Analyse Numérique CEA-EDF-INRIA (1983). English translation in "Topics in the Mathematical Modeling of Composite Materials"*, Progress in Nonlinear Differential Equations and their Applications, Boston, 1995. Birkhäuser.

[MT95b]   F. Murat and L. Tartar. *H*-convergence. In R. V. Kohn, editor, *Topics in the mathematical modeling of composite materials*, Progress in Nonlinear Differential Equations and their Applications, Boston, 1995. Birkhäuser.

[Mül87]   S. Müller. Homogenization of nonconvex integral functionals and cellular elastic materials. *Arch. Rat. Mech. Anal.*, 99:189–212, 1987.

[Mur78]   F. Murat. Compacité par compensation. *Ann. Scuola Norm. Sup. Pisa, Ser. 4*, 5:489–507, 1978.

[MV58]   W. V. R. Malkus and G. Veronis. Finite amplitude cellular convection. *J. Fluid Mech.*, 1:225–260, 1958.

[Ngu89]   G. Nguetseng. A general convergence result for a functional related to the theory of homogenization. *SIAM J. Math. Anal.*, 20:608–629, 1989.

[Nor85]   A. N. Norris. A differential scheme for the effective moduli of composites. *Mech. Mat.*, 4:1–16, 1985.

[Nun71]   J. W. Nunziato. On heat conduction in materials with memory. *Quarterly Appl. Math.*, 29:187–204, 1971.

[NvVE86] Th. M. Nieuwenhuizen, P. F. J. van Velthoven, and M. H. Ernst. Diffusion and long-time tails in a two-dimensional sit-percolation model. *Phys. Rev. Lett.*, 57:2477–2480, 1986.

[OSY92]   O. A. Oleinik, A. S. Shamaev, and G. A. Yosifian. *Mathematical Problems in Elasticity and Homogenization*. Elsevier, Amsterdam, 1992.

[Pan90]   M. B. Panfilov. Mean mode of porous flow in highly homogeneous media. *Sov. Phys. Dokl*, (35) 3:225–227, 1990.

[Pav90]   D. R. Pavone. A Darcy's law extension and a new capillary pressure equation for two phase flow in porous media. *Soc. Petrol. Eng. J.*, 20474, 1990.

[Pea76]   D. W. Peaceman. Convection in fractured reservoir. The effect of matrix fissure transfer on the instability of a density inversion in a vertical fissure. *Soc. Petrol. Eng. J.*, 261:269–280, 1976.

[Pes92]   M. Peszyńska. *Flow through fissured media. Mathematical analysis and numerical approach*. Ph. D. dissertation, Augsburg, 1992.

[Pes94]   M. Peszyńska. Finite element approximation of a model of nonisothermal flow through fissured media. In M. Křížek, P. Neittaanmäki, and R. Stenberg, editors, *Finite Element Methods: Fifty Years of the Courant Element*, volume 164 of *Lecture Notes in Pure and Applied Mathematics*, pages 357–366, New York, 1994. Dekker.

[Pes95a] M. Peszyńska. Analysis of an integro–differential equation arising from modelling of flows with fading memory through fissured media. *J. Partial Differential Equations*, 8:159–173, 1995.

[Pes95b]   M. Peszyńska. On a model for nonisothermal flow in fissured media. *Differential & Integral Equations*, 8:1497–1516, 1995.

[Pes96]   M. Peszyńska. Finite element approximation of diffusion equations with convolution terms. *Math. Comp.*, 1996. To appear.

[Pra61]   S. Prager. Viscous flow through porous media. *Phys. Fluids*, 4:1477–14822, 1961.

[PS94]   L. Packer and R. E. Showalter. Distributed capacitance microstructure in conductors. *Applicable Anal.*, 54:211–224, 1994.

[PS95]   L. Packer and R. E. Showalter. The regularized layered medium equation. *Applicable Anal.*, 58:137–155, 1995.

[PV82]   G. Papanicolaou and S. Varadhan. Boundary value problems with rapidly oscillating coefficients. In *Colloquia Mathematica Societatis János Bolyai, Random Fields (Esztergom, Hungary 1979)*, volume 27, page 835. North Holland–Elsevier Science Publishers, 1982.

[Rad92]   M. Radu. Homogenization techniques. Diploma thesis, Fakultät für Mathematik, Universität Heidelberg, 1992.

[Ray92]   J. W. Rayleigh. On the influence of obstacles arranged in rectangular order upon the properties of a medium. *Phil. Mag.*, pages 481–491, 1892.

[RB49]   W. D. Rose and W. A. Bruce. Evaluation of capillary characters in petroleum reservoir rock. *J. Petrol. Technol.*, 1:127, 1949.

[Rei80]   L. H. Reiss. *The Reservoir Engineering Aspects of Fractured Formations*. Editions Techniques, Paris, 1980.

[Ros52]   J. B. Rosen. Kinetics of a fixed bed system for solid diffusion into spherical particles. *J. Chem. Phys.*, 20:387–394, 1952.

[RS85]   J. N. Roberts and L. M. Schwartz. Grain consolidation and electrical conductivity in porous media. *Phys. Rev. B*, 31:5990–5997, 1985.

[RS95]   J. Rulla and R. E. Showalter. Diffusion in partially fissured media and implicit evolution equations. *Adv. Math. Sci. Appl.*, 5:163–191, 1995.

[Rub48]   L. I. Rubinstein. On the problem of the process of propagation of heat in heterogeneous media. *Izv. Akad. Nauk SSSR, Ser. Geogr.*, 1, 1948.

[Rub86]   J. Rubinstein. On the macroscopic description of slow viscous flow past a random array of spheres. *J. Statist. Phys.*, 44:849–863, 1986.

[RW50]   J. B. Rosen and W. E. Winshe. The admittance concept in the kinetics of chromatography. *J. Chem. Phys.*, 18:1587–1592, 1950.

[SA92]   D. Stauffer and A. Aharony. *Introduction to Percolation Theory*. Taylor and Francis, London, second edition, 1992.

[Sab92]   K. Sab. On the homogenization and the simulation of random materials. *Eur. J. Mech. A Solids*, 11:585–607, 1992.

[Saf71]   P. G. Saffman. On the boundary condition at the interface of a porous medium. *Studies Appl. Math.*, 1:93–101, 1971.

[Sah93]   M. Sahimi. Nonlinear transport processes in disordered media. *AIChE J.*, 39:369, 1993.

[Sah95]   M. Sahimi. *Flow and Transport in Porous Media and Fractured Rock*. VCH, Weinheim, 1995.

[Sch57]   A. E. Scheidegger. *The Physics of Flow Through Porous Media*. Univ. Toronto Press, 1957.

[Ser59]   J. Serrin. On the stability of viscous fluid motion. *Arch. Rat. Mech. Anal.*, 3:1–73, 1959.

[Sha77]   V. K. Shante. Hopping conduction in quasi-one-dimensional disordered compounds. *Phys. Rev. B*, 16:2597–2612, 1977.

[Sho89]   R. E. Showalter. Implicit evolution equations. In *Proc. Int. Conf. Theory and Applications of Differential Equation, Columbus, OH*, volume II of *Differential Equations and Applications*, pages 404–411, Athens, 1989. University Press.

[Sho91]   R. E. Showalter. Diffusion models with microstructure. *Transp. Porous Media*, 6:567–580, 1991.

[Sho93]   R. E. Showalter. Diffusion in a fissured medium with micro-structure. In J. M. Chadam and H. Rasmussen, editors, *Free Boundary Problems in Fluid Flow with Applications*, volume 282 of *Pitman Research Notes in Mathematics*, pages 136–141, New York, 1993. Longman.

[SK82]   P. Sheng and R. V. Kohn. Geometric effects in continuous-media percolation. *Phys. Rev. B*, 26:1331, 1982.

[SP80]   E. Sanchez-Palencia. *Non-Homogeneous Media and Vibration Theory*, volume 129 of *Lecture Notes in Physics*. Springer-Verlag, Berlin, 1980.

[SP82]   E. Sanchez-Palencia. On the asymptotics of the fluid flow past an array of fixed obstacles. *Int. J. Eng. Sci.*, 20:1291–1301, 1982.

[Spa68]   S. Spagnolo. Sulla convergenza delle soluzioni di equazioni paraboliche ed ellittiche. *Ann. Sc. Norm. Sup. Pisa Cl. Sci. (3)*, 22:571–597, 1968.

[Spa76]   S. Spagnolo. Convergence in energy for elliptic operators. In B. Hubbard, editor, *Proc. 3rd Symp. Numerical Solution Partial Differential Equations (College Park 1975)*, pages 469–498, San Diego, 1976. Academic Press.

[SS88]   K. Sattel-Schwind. *Untersuchung über Diffusionsvorgänge bei der Gelpermeations-Chromatographie von Poly-p-Methylstyrol*. Ph. D. dissertation, Fachbereich Chemie, Universität Heidelberg, 1988.

[ST91]   J. Saint Jean Paulin and K. Taous. *A generalized Darcy law Homogenization in Porous Medium*. Preprint Université de Metz, Dept. Mathématiques, 08/91, 1991.

[Sta77]   H. E. Stanley. Cluster shapes at the percolation threshold: An effective cluster dimensionality and its connection with critical-point exponents. *J. Phys. A*, 10:L211–L220, 1977.

[SW90]   R. E. Showalter and N. J. Walkington. A diffusion system for fluid in fractured media. *Differential & Integral Equations*, 3:219–236, 1990.

[SW91a]   R. E. Showalter and N. J. Walkington. Diffusion of fluid in a fissured medium with micro-structure. *SIAM J. Math. Anal.*, 22:1702–1722, 1991.

[SW91b]   R. E. Showalter and N. J. Walkington. Micro-structure models of diffusion in fissured media. *J. Math. Anal. Appl.*, 155:1–20, 1991.

[SW93]   R. E. Showalter and N. J. Walkington. Elliptic systems for a medium with micro-structure. In I. J. Bakelman, editor, *Geometric Inequalities and Convex Bodies*, pages 91–104, New York, 1993. Dekker.

[Swa76]   A. de Swaan. Analytic solutions for determining naturally fractured reservoir properties by well testing. *Soc. Petrol. Eng. J.*, 16:117–122, 1976.

[SY93]   Ch. Shah and Y. C. Yortsos. Aspects of non-Newtonian flow and displacement in porous media. Technical report, University of Southern California, 1993.

[Tar78]   L. Tartar. Quelques remarques sur l'homogénéisation. In H. Fujita, editor, *Proc. of the Japan-France Seminar 1976 "Functional Analysis and Numerical Analysis"*, pages 469–481. Japan Society for the Promotion of Sciences, 1978.

[Tar79]   L. Tartar. Compensated compactness and applications to partial differential equations. In *Nonlinear Analysis and Mechanics, Heriot-Watt Symposium IV*, volume 39 of *Research Notes in Mathematics*, pages 136–211, London, 1979. Pitman.

[Tar80]   L. Tartar. Incompressible fluid flow in a porous medium - convergence of the homogenization process. *in [SP80] 368-377*, 1980.

[Tar90]   L. Tartar. H–measures, a new approach for studying homogenization, oscillations and concentration effects in partial differential equation. *Proc. Roy. Soc. Edinburgh*, 115A:193–230, 1990.

[Tem79]   R. Temam. *Navier–Stokes Equations*. North Holland–Elsevier Science Publishers, Amsterdam, New York, Oxford, 1979.

[TH86]   C. T. Tan and G. M. Homsy. Stability of miscible displacements in porous media: Rectilinear flow. *Phys. Fluids*, 29:3549–3556, 1986.

[Tor87]   S. Torquato. Thermal conductivity of disordered heterogeneous media from the microstructure. *Rev. Chem. Eng.*, 4:151–204, 1987.

[Vog82]   C. Vogt. A homogenization theorem leading to a Volterra integro-differential equation for permeation chromatography. Technical report, SFB 123, Heidelberg, 1982.

[WB36]   R. D. Wyckoff and H. G. Botset. The flow of gas-liquid mixture through unconsolidated sands. *Physics*, 7:325, 1936.

[Whi86]   S. Whitaker. Flow in porous media II: The governing equations for immiscible, two-phase flow. *Transp. Porous Media*, 1:105–125, 1986.

[WL74]   B. P. Watson and P. L. Leath. Conductivity in the two-dimensional-site percolation problem. *Phys. Rev. B*, 9:4893–4896, 1974.

[WMS76]   R. A. Wooding and H. J. Morel-Seytoux. Multiphase fluid flow through porous media. *Ann. Rev. Fluid Mech.*, 8:233–274, 1976.

[Woo63]   R. A. Wooding. Convection in a saturated porous medium at large Rayleigh number or Péclet number. *J. Fluid Mech.*, 15:527–541, 1963.

[WPW91]   Y. S. Wu, K. Pruess, and A. Witherspoon. Displacement of a Newtonian fluid by a non-Newtonian fluid in a porous medium. *Transp. Porous Media*, 6:115–142, 1991.

[WR63]   J. Warren and P. Root. The behavior of naturally fractured reservoirs. *Soc. Petrol. Eng. J.*, 3:245–255, 1963.

[YL52]   C. N. Yang and T. D. Lee. Statistical theory of equations of state and phase transitions. I. Theory of condensation. *Phys. Rev.*, 87:404, 1952.

[Zhi89]   V. V. Zhikov. Effective conductivity of random homogeneous sets. *Math. Zametki*, 45:34, 1989.

Index

Interdisciplinary Applied Mathematics
Volume 6

Springer-Science+Business Media, LLC